Laboratory Manual for Exercise Physiology

Second Edition

G. Gregory Haff, PhD, CSCS,*D, FNSCA, ASCC

School of Medical and Health Sciences
Edith Cowan University

Charles Dumke, PhD, FACSM

Health and Human Performance
University of Montana

HUMAN KINETICS

Library of Congress Cataloging-in-Publication Data

Names: Haff, Greg, author. | Dumke, Charles, 1966- author.
Title: Laboratory manual for exercise physiology / G. Gregory Haff, Charles
  Dumke.
Description: Second edition. | Champaign, IL : Human Kinetics, [2019] |
  Includes bibliographical references.
Identifiers: LCCN 2017021596 (print) | LCCN 2017022349 (ebook) | ISBN
  9781492536949 (print) | ISBN 9781492536956 (e-book)
Subjects: | MESH: Exercise--physiology | Physical Fitness--physiology |
  Exercise Test | Laboratory Manuals
Classification: LCC QP301 (ebook) | LCC QP301 (print) | NLM QT 25 | DDC
  613.7/1--dc23
LC record available at https://lccn.loc.gov/2017021596

ISBN: 978-1-4925-3694-9 (print)

Senior Acquisitions Editor: Amy N. Tocco; **Senior Developmental Editor:** Amanda S. Ewing; **Managing Editor:** Anna Lan Seaman; **Copyeditor:** Alisha Jeddeloh; **Permissions Manager:** Dalene Reeder; **Senior Graphic Designer:** Joe Buck; **Cover Designer:** Keri Evans; **Photograph (cover):** © Human Kinetics; **Photographs (interior):** © Human Kinetics, unless otherwise noted; **Photo Asset Manager:** Laura Fitch, **Photo Production Manager:** Jason Allen; **Senior Art Manager:** Kelly Hendren; **Illustrations:** © Human Kinetics, unless otherwise noted; **Printer:** Walsworth

We thank the University of Montana in Missoula, Montana, for assistance in providing the location for the photo and video shoot for this book.

The video contents of this product are licensed for educational public performance for viewing by a traditional (live) audience, via closed circuit television, or via computerized local area networks within a single building or geographically unified campus. To request a license to broadcast these contents to a wider audience—for example, throughout a school district or state, or on a television station— please contact your sales representative (**www.HumanKinetics.com/SalesRepresentatives**).

Printed in the United States of America

10 9 8 7 6 5 4 3

The paper in this book was manufactured using responsible forestry methods.

Human Kinetics
P.O. Box 5076
Champaign, IL 61825-5076
Website: www.HumanKinetics.com

In the United States, e-mail info@hkusa.com or call 800-747-4457.
In Canada, e-mail info@hkcanada.com.
In the United Kingdom/Europe, e-mail hk@hkeurope.com.

For information about Human Kinetics' coverage in other areas of the world,
please visit our website: **www.HumanKinetics.com**

E6910

CONTENTS

Laboratory 1   Primary Data Collection   1

Laboratory 2   Pretest Screening   43

Laboratory 3   Flexibility Testing   89

Laboratory 16 | Electrocardiograph Measurements   383

LABORATORY ACTIVITY FINDER

Here is an alphabetical listing of the laboratory activities. The table of contents lists the laboratory activities by laboratory activity number.

PREFACE

Laboratory Manual for Exercise Physiology, Second Edition, is a detailed source of tests for an undergraduate or graduate exercise physiology laboratory course. The text covers a wide variety of tests typically performed in an exercise physiology laboratory when evaluating athletes, clinical clients, or other generally healthy individuals. The design allows instructors to choose activities that best suit their course needs. Specifically, each chapter offers a wide variety of laboratory activities that can be mixed and matched depending upon the instrumentation and time allotted for your course. The range of field and laboratory tests presented here gives students broad exposure to testing that can be applied in a wide variety of professional settings. Organized in a logical progression, the labs build in complexity as students progress through the book and develop their knowledge base. Ultimately, the text serves as a resource for basic testing procedures used in assessing human performance, health, and wellness.

UPDATES TO SECOND EDITION

The second edition of this text brings some new and exciting updates. Perhaps the most important is the inclusion of 10 virtual laboratory activities that provide an enhanced, immersive learning experience for students. These activities, in which video is used extensively, allow students to interact with virtual lab partners, observe the proper setup and use of equipment, and participate in following test protocols in the online environment. The virtual labs align with specific aspects of the lab activities introduced in the manual, often presenting some of the more complex lab concepts.

Other updates in this second edition of the manual include laboratory activities that introduce common intermittent fitness tests such as the Léger 20 m shuttle run test, the Yo-Yo Intermittent Recovery Test, and the 30-15 Intermittent Fitness Test. These types of tests are increasingly popular in the fitness world, and their addition to the manual allows students to learn how to perform these tests and interpret their results.

In addition to these new lab activities, updates have been made to every lab in the manual:

- Added new research and information pertaining to each laboratory topic
- Updated standards and norms with new published research
- Clarified instructions
- Added new case studies to illuminate laboratory concepts
- Included answers to the case studies
- Updated question sets to help students better understand lab concepts

SPECIAL LABORATORY FEATURES

Each laboratory chapter is a complete lesson, beginning with objectives, definitions of key terms, and background information that sets the stage for learning. For each of the laboratory activities, you will find step-by-step instructions, making it easier for those new to the lab setting to complete the procedures. Each laboratory activity has a data sheet to record individual findings, as well as question sets related to the data collected by students; these questions invite students to put their laboratory experience into context.

eBook
available at
your campus bookstore
or HumanKinetics.com

WEB STUDY GUIDE

The web study guide provides additional tools that can assist students in working through the 49 laboratory activities in the book. All laboratory activities in the book are supported by the following:

- Electronic versions of the individual data sheets from the book, which students can download to a computer or mobile device and print.

- Group data sheets, found only in the web study guide, that allow students to move beyond collecting individual data; with these group sheets, students can compile data from the entire class, calculate values such as mean and range, and compare findings to the normative data discussed in the lab.

- Downloadable versions of the question sets from the book, which can be submitted electronically, giving students an easy way to turn in answers after completing a laboratory activity.

- Practical case study questions, found only in the web study guide, that help students begin to critically analyze data collection and synthesize it with material they have learned in lecture and other courses. The case studies are provided as downloadable electronic files that students can complete and submit online.

In addition, 10 of the laboratory activities are provided as interactive labs, all of which include video, that give students an approximation of the real-world experience of performing the lab activities. These interactive labs complement the laboratory manual. This icon identifies the 10 interactive labs:

> This laboratory activity is supported by a virtual lab experience in the web study guide.
>
> WWW

The web study guide is available at www.HumanKinetics.com/ LaboratoryManualForExercisePhysiology.

INSTRUCTOR ANCILLARIES

An image bank is available for instructor use. The image bank includes most of the figures, tables, and photos from the book, saved as individual files. Instructors can use these items to create a PowerPoint presentation, enhance lecture notes, create student handouts, and so on. Instructors also have full access to the web study guide, and they have access to the case study answers.

NOTES FOR INSTRUCTORS

This manual is geared toward use in an exercise physiology laboratory course. It is designed to translate the scientific foundation developed in a core exercise physiology lecture course—using, for example, a text such as *Physiology of Sport and Exercise* by Kenney, Wilmore, and Costill (Human Kinetics, 2016)—into practical applications typically performed in a variety of settings. To accomplish this goal, the manual is divided into 16 laboratories that lead students through a series of activities: primary data collection, pretest screening, flexibility testing, blood pressure measurements, resting metabolic rate determinations, oxygen deficit and EPOC evaluations, submaximal exercise testing, aerobic power field assessments, high-intensity fitness testing, maximal oxygen consumption measurements, blood lactate threshold assessment, musculoskeletal fitness measurements, anaerobic fitness measurements, pulmonary function testing, body composition assessments, and electrocardiograph measurements.

Each laboratory provides background information and detailed step-by-step procedures for a variety of tests. In addition, because exercise physiology laboratories are equipped in various ways, the labs present multiple methods for introducing the testing concept. For example, laboratory 13 presents multiple methods for assessing vertical jump performance: jump and reach, Vertec, and switch mat. Equipment lists at the beginning of each activity make it easier to choose the labs that will work best in your facility. This versatility enables you to choose activities that best fit your facilities and best meet the needs of your students.

ACKNOWLEDGMENTS

I would like to thank Human Kinetics for their patience with Chuck and me as we slogged through the writing of the second edition of this text. Specifically, I would like to thank Roger Earle for his belief in our abilities and, more importantly, for his friendship. Additionally, I would like to thank Amy, Lisa, and Amanda for working tirelessly to assist us in updating this text to improve upon our first edition.

While the book was difficult to complete and, at times, very stressful, I would like to thank Chuck for his patience and dedicated efforts and, more importantly, for being one of my very best friends and favorite colleagues.

To my friends Michael Stone, Duncan French, Joel Cramer, Jay Dawes, and Travis Triplett, I am honored to call you my friends, and you each impact my life more than you know. I look forward to enjoying life in your company.

Finally, I would be remiss if I did not acknowledge the most important person in my life—my wife, Erin. You are the rock that supports me in all endeavors that I undertake. While things never seem to go smoothly or appear to be working out, your ability to ground me and to make me laugh, stop, and watch the waves is more than anyone could ever want in a life partner. I am blessed to have you in my life.

—Greg Haff

Thanks to Greg for being a great coauthor and friend. I wouldn't have wanted to do this by myself or with anybody else. Human Kinetics deserves a shout-out for their experience, professionalism, and patience with our ideas and extension of deadlines. Thank you for taking this book to product and being open to creative ideas for the second edition.

I would also like to thank my colleagues in the department of health and human performance at the University of Montana for their patience with me as I worked on this project.

Most importantly, thank you to my parents, Bob and Leah, and to my wife, Shannon; son, Carter; and dog, Rastro.

—Chuck Dumke

Primary Data Collection

Objectives

- Define basic terminology associated with testing.
- Learn metric conversions and the units recommended by the International System of Units (SI).
- Provide a rationale for collecting basic information during testing.
- Present the methods for evaluating temperature, barometric pressure, and relative humidity.
- Present basic statistical methods for evaluating test results.
- Describe types of graphics for presenting data.

DEFINITIONS

accuracy—Degree of a measurement's closeness to the actual value.

barometric pressure—Pressure exerted by ambient air.

central tendency—Score that best represents all scores collected for a group.

dependent variable—Effect or yield of the independent variable.

displacement—Length that an object moves in a straight line between two points.

distance—Total length that an object travels (may or may not be in a straight line).

effect size—Statistical method used to determine the magnitude of an effect.

energy—Capacity to do work, expressed as a joule (J).

field/laboratory test—Test that can be completed in either a field or a laboratory setting.

field test—Test completed in a field setting.

force—Mass multiplied by acceleration, expressed as a newton (N).

independent variable—Variable that is manipulated.

inferential statistics—Statistical methods that can be used to draw general conclusions about a population based on a population sample.

laboratory test—Sophisticated test that must be conducted in a laboratory setting.

magnitude statistics—Statistical methods that can be used to evaluate the magnitude of change, typically using the smallest worthwhile change or effect size.

mass—Measure of matter that constitutes an object; expressed in kilogram (kg).

mean—Average score of a sample.

median—Middle score of a sample.

mode—Most frequent score of a sample.

normative data—Placement within a population; also referred to as *norms* or *norm data*.

power—Rate at which work can be performed, represented as a watt (W).

precision—Degree to which a test is reproducible with nearly the same value.

range—Distance between end points in a group of scores.

relative humidity—Amount of water or percent saturation in ambient air.

reliability—Repeatability of a measure.

smallest worthwhile change—Smallest practically important change in some measure.

speed—Scalar quantity generally considered to be how fast a body is moving; calculated by dividing distance covered by time.

standard—Desirable or target score.

standard deviation (SD)—Measure of variability that shows variation or dispersion from the mean.

typical error (TE)—The most common measure of reliability; calculated as the standard deviation of the change scores between repeated measures divided by $\sqrt{2}$.

validity—Accuracy of a measure.

variability—Spread of a data set.

variable—A characteristic.

velocity—Vector quantity calculated by dividing displacement by time.

wet-bulb globe temperature (WBGT)—Measure of temperature that estimates cooling capacity of the surrounding environment.

work—Force times the distance through which it acts.

z-score—Standardized score indicating the distance of an individual score in standard deviation units from the mean of a group.

Testing human performance under exercise conditions allows for the evaluation of the human body's functional ability. This information can give us an understanding of the individual's overall health and wellness as well as athletic performance capacity. We can also garner information about the ability to tolerate and adapt to exercise by examining the individual's postexercise responses. This information can then be used to implement exercise programs designed to enhance health and wellness or sport performance. There are numerous tests that can be performed in the exercise physiology laboratory in order to evaluate health and wellness (1, 6, 21, 31) or examine athletic performance capacity (27, 38). Many of these tests fall into one of three classifications: field, field/laboratory, and laboratory.

Field tests allow us to assess specific fitness and performance variables in a real-world setting (16). These tests are generally practical and less expensive than their laboratory-based counterparts (32). Though not often used for research due to difficulty in controlling external variables (e.g., weather, terrain), these tests are extremely useful for screening and monitoring purposes (6). Because these tests are developed from their laboratory counterparts, they can offer a high degree of validity when conducted with attention to appropriate methodological controls. Examples in exercise physiology include the 1 to 1.5 mi (1.6-2.4 km) run test, the 1 mi (1.6 km) jogging test, the 12 min cycling test, sprints, the 30-15 Intermittent Fitness Test (11), and the quantification of the body mass index, or BMI (3, 10). Though typically done in field settings, some of these tests may also be conducted in laboratory settings (e.g., BMI, 12 min cycling test).

Field/laboratory tests can be conducted in either field or laboratory settings. Like field tests, they often require minimal equipment, but they are subjected to tighter controls (6), and a field/laboratory test in the field must be performed with the same tight controls that would be used in the laboratory (3). One example of a field/laboratory test is the step test (6). In the laboratory, this test can be performed using a step box, which limits the number of subjects to one. In the field, the step test can be performed on stadium bleachers with a large number of subjects at the same time. Regardless of location, the step test requires a metronome and stopwatch to appropriately conduct and control the test. Other examples of field/laboratory tests include the sit-and-reach test, skinfold assessments, vertical jump testing, and blood pressure (BP) measurements (6).

Laboratory tests are conducted with the highest level of control and often require expensive equipment that cannot be taken into the field; as a result, they are usually performed on one person at a time (6) and thus tend to be time consuming. In return, they offer a significantly higher degree of accuracy and precision (6). Examples include measurement of maximal oxygen consumption, quantification of resting metabolic rate (RMR), exercise electrocardiograms (ECGs), dual X-ray absorptiometry (DXA), underwater weighing (UWW), quantification of isometric or dynamic force-time curves, and anaerobic treadmill testing (4-6, 12).

TEST VARIABLES

A central goal in the exercise physiology laboratory is to quantify specific physiological or performance characteristics. A characteristic is usually termed a **variable**. Variables often quantified in an exercise physiology laboratory include maximal strength, body composition, anaerobic or aerobic power, and flexibility.

Variables are generally categorized as either independent or dependent. The **independent variable** is the one that is manipulated, whereas the **dependent variable** is the effect of, response to, or yield of the independent variable (39). In other words, the independent variable is controlled by the person administering the test, and the dependent variable is the physiological or performance response to the independent variable. In a treadmill test, for example, the independent variable is the speed of the treadmill; the dependent variable is the heart rate (HR) response or oxygen consumption rate. When graphically representing these variables, we place the independent variable on the *x*-axis (horizontal axis) and the dependent variable on the *y*-axis (vertical axis) (39). In the example of the treadmill test, then, the workload or speed would be the independent variable presented on the *x*-axis, and the HR response would be the dependent variable presented on the *y*-axis.

MEASUREMENT TERMINOLOGY

When examining physiological and performance characteristics, exercise physiologists use specific measurement terminology to discuss their data. This terminology is based on the guidelines set forth in the Système International d'Unités, or International System of Units (SI) (42). This system is designed to adhere to the core principles of being simple, precise, and accurate (table 1.1).

Following are some examples of common terms used in the exercise physiology laboratory.

The most common measurements of **mass** performed in the exercise physiology laboratory are measurements of body mass, lean body mass, and fat mass. Though the terms *body mass* and *body weight* are often used interchangeably in

Table 1.1 Sample SI Units

Unit	Name	Symbol
BASE SI UNITS		
Amount of substance	Mole	mol
Electric current	Ampere	A
Length	Meter	m
Mass	Kilogram	kg
Thermodynamic temperature	Kelvin	K
Time	Second	s
DERIVED UNITS		
Energy	Joule	J
Force	Newton	N
Frequency	Hertz	Hz
Power	Watt	W
Pressure	Pascal	Pa
Work	Joule	J

the United States, it is more accurate to use the term *body mass*. The SI base unit for mass is the kilogram (kg) (42).

Force is a vector quantity characterized by both magnitude and direction. It can be calculated according to Isaac Newton's second law, which states that force is equivalent to mass multiplied by acceleration (29):

$$\text{Force} = \text{mass} \times \text{acceleration}$$

For example, body mass is used often in this laboratory manual when calculating work or work rate, and in such instances body mass is converted to newtons and used in other calculations. Forces are generally expressed as newtons (N), which are calculated by multiplying the object's mass in kg by its downward acceleration due to gravity:

$$\text{Force (N)} = \text{mass} \times \text{acceleration} = \text{mass (kg)} \\ \times 9.81 \text{ m} \cdot \text{s}^{-2}$$

When thinking about force, it is not uncommon to consider the expression of strength. Generally, strength is the maximal force generated by a muscle or group of muscles at a specific velocity or the ability to generate external force (29). Such capability often constitutes a major concern. Force-generating capacity plays an important role in the ability to perform in sport (10) and activities of daily living (22). In addition, the ability to repetitively express submaximal forces is important in endurance activities. Thus strength, or force-generating capacity, exerts a major effect on athletic performance and overall fitness.

Both **displacement** and **distance** can be considered as lengths. Displacement is measured in a straight line from one point to another, whereas distance is the total length that an object travels and may or may not be limited to a straight line. Both are typically measured in centimeters or meters, but the meter is the SI base unit. Quantification of displacement and distance contributes to calculations such as work, velocity of movement, and expression of power.

To calculate the amount of **work** completed, multiply the amount of force exerted on an object by the distance that the object is moved:

$$\text{Work (J)} = \text{force (N)} \times \text{distance (m)}$$

The results of this equation are reported in joules (J; the SI unit for work), which is equal to newton × meters (N · m). As a whole, work is directly related to the amount of metabolic energy expended—the more work performed, the more kilocalories used (28, 37).

It is important to differentiate between speed and velocity. **Speed** is a scalar quantity generally considered to be how fast a body is moving, which is directly related to the total distance covered divided by time. Though similar to speed, **velocity** is an actual vector that has a magnitude and a direction. Therefore, velocity is based on both speed and direction (26). The difference between speed and velocity centers on the difference between distance and displacement. Mathematically, speed is calculated by dividing distance by time, whereas velocity is calculated by dividing displacement by time:

$$\text{Speed} = \text{distance} / \text{time}$$

$$\text{Velocity} = \text{displacement} / \text{time}$$

For example, a runner who completes a 400 m dash in 49 s would have an average velocity of 0 m · s^{-1} because the displacement would be zero. In this instance, then, it would be better to use average speed to represent how fast the runner is moving, and the result would be 8.16 m · s^{-1}.

Velocity is the linear speed of the object, distance represents how far the object has moved in a given direction, and time represents how long it took to cover that distance (26). For example, if a 100 m sprint was run in 9.58 s, the average velocity of the event would be 10.4 m · s^{-1}.

Power is the rate at which work can be performed (37, 41), and the ability to express high power outputs is one of the most important factors in sport performance (8). Power can be calculated in several ways:

$$\text{Power} = \text{work} / \text{time}$$
$$= \text{force} \times \text{velocity}$$
$$= \text{force} \times (\text{displacement} / \text{time})$$

Power is generally proportional to the amount of energy used (37)—thus, the higher the power output, the higher the work rate, which corresponds to a faster expenditure of energy. Because of this relationship, power is often used when discussing the transfer of metabolic energy to physical performance (24, 25). For example, power is used to describe this transformation when talking about aerobic power and anaerobic power.

In a general sense, **energy** is the capacity to do work (9), and it is often represented in joules. When examining metabolic energy release—the result of work done (energy used) and heat released (energy wasted)—the joule is the universally accepted unit (6). In the United States, however, it is more common to represent this release as kilocalories (kcal).

It is possible to estimate caloric or kilojoule (kJ) expenditure by determining oxygen uptake and cost. It can be assumed that 1 liter (L) of oxygen uptake corresponds to approximately 5 kcal or 21 kJ of energy expenditure. Thus, caloric expenditure can be calculated with one of the following formulas (3):

$$\text{Energy expenditure (kcal} \cdot \text{min}^{-1}) = \text{oxygen uptake (L} \cdot \text{min}^{-1}) \times 5 \text{ (kcal} \cdot \text{L}^{-1})$$

$$\text{Energy expenditure (kJ} \cdot \text{min}^{-1}) = \text{oxygen uptake (L} \cdot \text{min}^{-1}) \times 21 \text{ (kJ} \cdot \text{L}^{-1})$$

These figures are only approximations, and they can be influenced by the intensity of exercise. Specifically, higher intensities result in more energy being expended per liter of oxygen consumed (6).

METRIC CONVERSIONS

The use of metric units is standard in most exercise physiology laboratories. Though the metric system is not popular in the United States, it is the preferred system when conducting testing and research. Laboratory work in exercise physiology typically involves three categories of conversion: length, weight, and volume.

The standard metric and SI unit for length in scientific publications is the meter; it is generally used when reporting a person's height. The meter is easily converted to other metric units, such as the millimeter, centimeter, and kilometer (e.g., 0.01 km = 10 m = 1,000 cm = 10,000 mm) (see appendix A). To convert from the American system to the metric system, simply multiply

length in inches by 2.54 to figure the length in centimeters. A more complete list of conversions can be found in appendix A.

When referring to weight, the base metric and SI unit is the kilogram (kg), which is the preferred method for representing mass in scientific literature (40, 42). In the exercise physiology laboratory, it is common to represent body mass, lean body mass, and fat mass in terms of kilograms (3). The conversion of pounds to kilograms is easily accomplished by dividing the pound weight by 2.2046. A summary of basic conversions for measurements of mass can be found in appendix A.

The basic metric unit for volume is the liter, also known as a cubic decimeter (dm^3) (40). The SI unit for volume is the cubic meter (m^3) because it can be used to express the volumes of solids, liquids, and gases (40, 42). For example, 0.0015 m^3 is equivalent to 1,500 mL or 1.5 L (appendix A). The liter is commonly used when quantifying lung volume, oxygen consumption, cardiac output (CO), stroke volume (SV), and sweat loss. Length and weight measures can also be converted easily to volume measures. For example, if an athlete loses 1 kg of body mass during a training session, this is equivalent to 1 L of sweat loss and thus would require >1 L of water to restore fluid balance (6). A summary of volume conversions can be found in appendix A.

BACKGROUND AND ENVIRONMENTAL INFORMATION

Collecting basic information is an important organizational part of the testing process (3, 6). In planning a testing session, you should consider several distinct items as part of the basic information collected—for example, the subject's name or identification number, age, and sex. It is also important to note the date, time of day, and who conducted the testing session.

- *Name or identification number:* Typically, the subject's surname is noted first, followed by a comma and the subject's first name; however, if the data are being used for research, a subject number should be noted instead of the subject's name in order to ensure confidentiality and compliance with research procedures for human subjects (7). This information is generally placed at the top of each data sheet and on all forms related to the test.

- *Age and sex:* It is crucial to note the subject's age, especially when comparing the subject's data with those presented in normative data tables. The subject's age is typically recorded to the nearest year, though some instances may warrant reporting the year to the closest tenth of a year (3, 6). For example, if a subject is 18 years and 6 months old, the age should be recorded as 18.5 years. It is also necessary to document the subject's sex, generally by recording an *M* for male or an *F* for female on the data sheet.

- *Date:* The date should be recorded as either month/day/year or day/month/year. For example, the date of March 3, 2018, could be noted as 3/3/2018 in the appropriate location on the laboratory data sheet. This information should be clearly noted on any data sheet used in the testing process (3, 6).

- *Time:* It is especially important to note time when performing longitudinal testing because some biological and performance measures exhibit diurnal or circadian variations (10, 35, 36). As a result, longitudinal testing sessions should generally be conducted at the same time each day in order to minimize the possibility of diurnal- or circadian-induced variations in performance.

- *Tester's initials:* The tester should initial the data sheet to create a record of who conducted the testing session (6). This information identifies a contact person to whom one can direct questions about the session. It also enables you to match the subject and tester if differences arise between testers in the facility.

In testing procedures in the exercise physiology laboratory, the two basic variables most often assessed are height and weight. These measures can serve simply as descriptors or as integral parts of a testing program.

Height, technically referred to as *stature*, is routinely measured in most exercise physiology laboratories. Stature is generally measured with a physician's scale, a stadiometer, or a metric scale attached to a wall (6). The measurement should be made to the nearest tenth of a centimeter (0.1 cm) or to the nearest quarter or half of an inch (.25-.5 in.) and then converted to meters. For example, if stature is determined to be 5 ft 11 in., the following conversion would be made:

$$m = in. \times 0.0254$$

$$Stature\ in\ m = 71\ in. \times 0.0254 = 1.8034\ m$$

Rounding to the nearest hundredth of a meter, 1.8034 m would be reported as a value of 1.80 m. Similarly, if the height had been calculated as 1.8288 m, the reported value would be rounded to 1.83 m.

The assessment of weight is probably the most common test in the exercise physiology laboratory because it plays a role in many calculations. Body weight, which is equivalent to body mass under normal gravitational forces (6), is represented as a kilogram value in the scientific literature. Most Americans are familiar with pound measurements, but this representation of weight does not meet SI standards. To convert body mass from pounds to kilograms, use the following equation:

$$Mass\ (kg) = mass\ (lb) / 2.2046$$

Every 1 kg is equal to 2.2046 lb, a figure that is more commonly expressed as 2.2 lb. If a person weighs 225 lb, body mass would be calculated as follows:

$$Mass\ (kg) = 225\ lb / 2.2046 = 102.1\ kg$$

If this equation were used with 2.2 lb in the denominator, then the individual's body mass would be recorded as 102.3 kg. Body mass results should be rounded to the nearest tenth of a kilogram.

Once you have recorded the background information, the next step is to measure and record meteorological information about the testing environment. Typically, you will assess temperature, barometric pressure, and relative humidity (6) because they can profoundly affect the results of certain physiological and performance tests (14, 28, 33). It is well documented that high temperature can exert a significant physiological effect on test results (14, 33). In fact, it appears that HR increases 1 beat per minute (beats · min⁻¹) for every increase of 1 °C above 24 °C (33). In contrast, cold environments can increase the respiratory rate, which can negatively affect performance by increasing the risk of dehydration.

Temperature is commonly represented in units of Fahrenheit, Celsius, or Kelvin (40). Most Americans are familiar with the Fahrenheit temperature scale, in which 32 °F represents the melting point of ice (6), but this measure is not recommended by the international system (SI) that sets the standard in research settings (40).

Another common method for representing temperature is the Celsius scale, in which 0 °C represents the freezing point of water and 100 °C represents the boiling point (6, 16). To convert from Fahrenheit to Celsius, use either of the following formulas:

$$°C = (°F - 32) / 1.8$$

$$°C = 0.56 \times (°F - 32)$$

This measure of temperature is commonly seen in the scientific literature, but, like Fahrenheit, it is not the recommended SI unit (42). Instead, the SI thermal unit is the Kelvin (K), which contains no negative or below-zero temperatures. The conversion from Celsius to the Kelvin scale is accomplished by the following formula:

$$K = 273.15 + °C$$

This system represents the coldest possible temperature as 0 K (6).

The pressure of ambient air is represented as **barometric pressure** and can fluctuate with changes in altitude (23) and weather pattern (6). As barometric pressure changes, so do the partial pressures of the gases that make up ambient air (oxygen, carbon dioxide, and nitrogen). Regardless of the change in overall barometric pressure, the percentage of the gases contained in the ambient air remain constant. Ambient air contains 79.04% nitrogen (N_2), 20.93% oxygen (O_2), and 0.03% carbon dioxide (CO_2). In order to determine the partial pressures of these gases, simply multiply the total barometric pressure by the percent contribution of the gases.

Barometric pressure is measured with an aneroid or mercury barometer in units of millimeters of mercury (mmHg). For example, at sea level, barometric pressure is about 760 mmHg, and the partial pressure of oxygen is around 159 mmHg. At an elevation of 2,000 m above sea level, the barometric pressure would fall to about 596 mmHg, whereas the partial pressure of oxygen would decrease to 125 mmHg. This reduction in barometric pressure and concomitant decrease in the partial pressure of oxygen can result in a significant decrease in the ability to perform aerobic exercise (18, 28). Because of the impact of changes in barometric pressure on both pulmonary and cardiovascular function, both respiratory ventilation and metabolic volumes are often corrected for these changes (6).

In the scientific literature, barometric pressure is typically reported in the following units: mmHg, torr, hectopascal (hPa), or kilopascal (kPa). Generally, the following formulas can be used to convert the various units:

$$1 \text{ torr} = 1 \text{ mmHg}$$

$$kPa = torr \times 0.1333 = torr / 7.50$$

$$hPa = torr \times 1.333 = torr / 0.750$$

Therefore, a barometric pressure of 674 mmHg or 674 torr would be converted as follows:

$$674 \text{ torr} \times 0.1333 = 89.8 \text{ kPa}$$

$$674 \text{ torr} \times 1.333 = 898.4 \text{ hPa}$$

The **relative humidity** of the environment is the amount of water or percent saturation in the ambient air (6, 23). To quantify relative humidity in the exercise physiology laboratory, use an instrument called a *hygrometer*, which yields a percent relative humidity value (e.g., 60% relative humidity). If, for example, the ambient air is completely saturated with water vapor, then the relative humidity is represented as 100% at that temperature. Generally, the amount of water that can be contained in ambient air increases with temperature.

Relative humidity values between 20% and 60% generally do not affect exercise, but values above or below this range can influence physical performance (2). Specifically, high humidity limits the evaporative capacity of sweat, which can significantly reduce blood plasma volume and thus increase cardiovascular stress. High relative humidity can also affect thermoregulation, which can increase the effects of temperature on cardiovascular function (28). Instead of the 1 beat · min^{-1} increase in HR typically associated with a 1 °C increase in temperature above 24 °C, an increase in relative humidity can result in a 2 to 4 beats · min^{-1} increase in HR (6, 33). Because of these potential effects on physical performance, relative humidity is commonly assessed in the exercise physiology laboratory.

Heat stress can be estimated in terms of the **wet-bulb globe temperature (WBGT)**. This measure simultaneously accounts for three thermometer readings and provides a single temperature reading to estimate the cooling capacity of the surrounding environment (23). The first thermometer measurement is the dry-bulb temperature (T_{db}), which is taken with a standard thermometer and evaluates the actual air temperature. The second thermometer measurement is taken with a wet-bulb thermometer, which reflects the effect of sweat evaporating from the skin (23). With this measure, water evaporates from the bulb, which effectively lowers the temperature below that represented by the dry bulb, thus yielding the wet-bulb temperature (T_{wb}). The difference between the wet- and dry-bulb temperatures represents the environment's capacity for cooling. The third thermometer is placed in a black globe and generally has a higher temperature than that indicated by the dry bulb because the black globe absorbs heat. This globe temperature (T_g) is used to estimate the radiant heat load of the environment (23).

Once the temperature has been measured with the three thermometers, the three results can be combined to estimate the overall atmospheric challenge to body temperature in outdoor environments using the following equation:

$$WBGT = 0.1T_{db} + 0.7T_{wb} + 0.2T_g$$

Careful examination of this equation reveals that the T_{wb} reflects the importance of sweat evaporation in the physiology of heat exchange. If the relative humidity is high, this measure reflects impairment in the ability to evaporate

sweat, which in turn increases the heat load encountered by the body (23). As a general rule, if the WBGT is >28 °C (>82-83 °F), then modifications to exercise or practice should be considered—including canceling practice, moving to indoor facilities, or reducing the training intensity.

DESCRIPTIVE STATISTICS

Statistics are a mathematical method for describing and analyzing numerical data (39). Exercise physiology laboratories typically involve calculating descriptive statistics such as measures of **central tendency** and **variability**.

Central Tendency

Measures of central tendency are commonly used to present a score that best represents all of the scores collected for a group (39). The most common statistics calculated when representing central tendency are the mean, median, and mode. The **mean** is calculated by summing all the scores and dividing the result by the number of scores. The following formula represents the calculation of the mean:

$$Mean = \frac{\Sigma X}{N}$$

In this equation, X is the individual scores, and N equals the number of scores. For example, if you measured the body mass of five people as 53, 55, 65, 48, and 60 kg, the mean would be calculated as follows:

$$Mean = \frac{53+55+65+48+60}{5} = \frac{281}{5} = 56.2 \text{ kg}$$

Therefore, for this example, the mean weight for the five individual tests is 56.2 kg.

The **median** represents the middle score in a series of data. It is generally calculated using the following equation when the sample is placed in order:

$$Median = [(N + 1) / 2]\text{th score}$$

In this equation, N represents the total number of scores in the sample. For the preceding example involving five body weights—now placed in order as 48, 53, 55, 60, and 65 kg—the median would be calculated as follows:

$$\text{Median} = (5 + 1) / 2 = \text{3rd score}$$

You would then count 3 places from the first score, thus revealing the median as 55 kg. If trying to calculate the median with an even number of values, you need to first find the middle pair of numbers. For example, if the six weights are 48, 53, 55, 58, 60, and 65 kg, the preceding equation would give (6 + 1) / 2 = 3.5. You would then move in 3.5 positions and find the two middle values, 55 and 58. These would be added together, 55 + 58 = 113, and then divided by 2, which yields a median value of 56.5.

The **mode** is the most frequent score in a sample, and it is possible to have more than one mode in a group of scores. If the body masses of 10 people were measured as 48, 49, 53, 53, 55, 59, 60, 60, 60, and 62 kg, the most frequently occurring value would be 60 kg, which would thus be identified as the mode of this sample population. If each score in a sample appears with the same frequency, then the mode is undefined for that sample.

Variability

The variability of a data set allows the spread of the data to be depicted. In examining a group of data in the exercise physiology laboratory, variability is commonly examined with the use of the standard deviation and the range.

The **standard deviation (SD)** of a data set is easily calculated by many spreadsheet programs, but it can also be calculated by hand with the following equation:

$$\text{Standard deviation} = \sqrt{\frac{\Sigma(X - M)^2}{(N - 1)}}$$

In this equation, the mean (M) is subtracted from each score (X) in the group. The result for each score is then squared, and these figures are summed to provide a total, which is then divided by the number (N) of scores in the group minus 1. The square root of this is the standard deviation. An example of how one might calculate the standard deviation is presented in the accompanying highlight box; the example involves body mass measurements of five subjects.

Sample Calculation of Standard Deviation

Sample Laboratory Data Table

Sample	X	X – M	(X – M)²
1	53	–3.2	10.2
2	65	8.8	77.4
3	48	–8.2	67.2
4	60	3.8	14.4
5	55	–1.2	1.4
Σ =	281	0.0	170.8

$$\text{Mean} = \frac{\Sigma X}{N} = 281 / 5 = 56.2 \qquad SD = \sqrt{\frac{\Sigma(X - M)^2}{(N - 1)}} = \sqrt{\frac{170.8}{4}} = \sqrt{42.7} = 6.53$$

Note: SD = standard deviation; X = sample score; M = mean; N = number of samples.

Thus the mean body mass was 56.2 kg, and the standard deviation was 6.53 kg. In a results section of a manuscript or laboratory report, we would then represent these data as 56.2 ± 6.5 kg.

If the standard deviation and the mean of a group are calculated, a **z-score** can be determined in order to express the distance of any individual score in standard units from the mean (30):

$$z = \frac{\text{sample score} - \text{group mean}}{\text{group standard deviation}}$$

For example, if a group of weightlifters were tested in the isometric midthigh pull, and the group average peak force was 3013.9 N and the group standard deviation was 360.7 N, then the z-score for an athlete who had a peak force of 2679.7 would be calculated as follows:

$$z = \frac{2679.7 - 3013.9}{360.7} = \frac{-334.2}{360.7} = -0.93$$

This athlete's performance on the test would then be −0.93 standard deviations below (i.e., weaker) than the group tested.

The second measure of variability often used in the exercise physiology laboratory is **range**—the distance between the end points in a group of scores (7). The range is easily calculated with the following formula:

$$\text{Range} = (\text{high score} - \text{low score}) + 1$$

Thus, for our previous example of body mass scores, the range would be calculated as follows:

$$\text{Range} = (65 - 48) + 1 = 17 + 1 = 18$$

The high score can also be considered as the *maximum,* and the low score would be considered as the *minimum.* These variables are also sometimes reported in the results generated by performance testing in the exercise physiology laboratory.

Reliability and Validity

Whether you are conducting basic testing or using test results for research purposes, it is essential to determine the reliability and validity of the testing process. If the tests chosen do not meet both of these criteria, you may produce false information about the subject's physiological or physical performance capacity. Therefore, before using any testing procedure—whether in the field or laboratory—you must determine its reliability and validity. Typically, you can do so by using established testing protocols that have been previously validated and determined to be reliable.

Reliability refers to the repeatability or consistency of a measure (7, 19, 39). The reliability of a measure is often affected by experimental or biological errors (6). Experimental errors may include technical errors such as miscalibration of instruments or changes to the testing environment. Biological errors may include natural shifts in performance that occur in response to the time of day when testing is undertaken (35, 36) or in response to accumulated fatigue. The subject's familiarity with the testing process can also affect reliability (6). Therefore, when introducing new testing procedures, it is important that athletes and clients are adequately familiarized with the tests. This can be done via several practice trials before the start of the actual testing session.

Statistically, the primary methods for evaluating reliability are the change in the mean between measurements, the standard error of the measurement, and the intraclass correlation coefficient (ICC) (34). Highly reliable field, field/laboratory, and laboratory tests generate high ICCs and low coefficients of variation (CVs) when repeated trial data are compared (6, 19). Acceptable ICCs range between 0 and 1 or between 0 and −1. Perfect correlations are either 1 or −1 (4); they rarely occur with the variables analyzed in the exercise physiology laboratory. Generally, correlations greater than 0.90 demonstrate high reliability, whereas correlations less than 0.70 exhibit poor reliability (6). The most common reliability measure is the **typical error (TE)** of the measurement. The typical error is calculated by dividing the standard deviation of the change scores between repeated measures divided by the square root of 2 (34).

Generally, the smaller the TE, the more reliable the measure. With performance-based measures, the general marker of reliability is an ICC >0.80 and a CV (%) <10.0 (15).

Even if a measure or test is reliable, it may not be valid; in other words, rather than providing

Sample Reliability and Error Calculations

Subject	Trial 1	Trial 2	ΔT2-T1	100*LN(T1)	100*LN(T2)	ΔT2-T1
1	2679.7	2684.0	4.3	789.3	789.5	0.2
2	2775.1	2752.3	−22.8	792.8	792.0	−0.8
3	3600.5	3609.2	8.6	818.9	819.1	0.2
4	2751.7	2979.0	227.3	792.0	799.9	7.9
5	3258.3	3170.7	−87.6	808.9	806.2	−2.7
6	3297.2	3222.8	−74.4	810.1	807.8	−2.3
7	2897.2	3093.2	196.0	797.2	803.7	6.5
8	2452.9	2453.9	0.9	780.5	780.5	0.0
9	3069.2	3043.6	−25.7	802.9	802.1	−0.8
10	3357.1	3356.3	−0.8	811.9	811.9	0.0
Mean	3013.9	3036.5	22.6	800.5	801.3	0.8
SD	360.7	339.0	105.1	12.0	11.4	3.5
Reliability in raw values				Reliability as a percentage of log transformation		
Change in mean	22.6			Change in mean %	0.8	
TE	74.3			TE as a CV (%)	2.5	
ICC	0.96			ICC	0.97	

Note: TE is calculated as the SD of the change score divided by $\sqrt{2}$. In this example, TE = 105.1 / $\sqrt{2}$ = 105.1 / 1.414 = 74.3. LN = natural logarithm; SD = standard deviation; TE = typical error; ICC = intraclass correlation coefficient; CV = coefficient of variation.

a suitable assessment of what you are trying to measure, your measures might be dependable but consistently wrong. Therefore, you must determine that a test is both reliable and valid before using it in a clinical or sport testing program.

Validity is the degree to which a test or instrument measures what it is supposed to measure (39). In the exercise physiology laboratory, the main form of validity is criterion validity—the ability of a test to be related to a recognized standard or criterion measure (39). For example, when examining body composition, UWW is considered the gold standard or criterion measure. When considering another method of assessing body composition—say, skinfold measurements—validity is determined by correlating the results obtained from the skinfold estimate to the body fat as determined by underwater weighing. Typically, this correlation is performed with the use of a Pearson product-moment correlation coefficient, and a high correlation (*r*) between the criterion variable and the test represents high validity.

A measurement is considered valid if it is both accurate and precise. The closer a measurement is to an accepted value, the higher the **accuracy** of the measurement. For example, if multiple arrows are shot at a target, the closeness of the arrows to the bull's-eye indicates the accuracy of the measure. If the arrows are clustered close to one another, then there would be a high degree of **precision**. Thus, precision is considered to be the reproducibility of the measurement.

Ultimately, a measure can be highly accurate but not precise, highly precise but not accurate, highly accurate and highly precise, or neither highly precise nor highly accurate (see figure 1.1).

PRESENTATION OF RESULTS

Data collected on a group of subjects can be presented in many ways in a publication or laboratory report. One of the first steps in determining how to present data is to decide whether it should be placed in a table, a figure, or the text of the document. Consider whether you can best represent the information in the form of the actual numbers collected or as a picture of the results. Once you make this decision, you can create a table or figure to best represent the data.

Though it is relatively easy to create a table with modern software, it remains important to consider some basic rules. First, tables provide a means for communicating information about data—not storing them (39). For example, a good use of tables is creating normative data tables for evaluating testing. You can use these tables for communicating information about typical results for any given field or laboratory test. When creating a table, follow these rules suggested by Thomas, Nelsen, and Silverman (39):

- Table headings should be clearly labeled and easy to follow. To improve readability of headings, avoid excessive abbreviation and make formatting easy to follow.
- Like characteristics should be presented vertically. The columns and rows contained in the table should make sense. For example, columns could contain variables, means, and standard deviations, and each row could represent a specific variable.
- Readers should not have to refer to the text to understand the content of a table. As a rule, a table should stand alone and not require the reader to search for abbreviations or text to explain what the table is presenting.

Table 1.2 shows an example that meets these rules and presents basic data collected in an exercise physiology laboratory. This tabular format can be modified to fit many data sets.

Another way to present data is by using a figure. Deciding whether to place data in table or figure form depends largely on whether a picture of the results (i.e., a figure) will work better than the actual numbers (presented in a table). If the data are better represented as a figure, consider the following rules (39):

- Make sure the figure is clearly labeled and easy to read.
- Present important information so that it is easily evaluated.

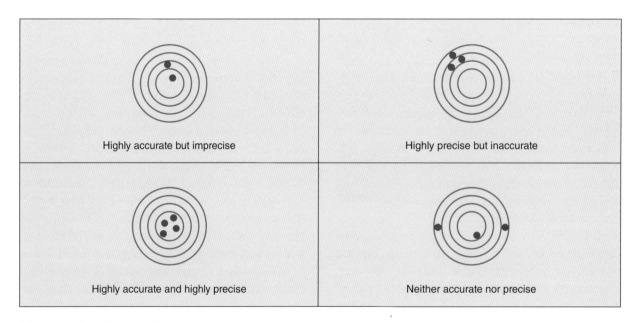

Figure 1.1 The relationship between accuracy and precision.

Table 1.2 Sample Laboratory Table

Characteristics	Mean	SD
Age (y)	22.2	±1.0
Height (m)	1.72	±0.02
Weight (kg)	83.5	±5.6
Percent body fat (%)	8.2	±3.1
$\dot{V}O_2$max (ml · kg^{-1} · min^{-1})	45.2	±2.2
1RM back squat (kg)	110.2	±10.5
1RM bench press (kg)	85.4	±4.6

1RM = 1-repetition max; SD = standard deviation.

- Create a figure that is free from visual distractions.

Determine what type of figure will best represent the data. Types include bar and column charts, line graphs, scatter plots, flow charts, and pie charts.

- *Bar and column charts:* These figures are useful when comparing single responses—often a mean or a single most important time point—between treatment groups. Though not the best method for determining trends over time, column charts can be used for comparing amounts over time so that trends can be seen or for comparing variables between groups. You can often use shading or coloring to distinguish between columns and thus help your audience easily interpret the data. You can also depict standard deviations and standard errors of the mean in order to make the chart more descriptive.

- *Line graphs:* This type of graphical representation is often used to present longitudinal data and show how things change over time. When creating charts of this type, time is often placed on the *x*-axis and the variable being measured or quantified on the *y*-axis. When multiple lines are placed on these charts, it is prudent to use shading, differing symbols, and broken or colored lines to allow the reader to easily distinguish between variables.

- *Scatter plots:* In this type of chart, each point on the plot represents a data point on the *x*-axis and the *y*-axis. As a result, it is easy to see the patterns of indi- vidual scores. Because scatter plots illustrate individual data points, researchers can use them to gain a sense of the spatial distribution of the results and to determine linearity, outliers, or clumps of data. Researchers often facilitate interpretation of the data by using multiple colors or symbol shapes to distinguish between groups.

- *Flow charts:* These charts are useful in depicting procedures or steps within a process. This type of representation can often be used in methods sections, especially when depicting decisions and concomitant steps.

- *Pie charts:* This type of graphic representation can show proportions within a whole. The entire circle represents the whole, and each segment represents a percentage of the whole. For example, a pie chart might be used to show the leading causes of death from cardiovascular disease, with each segment representing a percentage of the overall death rate. Pie charts should generally contain a maximum of five segments or variables. Each segment should be shaded so that the segments can be easily distinguished from one another.

INTERPRETATION OF DATA

Once you have collected the data, you need to interpret them in order to develop meaning from them. In clinical settings, the data collected in the exercise physiology laboratory are generally compared with normative data or standards (6).

When examining data, **inferential statistics** can be used to draw conclusions about a population from a sample population. For example, if a football team is put through a battery of tests, and it is assumed that this team (sample) is representative of all football players (population), then the test results can be used to make inferences about football players (population). A key aspect of inferential statistics is that the sample is representative of the population (30).

When a population is normally distributed, the mean, median, and mode are all the same, and the data will look similar to figure 1.2 when graphed. **Normative data** are used to derive percentile ranks and standard deviations of the data, which are then associated with a descriptive category such as poor, average, or excellent. In figure 1.2, normative data are presented as a normal curve, wherein the middle portion contains the majority of the population (68%), which would correspond to the 50th percentile. Thus, subjects whose results fell within this area of the curve would be considered average. For example, if the average bench press (50th percentile) is 1.06 (1-repetition maximum [1RM] / body mass) and our subject achieves a result of 1.22, that individual would be above average or within the 70th percentile (see table 12.2). Another method that could be used to interpret subjects' test results is to convert their score to a z-score based on the population they are in and report the distance of their results from the mean.

Standards and *normative data* are often wrongly used interchangeably (6). A **standard** is better used to describe a desirable or target score (3), whereas normative data represent a subject's placement within a population (6). Standards are often set for items such as the recommended quantity and quality of exercise for maintaining a minimal level of physical

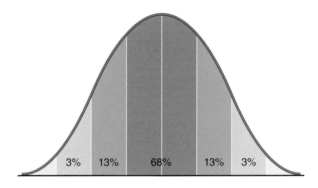

Figure 1.2 The normal curve.

fitness (3). In many cases, both standards and normative data are presented for a given population in order to facilitate interpretation of the subject's results (6).

Another way to interpret test results is to use **magnitude statistics** (30). These statistics are often more useful to practitioners because they allow for interpretation of the clinical significance of a test (20, 30). Typically, the smallest worthwhile change and the effect size are used to make magnitude-based inferences about test data.

If the tests used in the exercise physiology laboratory are reliable and valid, meaningful changes in performance can easily be tracked. One method for tracking meaningful changes in performance is to calculate the smallest worthwhile change. The **smallest worthwhile change** refers to the ability of a test to detect the smallest practically important change in the performance assessed (30). Though there are many ways to calculate the smallest worthwhile change, the typical method is to multiply the between-subject standard deviation by 0.20 (20). For example, if the standard deviation in our group of weightlifters' isometric midthigh pull data is 360.7 N, then the smallest worthwhile change for this group of athletes is 72 N (0.2 × 360.7). Thus, if our athlete is tested after a period of training and his or her performance increases by 120 N, that would be considered a meaningful training adaptation.

The **effect size** is useful for analyzing group performance following a training program or for making comparisons between groups of athletes (13). For example, if a group of subjects undertook a 12 wk resistance training program, the effect of the program would be calculated as the difference or change between the mean performance scores of the pre- and posttest divided by the standard deviation of the pretest:

Effect size = (mean posttest − mean pretest) / SD pretest

In our example of the athletes who undertook a 12 wk resistance training program, their pre-training bench press 1RM average was 103 kg (SD = 5 kg), and their posttraining bench press 1RM increased to 110 kg.

Based on these results, the effect size would be 1.4 (i.e., [110 − 103] / 5). Several scales have been developed to interpret effect size statistics (13, 20), but reference values for small (0.2), moderate (0.6), large (1.2), and very large (2.0) are generally used when interpreting data (13, 20). When interpreting the effect size of our example, the practitioner could thus conclude that the training program had a large effect on bench press strength.

Go to the web study guide to access electronic versions of individual and group data sheets, the question sets from each laboratory activity, and case studies for each laboratory activity.

Basic Data

Equipment

- Platform scale (or digital or other scale)
- Certified weights for calibrating scale
- Stadiometer (wall or freestanding) or physician's scale with attached anthropometer
- Mercury or aneroid barometer
- Hygrometer and hygrometer chart
- Laboratory thermometer
- Sliding caliper
- Screwdriver
- Individual data sheet
- Group data sheet

Background Information

In most testing situations, you can follow these steps when collecting basic informational data.

Step 1: Note the subject's name or identification number, as well as the time and date of the testing session, in the appropriate locations on the individual data sheet.

Step 2: Indicate the sex and age of the subject on the individual data sheet, and transfer the subject's age to the group data sheet.

Step 3: Initial the individual data sheet in the appropriate location.

Height

Step 1: Verify the accuracy of the stadiometer by measuring the distance from the floor to the base to the horizontal headboard. If using a physician's scale, measure the distance from the scale platform to the hinged horizontal level.

Step 2: Have the subject remove his or her shoes and then stand with his or her back to the stadiometer and heels placed evenly apart. If using a wall-mounted stadiometer, have the subject stand with the buttocks, scapulae, and heels in contact with the wall.

Step 3: Instruct the subject to stand as tall as possible, inhale, and hold his or her breath while looking straight ahead.

Step 4: Lower the headboard so that it rests atop the subject's head and is perpendicular to the measuring scale. If using a physician's scale, move the hinged lever so it is perpendicular to the graduated vertical rod while resting evenly on top of the subject's head.

Step 5: Measure the subject's height to the nearest tenth of a centimeter (0.1 cm) or to the nearest eighth or quarter of an inch, convert it to meters, and record this value on the individual and group data sheets.

Step 6: Repeat steps 2 through 5 twice for a total of three measurements.

Step 7: To begin interpreting the data, convert (if necessary) the measured height from inches to centimeters and then calculate the average of the three trial measures.

Step 8: Compare the results with the normative data presented in table 1.3 for average height of Americans, and record the result on the individual data sheet.

Step 9: Calculate a difference score by subtracting the average height from the measured height and transfer to the individual data sheet.

Step 10: Calculate the percent difference with the following equation and transfer to the individual data sheet.

$$\text{Percent difference} = \frac{\text{actual height (m)} - \text{average height (m)}}{\text{average height (m)}} \times 100$$

Table 1.3 Average Height of Americans

	Age (y)	MALES		FEMALES	
		in.	m	in.	m
Young children	2	35.9	0.912	35.5	0.902
	3	38.8	0.986	38.4	0.975
	4	41.9	1.064	41.7	1.059
	5	44.5	1.130	44.3	1.125
	6	46.9	1.191	46.1	1.171
	7	49.7	1.262	49.0	1.245
	8	52.2	1.326	51.5	1.308
Preadolescents	9	54.4	1.382	53.9	1.369
	10	55.7	1.415	56.4	1.433
	11	58.5	1.486	59.6	1.514
	12	60.9	1.547	61.4	1.560
Adolescents	13	63.1	1.603	62.6	1.590
	14	66.3	1.685	63.7	1.618
	15	68.4	1.738	63.8	1.621
	16	69.0	1.753	63.8	1.620
	17	69.0	1.753	64.2	1.631
Adults	18	69.5	1.765	64.2	1.631
	19	69.6	1.768	64.2	1.631
	20-29	69.6	1.768	64.2	1.631
	30-39	69.5	1.765	64.2	1.631
	40-49	69.7	1.770	64.3	1.633
	50-59	69.2	1.758	63.9	1.623
	60-74	68.6	1.742	63.0	1.600
	≥75	67.4	1.712	62.0	1.575

Reprinted, by permission, from J. Hoffman, 2006, *Norms for fitness, performance, and health* (Champaign, IL: Human Kinetics), 82; Data from C.L. Ogden et al., 2004, *Mean body weight, height, and body mass index, United States 1960-2002*, Advance data from vital and health statistics, no. 347. (Hyattsville, MD: National Center for Health Statistics).

Weight Scale Calibration

Scales come in several forms, including digital scales, which contain a load cell; spring resistance scales, such as a Chatillon scale, which measures weight by means of pressure resistance on an internal spring; and balance-beam scales, typically referred to as *physician's scales* (1). Calibration procedures depend largely on the type of scale. Every digital scale is slightly different, but in most cases calibration is initiated by turning the device on and off, by pressing the reset button, or by pressing the calibration button. For a Chatillon scale, to tare or set the zero point, a screw or knob is turned until the unweighted device displays a zero. Once the scale is zeroed, the tester can verify that it is calibrated using the procedures described in the following.

Simply zeroing a scale is not enough to ensure that the scale is appropriately calibrated. It is also important to verify the accuracy of the scale at three points—zero, midpoint, and high point—to ensure its accuracy across a variety of weights. The steps for calibrating a scale are as follows (4, 6):

Step 1: To set the zero point, move the beam weights to the zero points.

Step 2: Observe the position of the pointer. If the zero point is set, the pointer at the end of the beam should sit midway between the top and bottom of the pointer window. If the pointer is centered, perform steps 5 through 7. If the pointer is not centered, follow the next calibration step.

Step 3: Adjust the pointer by manipulating the tare screw weight with a screwdriver. If the pointer is below the middle position, turn the tare screw clockwise to move the pointer upward; if the pointer is above the middle position, turn the tare screw counterclockwise to move the pointer downward. Fine-tune the position of the pointer by adjusting the tare screw until the pointer rests in the middle of the pointer window.

Step 4: Repeat steps 1 through 3 twice for a total of three times. Then perform the verification-of-calibration procedure described in steps 5 through 7.

Step 5: To perform the high-point calibration, place the heaviest certified weight on the scale (with some scales, you may have to hang the weights). In most instances, this weight should correspond to the heaviest person you might weigh.

Step 6: Record the actual weight of the certified weight and the weight measured by the scale in the appropriate locations on the individual data sheet.

Step 7: If the actual weight and the weight measured are the same, the scale is accurate at the high-point range. If not, adjust the scale so that it reads correctly.

Step 8: Calculate a difference score between the actual and measured weights and use this score to create a correction factor. For example, if the scale consistently weighs 2 kg heavier at weights over 120 kg, note that people who weigh more than 120 kg should have 2 kg subtracted from their weight.

Step 9: To perform the midpoint calibration, estimate the average weight of the subjects you will measure, and then place this amount of certified weight on the scale.

Step 10: Record the actual weight of the certified weight and the weight measured by the scale in the appropriate locations on the individual data sheet.

Step 11: If the actual weight and the weight measured are the same, the scale is accurate at the midpoint range. If not, adjust the scale so it reads correctly.

Step 12: Record the actual weight of the certified weight and the weight measured by the scale in the appropriate locations on the individual data sheet.

Step 13: Calculate a difference score between the actual and measured weights and use it to create a correction factor.

Step 14: Repeat steps 5 through 13 twice for a total of three times.

Body Weight

Step 1: Determine whether the scale is calibrated; if it is not, follow the steps listed in the Weight Scale Calibration section.

Step 2: After completing the calibration, have the subject remove as much clothing as possible, including shoes and jewelry, in order to obtain the most accurate assessment. It is best to weigh the subject without clothing or have the subject wear a paper gown, which is considered to be weightless. If this is not possible, another option is to weigh the subject's weigh-in clothing to the nearest 0.25 lb (0.02 kg) and subtract this value from the total weight measured when the subject is wearing the items. If it is not possible to weigh the subject's clothing, then the subject should be weighed while wearing as little clothing as possible.

Step 3: Have the subject stand on the scale while nude, wearing a minimal amount of clothing, or wearing the weigh-in clothing. Stand behind the platform scale to adjust the beam weights and accurately read the measurement without touching the subject.

Step 4: If using a beam scale, adjust the beam weights until they are balanced and an accurate reading can be made of the subject's body weight. Record this weight in the appropriate location on the individual and group data sheets.

Step 5: Repeat steps 2 through 4 twice for a total of three measurements.

Step 6: Determine the subject's frame size by using the sliding caliper to measure elbow breadth. Have the subject extend the right arm forward, perpendicular to the body, and bend the arm to 90° so that the fingers point upward and the hand points away from the body. Determine the breadth of the elbow by measuring the distance between the two prominent bones on either side of the elbow (i.e., measuring the widest point). Record the inch value for the elbow breadth in the appropriate location on the individual data sheet.

Step 7: To begin interpreting the body mass data, determine the subject's frame size by comparing the elbow breadth measurement with the values presented in table 1.4. Indicate the appropriate size on the individual data sheet.

Step 8: Compare the actual weight measured with the weight recommendations from the National Institutes of Health (NIH) shown in table 1.5. Record the range and determine whether the subject falls into, below, or above the recommended range.

Step 9: Using the frame size estimation, compare the measured weight with the ranges presented in table 1.6 (for men) or 1.7 (for women), and record the corresponding range on the individual data sheet. Indicate whether the subject falls into, below, or above the recommended range.

Table 1.4 Metropolitan Life Insurance Frame Size Chart

| | HEIGHT | | ELBOW BREADTH | | | | | |
| | | | SMALL FRAME | | MEDIUM FRAME | | LARGE FRAME | |
	in.	m	in.	cm	in.	cm	in.	cm
Men	61-62	1.549-1.575	<2.5	<6.4	2.5-2.88	6.4-7.3	>2.88	>7.3
	63-66	1.600-1.676	<2.63	<6.7	2.63-2.88	6.7-7.3	>2.88	>7.3
	67-70	1.702-1.778	<2.75	<7.0	2.75-3.0	7.0-7.6	>3.0	>7.6
	71-74	1.803-1.880	<2.75	<7.0	2.75-3.13	7.0-8.0	>3.13	>8.0
	≥75	≥1.905	<2.88	<7.3	2.88-3.25	7.3-8.3	>3.25	>8.3
Women	57-58	1.448-1.473	<2.25	<5.7	2.25-2.5	5.7-6.4	>2.5	>6.4
	59-62	1.499-1.575	<2.25	<5.7	2.25-2.5	5.7-6.4	>2.5	>6.4
	63-66	1.600-1.676	<2.38	<6.0	2.38-2.63	6.0-6.7	>2.63	>6.7
	67-70	1.702-1.778	<2.38	<6.0	2.38-2.63	6.0-6.7	>2.63	>6.7
	≥71	≥1.803	<2.5	<6.4	2.5-2.75	6.4-7.0	>2.75	>7.0

Height is measured without shoes. Elbow breadth is measured by extending the right arm forward, perpendicular to the body, and bending the arm to 90° so that the fingers point upward and the hand points away from the body. The breadth of the elbow is determined by measuring the distance between the two prominent bones on either side of the elbow (i.e., measuring the widest point).

Adapted from the Metropolitan Life Insurance Company 1983.

Table 1.5 NIH Weight Recommendations

| HEIGHT | | | AGE 19 TO 34 y | | AGE 35 y OR OLDER | |
ft, in.	in.	m	lb	kg	lb	kg
5'0"	60	1.52	97-128	44-58	108-138	49-63
5'1"	61	1.55	101-132	46-60	111-143	51-65
5'2"	62	1.58	104-137	47-62	115-148	52-67
5'3"	63	1.60	107-141	49-64	119-152	54-69
5'4"	64	1.63	111-146	50-66	122-157	56-71
5'5"	65	1.65	114-150	52-69	126-162	57-74
5'6"	66	1.68	118-155	54-71	130-167	59-76
5'7"	67	1.70	121-160	55-73	134-172	61-78
5'8"	68	1.73	125-164	57-75	138-178	63-81
5'9"	69	1.75	129-169	59-77	142-183	65-83
5'10"	70	1.78	132-174	60-79	146-188	66-86
5'11"	71	1.80	136-179	62-81	151-194	69-88
6'0"	72	1.83	140-184	64-84	155-199	71-91
6'1"	73	1.85	144-189	66-86	159-205	72-93
6'2"	74	1.88	148-195	67-89	164-210	75-96
6'3"	75	1.91	152-200	69-91	168-216	76-98
6'4"	76	1.93	156-205	71-93	173-222	79-101
6'5"	77	1.96	160-211	73-96	177-228	81-104
6'6"	78	1.98	164-216	75-98	182-234	83-106

Height and weight are measured without shoes.

Data from American College of Sports Medicine 2010; US Department of Health and Human Services.

Table 1.6 Metropolitan Life Insurance Height and Weight Table for Men

HEIGHT			SMALL FRAME		MEDIUM FRAME		LARGE FRAME	
ft, in.	in.	m	lb	kg	lb	kg	lb	kg
5'2"	62	1.58	128-134	58-61	131-141	60-64	138-150	63-68
5'3"	63	1.60	130-136	59-62	133-143	61-65	140-153	64-69
5'4"	64	1.63	132-138	60-63	135-145	61-66	142-156	65-71
5'5"	65	1.65	134-140	61-64	137-148	62-67	144-160	66-73
5'6"	66	1.68	136-142	62-65	139-151	63-69	146-164	66-75
5'7"	67	1.70	138-145	63-66	142-154	65-70	149-168	68-76
5'8"	68	1.73	140-148	64-67	145-157	66-71	152-172	69-78
5'9"	69	1.75	142-151	65-69	148-160	67-73	155-176	71-80
5'10"	70	1.78	144-154	66-70	151-163	68-74	158-180	72-82
5'11"	71	1.80	146-157	66-71	154-166	70-76	161-184	73-84
6'0"	72	1.83	149-160	68-73	157-170	71-77	164-188	75-86
6'1"	73	1.85	152-164	69-75	160-174	73-79	168-192	76-87
6'2"	74	1.88	155-168	71-76	164-178	75-81	172-197	78-90
6'3"	75	1.91	158-172	72-78	167-182	76-83	176-202	80-92
6'4"	76	1.93	162-176	74-80	171-187	78-85	181-207	82-94

Weights are based on height and frame size for men wearing indoor clothing of 5 lb (2.3 kg) and 2 in. (5.1 cm) heels.

Adapted from the Metropolitan Life Insurance Company 1983.

Table 1.7 Metropolitan Life Insurance Height and Weight Table for Women

HEIGHT			SMALL FRAME		MEDIUM FRAME		LARGE FRAME	
ft, in.	in.	m	lb	kg	lb	kg	lb	kg
4'10"	58	1.47	102-111	46-51	109-121	50-55	118-131	54-60
4'11"	59	1.50	103-113	47-51	111-123	51-56	120-134	55-61
5'0"	60	1.52	104-115	47-52	113-126	51-57	122-137	56-62
5'1"	61	1.55	106-118	48-54	115-129	52-59	125-140	57-64
5'2"	62	1.58	108-121	49-55	118-132	54-60	128-143	58-65
5'3"	63	1.60	111-124	51-56	121-135	55-61	131-147	60-67
5'4"	64	1.63	114-127	52-58	124-138	56-63	134-151	61-69
5'5"	65	1.65	117-130	53-59	127-141	58-64	137-155	62-71
5'6"	66	1.68	120-133	55-61	130-144	59-66	140-159	64-72
5'7"	67	1.70	123-136	56-62	133-147	61-67	143-163	65-75
5'8"	68	1.73	126-139	57-63	136-150	62-68	146-167	66-76
5'9"	69	1.75	129-142	59-65	139-153	63-70	149-170	68-77
5'10"	70	1.78	132-145	60-66	142-156	65-71	152-173	69-79
5'11"	71	1.80	135-148	61-67	145-159	66-72	155-176	71-80
6'0"	72	1.83	138-151	62-69	148-162	67-74	158-179	72-81

Weights are based on height and frame size for women wearing indoor clothing of 3 lb (1.4 kg) and 2 in. (5.1 cm) heels.

Adapted from the Metropolitan Life Insurance Company 1983.

Environmental Information

One of the first things to do in the testing environment is collect basic meteorological data.

Temperature

Step 1: Use a thermometer to measure the temperature of the laboratory in degrees Fahrenheit, and record this value on the individual data sheet.

Step 2: Convert the temperature from degrees Fahrenheit to degrees Celsius and note the result on the individual data sheet. Next, convert the Celsius value to Kelvin and note the result in the appropriate location on the individual data sheet.

Barometric Pressure

Step 1: Use a mercury or aneroid barometer to determine the barometric pressure in mmHg, and record this value on the individual data sheet.

Step 2: Convert the barometric pressure in mmHg to torr, kPA, and hPa values and note the results on the individual data sheet.

Relative Humidity

Step 1: Use a hygrometer to measure the relative humidity of the laboratory. To use this device, fill the water bottle and attach it to the device. Hang it on the wall and wait 5 to 10 min.

Step 2: Read and record both the wet- and dry-bulb temperatures in Celsius on the individual data sheet.

Step 3: Subtract the wet-bulb temperature from the dry-bulb temperature to get the wet-bulb depression value. Then determine the relative humidity from the hygrometer chart by finding the intersection between the wet-bulb depression and the dry-bulb temperature.

Step 4: Record the relative humidity on the individual data sheet.

Question Set 1.1

1. Explain the calibration process and discuss the results achieved during this process.
2. What are the SI units for length, mass, and volume? How does one convert to these values from the American system?
3. Explain the process for assessing body mass. How can the method used affect the accuracy of the measurement?
4. How does the average height of the group compare with the normative data tables presented in this text?
5. What does the percent difference score indicate?
6. How do the average body weight results for the males and females you measured in this laboratory activity compare with the Metropolitan Life Insurance ranges? With the NIH recommendations?

LABORATORY ACTIVITY 1.1

INDIVIDUAL DATA SHEET

Name or ID number: _____ Date: _____

Tester: _____ Time: _____

Sex: M / F (circle one)   Age: _____ y

Height: _____ in. _____ m

Weight: _____ lb _____ kg

Temperature, Barometric Pressure, and Relative Humidity

Temperature: _____ °F / 1.8 = _____ °C + 273.15 = _____ K

Barometric pressure: _____ mmHg = _____ torr × 1.33 = _____ hPA = _____ kPA

Relative humidity: $\underset{\text{dry-bulb temp °C}}{\underline{\hspace{2cm}}}$ – $\underset{\text{wet-bulb temp °C}}{\underline{\hspace{2cm}}}$ – $\underset{\text{wet-bulb depression °C}}{\underline{\hspace{2cm}}}$ = _____ % humidity

Scale Calibration

Zero-Point Calibration Verification

Trial 1 known mass: _____ lb _____ kg   Measured mass: _____ lb _____ kg

Trial 2 known mass: _____ lb _____ kg   Measured mass: _____ lb _____ kg

Trial 3 known mass: _____ lb _____ kg   Measured mass: _____ lb _____ kg

Average weight: _____ kg

Difference score: $\underline{\hspace{3cm}}$ = $\underset{\text{known mass}}{\underline{\hspace{3cm}}}$ – $\underset{\text{average mass}}{\underline{\hspace{3cm}}}$

High-Point Calibration Verification

Trial 1 known mass: _____ lb _____ kg   Measured mass: _____ lb _____ kg

Trial 2 known mass: _____ lb _____ kg   Measured mass: _____ lb _____ kg

Trial 3 known mass: _____ lb _____ kg   Measured mass: _____ lb _____ kg

Average weight: _____ kg

Difference score: $\underline{\hspace{3cm}}$ = $\underset{\text{known mass}}{\underline{\hspace{3cm}}}$ – $\underset{\text{average mass}}{\underline{\hspace{3cm}}}$

Midpoint Calibration Verification

Trial 1 known mass: _____ lb _____ kg   Measured mass: _____ lb _____ kg

Trial 2 known mass: _____ lb _____ kg   Measured mass: _____ lb _____ kg

Trial 3 known mass: _____ lb _____ kg   Measured mass: _____ lb _____ kg

Average weight: _____ kg

Difference score: _____ = _____ − _____
                                            known mass   average mass

Height Assessment

Trial 1: _____ in. × 0.0254 = _____ m

Trial 2: _____ in. × 0.0254 = _____ m

Trial 3: _____ in. × 0.0254 = _____ m

Average: _____ in.   _____ m

Average height from table 1.3 for _____ y old = _____ in. _____ m

Difference score calculation: _____ − _____ = _____
                                          actual height (m)   average height (m)   difference score

Percent difference: = (_____ / _____) × 100 = _____
                               difference score   average height (m)

Weight Assessment

Trial 1:   (_____ lb − _____ lb) / 2.2205 = _____ kg
                                         weigh-in clothes*

Trial 2:   (_____ lb − _____ lb) / 2.2205 = _____ kg
                                         weigh-in clothes*

Trial 3:   (_____ lb − _____ lb) / 2.2205 = _____ kg
                                         weigh-in clothes*

Average: _____ lb   _____ kg

*If weigh-in clothes are not measured, simply use a zero here.

Elbow breadth: _____ in. × 2.54 = _____ cm

Frame size: _____

NIH weight recommendation range: _____ (actual weight) is _____ range.

Metropolitan Life Insurance height and weight range: _____ (actual height) is _____ range.

Statistical Procedures

Equipment

- Data collected in laboratory activity 1.1
- Calculator
- Individual data sheet
- Group data sheet
- Microsoft Excel or equivalent spreadsheet software

Calculating the Mean

Once the basic data have been collected in laboratory 1.1, they can be analyzed. Start this lab activity by gathering basic data (i.e., age, height, weight, temperature, barometric pressure, relative humidity) for the individual data sheet as described in laboratory activity 1.1.

Hand Calculation

Step 1: Note the number of subjects for whom data were collected in the appropriate location on the individual data sheet.

Step 2: Sum all scores recorded in laboratory 1.1 and note the result as variable X on the individual data sheet.

Step 3: Calculate the mean for these scores and mark the result on the individual data sheets for laboratory 1.2. The following formula should be used:

$$\text{Mean} = \frac{\Sigma X}{N}$$

Step 4: If not performing the Excel collection method, then transfer the mean score to the group data sheet.

Excel Calculation

Step 1: Enter the data presented in table 1.2 into an Excel spreadsheet. The rows should represent subjects, and the columns should represent variables. Place the Subject label in cell A2, highlight cells B1 and C1, and select the Merge Cells function located on the toolbar under the Alignment tab. Use the same process to merge cells D1 and E1. Now, label cells B2 and D2 as Male, C2 and E2 as Female, B1 and C1 as Height, and D1 and E1 as Weight. See table 1.8 (rows 1-11) for an example of what your data table should look like. (The other rows are described in the steps that follow.)

Step 2: In a cell under the last subject's variables, enter the following formula:

=AVERAGE(number1:number2)

Number1 indicates the first number of the range, and number2 indicates the last number of the range. In the sample shown in table 1.8, the formula appears in cells B11, C11, D11, and E11.

Step 3: Transfer the mean scores to the group data sheet.

Table 1.8 Sample Format and Formula Placements in a Spreadsheet

	A	B	C	D	E
1		HEIGHT (m)		WEIGHT (kg)	
2	Subject	Male	Female	Male	Female
3	1				
4	2				
5	3				
6	4				
7	5				
8	6				
9	7				
10	8				
11	Mean	=AVERAGE(B3:B10)	=AVERAGE(C3:C10)	=AVERAGE(D3:D10)	=AVERAGE(E3:E10)
12	Median	=MEDIAN(B3:B10)	=MEDIAN(C3:C10)	=MEDIAN(D3:D10)	=MEDIAN(E3:E10)
13	Mode	=MODE(B3:B10)	=MODE(C3:C10)	=MODE(D3:D10)	=MODE(E3:E10)
14	SD	=STDEV(B3:B10)	=STDEV(C3:C10)	=STDEV(D3:D10)	=STDEV(E3:E10)

Once a formula is entered into a cell, it will disappear, and the calculated values will appear in its place.

Calculating the Median and the Mode

The median and mode offer two other methods besides the mean for examining the central tendency of a data set.

Hand Calculation

Step 1: Place all of the scores for height and weight in order from lowest to highest for each group of data (male and female).

Step 2: Calculate the location of the median score for each group using the following formula:

$$\text{Median} = [(N + 1) / 2]\text{th term}$$

Step 3: Determine the median score for males and the median score for females, and record them in the appropriate locations on the individual and group data sheets.

Step 4: Calculate the mode by determining which value occurs most frequently. Note the result on the individual and group data sheets.

Excel Calculation

Step 1: Use the following equations:

$$=\text{MEDIAN(number1:number2)}$$

$$=\text{MODE(number1:number2)}$$

Number1 indicates the first number in the range, and number2 indicates the last number. Place the median formula in cells B12, C12, D12, and E12. Place the mode formula in cells B13, C13, D13, and E13. Your spreadsheet will be formatted as shown in table 1.8 (rows 12 and 13).

Step 2: Transfer the results to the individual and group data sheets.

Calculating the Standard Deviation

The standard deviation (SD) is one of the most commonly reported indicators of variability in a data set.

Hand Calculation

Step 1: Sum all the results for the variable being evaluated (X).

Step 2: Subtract the mean, calculated previously, from each variable collected ($X - M$). Hint: This must be done for each subject. If summed, the value should be 0.

Step 3: Square the value calculated in step 2 [$(X - M)^2$]. Then sum all the values.

Step 4: Use the following equation to calculate the standard deviation:

$$\text{Standard deviation} = \sqrt{\frac{\Sigma(X-M)^2}{(N-1)}}$$

Step 5: Record the results in the appropriate location on the individual and group data sheets.

Excel Calculation

Place the following formula in the cell underneath the data column:

$$=\text{STDEV(number1:number2)}$$

See row 14 of the sample spreadsheet in table 1.8.

Calculating the Range

The range of scores in a data set can also be used to depict the variability of the set.

Step 1: Determine the high and low scores within the data set for all the height and weight data: height for men, weight for men, height for women, and weight for women.

Step 2: For each data set, subtract the low score from the high score.

Step 3: Add 1 to each of the scores calculated in step 2.

Step 4: Record the range for each set in the appropriate location on the individual and group data sheets.

Calculating the Typical Error

The typical error (TE) can be used to determine the reliability of your data set.

Hand Calculation

Step 1: Determine the standard deviation for the change scores (trial 2 – trial 1) between trial 1 and 2 for both height and weight measurements as outlined in the previous section.

Step 2: Determine the typical error for height and weight measurements with the following formula:

$$\text{Typical error} = \frac{\text{standard deviation (trial 2 – trial 1)}}{\sqrt{2}}$$

Step 3: Record the calculated typical error values in the appropriate location on the individual and group data sheets.

Excel Calculation

To calculate the typical error in Excel, use the following equation:

$$=(\text{STDEV(number1:number2)/SQRT(2))}$$

Question Set 1.2

1. Are the height and weight data collected in this laboratory normally distributed? How can you tell?

2. When working with Excel, what formulas are used for mean, median, mode, and standard deviation?

3. Do the men or the women in the class have higher body mass? Was this expected? How do their heights compare? Explain the possible reasons for these results.

LABORATORY ACTIVITY 1.2

INDIVIDUAL DATA SHEET

Name or ID number: _____ Date: _____

Tester: _____ Time: _____

Sex: M / F (circle one) Age: _____ y Height: _____ in. _____ m

Weight: _____ lb _____ kg Temperature: _____ °F _____ °C

Barometric pressure: _____ mmHg Relative humidity: _____ %

Hand Calculation of the Mean

Male Data

Number of subjects: _____ = N

Height (m)

Trial 1: _____ = X Mean = $\dfrac{\underline{\hspace{2cm}}}{\text{sum of trials}} \Big/ \dfrac{\underline{\hspace{1cm}}}{N} = \dfrac{\underline{\hspace{1.5cm}}}{\bar{x}}$

Trial 2: _____ = X Mean = $\dfrac{\underline{\hspace{2cm}}}{\text{sum of trials}} \Big/ \dfrac{\underline{\hspace{1cm}}}{N} = \dfrac{\underline{\hspace{1.5cm}}}{\bar{x}}$

Trial 3: _____ = X Mean = $\dfrac{\underline{\hspace{2cm}}}{\text{sum of trials}} \Big/ \dfrac{\underline{\hspace{1cm}}}{N} = \dfrac{\underline{\hspace{1.5cm}}}{\bar{x}}$

Weight (kg)

Trial 1: _____ = X Mean = $\dfrac{\underline{\hspace{2cm}}}{\text{sum of trials}} \Big/ \dfrac{\underline{\hspace{1cm}}}{N} = \dfrac{\underline{\hspace{1.5cm}}}{\bar{x}}$

Trial 2: _____ = X Mean = $\dfrac{\underline{\hspace{2cm}}}{\text{sum of trials}} \Big/ \dfrac{\underline{\hspace{1cm}}}{N} = \dfrac{\underline{\hspace{1.5cm}}}{\bar{x}}$

Trial 3: _____ = X Mean = $\dfrac{\underline{\hspace{2cm}}}{\text{sum of trials}} \Big/ \dfrac{\underline{\hspace{1cm}}}{N} = \dfrac{\underline{\hspace{1.5cm}}}{\bar{x}}$

Female Data

Number of subjects: _____ = N

Height (m)

Trial 1: _____ = X   Mean = _____ / _____ = _____
                                            sum of trials   N $\bar{x}$

Trial 2: _____ = X   Mean = _____ / _____ = _____
                                            sum of trials   N $\bar{x}$

Trial 3: _____ = X   Mean = _____ / _____ = _____
                                            sum of trials   N $\bar{x}$

Weight (kg)

Trial 1: _____ = X   Mean = _____ / _____ = _____
                                            sum of trials   N $\bar{x}$

Trial 2: _____ = X   Mean = _____ / _____ = _____
                                            sum of trials   N $\bar{x}$

Trial 3: _____ = X   Mean = _____ / _____ = _____
                                            sum of trials   N $\bar{x}$

Excel Spreadsheet Mean

Male Data

Trial 1: Height: _____ m   Weight: _____ kg
Trial 2: Height: _____ m   Weight: _____ kg
Trial 3: Height: _____ m   Weight: _____ kg

Female Data

Trial 1: Height: _____ m   Weight: _____ kg
Trial 2: Height: _____ m   Weight: _____ kg
Trial 3: Height: _____ m   Weight: _____ kg

Hand Calculation of the Median and Mode

Male Data

Trial 1: Median: (_____ +1) / 2 = _____ m
                       number of samples

Trial 2: Median: (_____ +1) / 2 = _____ m
                       number of samples

Trial 3: Median: (_____ +1) / 2 = _____ m
                       number of samples

Female Data

Trial 1: Median: $(\underline{\hspace{3cm}}_{\text{number of samples}} + 1) / 2 = \underline{\hspace{3cm}}$ m

Trial 2: Median: $(\underline{\hspace{3cm}}_{\text{number of samples}} + 1) / 2 = \underline{\hspace{3cm}}$ m

Trial 3: Median: $(\underline{\hspace{3cm}}_{\text{number of samples}} + 1) / 2 = \underline{\hspace{3cm}}$ m

Determination of the Mode

Male Data

Trial 1: Height: _____ m   Weight: _____ kg

Trial 2: Height: _____ m   Weight: _____ kg

Trial 3: Height: _____ m   Weight: _____ kg

Female Data

Trial 1: Height: _____ m   Weight: _____ kg

Trial 2: Height: _____ m   Weight: _____ kg

Trial 3: Height: _____ m   Weight: _____ kg

Excel Spreadsheet Median and Mode

Male Median

Trial 1: Height: _____ m   Weight: _____ kg

Trial 2: Height: _____ m   Weight: _____ kg

Trial 3: Height: _____ m   Weight: _____ kg

Female Median

Trial 1: Height: _____ m   Weight: _____ kg

Trial 2: Height: _____ m   Weight: _____ kg

Trial 3: Height: _____ m   Weight: _____ kg

Male Mode

Trial 1: Height: _____ m   Weight: _____ kg

Trial 2: Height: _____ m   Weight: _____ kg

Trial 3: Height: _____ m   Weight: _____ kg

Female Mode

Trial 1: Height: _____ m   Weight: _____ kg

Trial 2: Height: _____ m   Weight: _____ kg

Trial 3: Height: _____ m   Weight: _____ kg

Excel Spreadsheet Calculation of the Standard Deviation

Male Data

Trial 1: Height: _____ m   Weight: _____ kg

Trial 2: Height: _____ m   Weight: _____ kg

Trial 3: Height: _____ m   Weight: _____ kg

Female Data

Trial 1: Height: _____ m   Weight: _____ kg

Trial 2: Height: _____ m   Weight: _____ kg

Trial 3: Height: _____ m   Weight: _____ kg

MALES	TRIAL 1 HEIGHT			TRIAL 2 HEIGHT			TRIAL 3 HEIGHT		
Subject	X	$X - M$	$(X - M)^2$	X	$X - M$	$(X - M)^2$	X	$X - M$	$(X - M)^2$
1									
2									
3									
4									
5									
6									
7									
8									
9									
10									
11									
12									
13									
14									
15									
16									
17									
18									
19									
20									
Sum =									
Mean =									
SD =		SD =			SD =			SD =	

FEMALES	TRIAL 1 HEIGHT			TRIAL 2 HEIGHT			TRIAL 3 HEIGHT		
Subject	X	$X - M$	$(X - M)^2$	X	$X - M$	$(X - M)^2$	X	$X - M$	$(X - M)^2$
1									
2									
3									
4									
5									
6									
7									
8									
9									
10									
11									
12									
13									
14									
15									
16									
17									
18									
19									
20									
Sum =									
Mean =									
SD =		SD =			SD =			SD =	

MALES	TRIAL 1 WEIGHT			TRIAL 2 WEIGHT			TRIAL 3 WEIGHT		
Subject	X	$X - M$	$(X - M)^2$	X	$X - M$	$(X - M)^2$	X	$X - M$	$(X - M)^2$
1									
2									
3									
4									
5									
6									
7									
8									
9									
10									
11									
12									
13									
14									
15									
16									
17									
18									
19									
20									
Sum =									
Mean =									
SD =		SD =			SD =			SD =	

FEMALES	TRIAL 1 WEIGHT			TRIAL 2 WEIGHT			TRIAL 3 WEIGHT		
Subject	X	$X - M$	$(X - M)^2$	X	$X - M$	$(X - M)^2$	X	$X - M$	$(X - M)^2$
1									
2									
3									
4									
5									
6									
7									
8									
9									
10									
11									
12									
13									
14									
15									
16									
17									
18									
19									
20									
Sum =									
Mean =									
SD =		SD =			SD =			SD =	

Calculation of the Range

Calculate the range by hand (note that these are the high and low scores for the group data):

Male Data

Trial 1: Height: (_____ – _____) + 1 = _____ = range
                  high score   low score

Trial 2: Height: (_____ – _____) + 1 = _____ = range
                  high score   low score

Trial 3: Height: (_____ – _____) + 1 = _____ = range
                  high score   low score

Trial 1: Weight: (_____ – _____) + 1 = _____ = range
                  high score   low score

Trial 2: Weight: (_____ – _____) + 1 = _____ = range
                  high score   low score

Trial 3: Weight: (_____ – _____) + 1 = _____ = range
                  high score   low score

Female Data

Trial 1: Height: (_____ – _____) + 1 = _____ = range
                  high score   low score

Trial 2: Height: (_____ – _____) + 1 = _____ = range
                  high score   low score

Trial 3: Height: (_____ – _____) + 1 = _____ = range
                  high score   low score

Trial 1: Weight: (_____ – _____) + 1 = _____ = range
                  high score   low score

Trial 2: Weight: (_____ – _____) + 1 = _____ = range
                  high score   low score

Trial 3: Weight: (_____ – _____) + 1 = _____ = range
                  high score   low score

MALES	WEIGHT			HEIGHT		
Subject	Trial 1	Trial 2	(Trial 2 – Trial 1)	Trial 1	Trial 2	(Trial 2 – Trial 1)
1						
2						
3						
4						
5						
6						
7						
8						
9						
10						
11						
12						
13						
14						
15						
16						
17						
18						
19						
20						
Sum =				Sum =		
Mean =				Mean =		
SD =				SD =		
TE =				TE =		

FEMALES	WEIGHT			HEIGHT		
Subject	Trial 1	Trial 2	(Trial 2 – Trial 1)	Trial 1	Trial 2	(Trial 2 – Trial 1)
1						
2						
3						
4						
5						
6						
7						
8						
9						
10						
11						
12						
13						
14						
15						
16						
17						
18						
19						
20						
Sum =				Sum =		
Mean =				Mean =		
SD =				SD =		
TE =				TE =		

Tables and Graphs

Equipment

- Data from laboratories 1.1 and 1.2
- Microsoft Word or equivalent word-processing software
- Microsoft Excel or equivalent spreadsheet software

Creating Tables

Step 1: Use either Microsoft Word or Microsoft Excel. If using Word, select the Insert Table command. Then select the number of columns and rows the table should have. For this lab, choose five rows and seven columns (see the following table for an example).

Step 2: Insert labels into the table. The first column should contain the variables Age (y), Height (m), and Weight (kg).

Step 3: Merge the second, third, and fourth columns. To merge the columns, highlight the top cell of each column and select the Merge Cells function. Label the resulting merged column as Males (click on the Center option to center the label). Repeat this process for columns 5 through 7, but use the label Females.

Step 4: Below the Males label and the Females label, insert the Mean, ±, and SD labels for each grouping. To insert the ± symbol, find the Insert Symbol link; if necessary (i.e., if the ± symbol is not included in the frequently used list), select the More Symbols option. To find the ± symbol, change the font to Symbol and scroll through the available symbols.

Step 5: Center the ± symbol by highlighting the columns in which it appears and selecting the Center option located under the Home tab. Right justify the Mean and SD labels by highlighting the columns in which they appear and selecting the Right Justify option under the Home tab.

Step 6: Resize the columns containing the ± symbol to make the table smaller and thus easier to interpret. Do so by double-clicking on the right column line to prompt the software to automatically resize the column. In addition, insert the ± symbol in each row under the ± label.

Step 7: Remove the lines on the table by highlighting the entire table, selecting the Borders option, and choosing No Border.

Step 8: Insert a border at the top and at the bottom of the table and below the row containing the column labels. If done correctly, the table should be formatted as shown in the following table.

Step 9: Enter into the table the pertinent data collected in laboratory activities 1.1 and 1.2. Once the data are entered, you can adjust the formatting of the table to make it easier to read as shown here:

	MALES			FEMALES		
Variables	Mean	±	SD	Mean	±	SD
Age (y)		±			±	
Height (m)		±			±	
Weight (kg)		±			±	

Creating Column Graphs

Bar graphs offer a great way to represent data and are relatively easy to create with software such as Microsoft Excel. Here, you will create a column graph based on data collected in laboratory activity 1.1 and analyzed in laboratory activity 1.2.

Step 1:   Locate the mean and standard deviation data recorded on the group data sheet for laboratory 1.2. If you analyzed these data using Excel, use the data set created in Excel to do this new analysis.

Step 2:   If you did not use Excel for your analysis in laboratory activity 1.2, enter the mean and standard deviation data into Excel now.

Step 3:   To create a bar graph for the height data, highlight cells B11 and C11. Then find the Insert Clustered Column Graph option.

Step 4:   A basic column graph should be created. To clean up the graph, click on the horizontal gridlines to effectively highlight them, and then delete them.

Step 5:   To add a y-axis label, find your software's option for editing the axis titles and the option for editing the primary vertical axis. Rotate the title so that it stretches along the y-axis. Then type in the units that correspond to the y-axis. In this case, use height (m).

Step 6:   Delete the Series 1 legend entry by clicking on it and then pressing the delete key.

Step 7:   Next, edit the x-axis. Find the option for editing the horizontal axis, and then highlight the range. Find the function to label the axis so that you can highlight the Male and Female labels located in cells B2 and C2, respectively. At this point, the columns should be labeled according to their contents.

Step 8:   To add standard deviation bars, click on the column chart, find your software's option for inserting error bars, and find the option for formatting error bars. Now find and select the custom option and insert a custom positive and negative range. In this example, highlight cells B14 and C14, and then click the box with the red arrow pointing down to return to the Custom Error Bars selection screen. Select the Negative Error Value Range option and then highlight B14 and C14. After both negative and positive error values are selected, click OK and then close, which will place the appropriate error bars in the column graph.

Step 9:   Add a title above the chart.

Step 10: Repeat these steps again to create a column graph for the weight variable.

Question Set 1.3

1. If you were creating a graphical depiction of a series of anthropometric data, what important variables would you need to present?

2. Is it important to include the standard deviation of the data in the graphical depiction of the data set? Why or why not?

3. Describe how you might set up your spreadsheet to create a graphical depiction of your data. Be specific.

4. How would you determine whether a graphical depiction is necessary for your data set?

Pretest Screening

Objectives

- Introduce the components of a pretesting screening process.
- Explain the informed consent process and outline what is contained in an appropriately structured informed consent form.
- Present the components of a medical and health history questionnaire.
- Practice the performance of pretesting screening.
- Describe the American College of Sports Medicine (ACSM) preparticipation health screening method.
- Explain the classic risk stratifications and guidelines for physician involvement in the testing process.

DEFINITIONS

coronary heart disease (CHD)—Condition in which blood flow through the coronary arteries is reduced, typically as a result of atherosclerosis; also known as *coronary artery disease*.

high risk—Presence of one or more signs or symptoms of cardiovascular or pulmonary disease or presence of known risk factors for cardiovascular, pulmonary, or metabolic disease.

hypertension—Chronically elevated BP; also referred to as *high blood pressure*.

informed consent—Process in which subjects are informed of the purposes, methods, risks, and benefits of a study or test and then sign a form to acknowledge that they have been informed.

institutional review board (IRB)—An ethics committee that is formally designated to approve, monitor, and review research protocols in order to protect human subjects from physical or psychological harm.

low risk—Absence of relevant symptoms of cardiovascular or pulmonary disease, along with the presence of no more than one CHD risk factor; characteristic of younger individuals (men under age 45, women under age 55).

moderate risk—Risk status typically assigned to older individuals (men older than 45, women older than 55), as well as to individuals of any age who have two or more CHD risk factors.

myocardial infarction—Heart attack.

Physical Activity Readiness Questionnaire for Everyone (PAR-Q+)—Questionnaire used to establish an individual's fitness for physical activity.

Before performing any type of exercise testing, it is essential to use a prescreening process in order to identify individuals who may be at an elevated risk for exercise-induced sudden cardiac death or myocardial infarction (2, 4, 6). The classic prescreening approach has the following goals (3):

- Determine whether the individual has any medical condition that contraindicates performing certain health-related fitness assessments.
- Determine whether the individual should have a medical evaluation before undergoing a health-related fitness assessment, and consult a physician.

- Establish whether the individual should undergo a medically supervised fitness assessment.
- Determine whether the individual has other health or medical concerns (e.g., diabetes mellitus, orthopedic injuries).

Based on these goals, any classic prescreening process has typically recommended the inclusion of a minimum of three components (6):

1. Health history questionnaire
2. Physical Activity Readiness Questionnaire for Everyone (PAR-Q+)
3. Medical or health exam

However, performing a more comprehensive, seven-step screening process is generally warranted to decrease the subject's potential risk for an adverse effect during the testing process (4). See table 2.1 for a description of this process.

Although the classic method of prescreening has been commonly used, a roundtable of experts concluded that in order to avoid excessive physician referrals and promote the adoption of an exercise prescription, a less conservative approach to preparticipation health screening should be employed by exercise

professionals (2, 8, 10). The goals of these new prescreening guidelines are to identify individuals who

- need to receive medical clearance before starting an exercise program or before increasing the intensity, volume, or frequency of their current exercise regime;
- have a clinically significant disease (or diseases) and therefore may benefit from a medically supervised exercise intervention; or
- have a medical condition (or conditions) that contraindicates an exercise program until the condition is better controlled or abated.

In order to accomplish these goals, an evidence model for exercise preparticipation health screening has been established based upon the following (2, 8, 10):

1. Identification of the individual's current physical activity level.
2. Determination of the existence of cardiovascular, metabolic, or renal disease or signs and symptoms of such disease.
3. Use of signs and symptoms, as well as the individual's disease history and current exercise participation history, coupled with

Table 2.1 Seven-Step Comprehensive Screening Process

Questionnaire or screening form	Purpose
Informed consent	Explains the purpose, risks, and benefits of participating in the testing process
PAR-Q+	Helps determine the individual's readiness to perform physical activity
Medical and health history	Collects information about the individual's past and present personal health history and family history; the focus is on conditions that require medical referrals and clearance
Signs and symptoms of disease and medical clearance	Identifies individuals who need a physician's approval, via a medical referral, to perform exercise testing
Coronary risk factor analysis	Establishes risk by quantifying the individual's number of coronary heart disease risk factors
Lifestyle evaluation	Gives insight into the individual's living habits
Disease risk classification	Categorizes the individual as high, medium, or low risk

Adapted, by permission, from V. Heyward and A.L. Gibson, 2014, *Advanced fitness assessment and exercise prescription*, 7th ed. (Champaign, IL: Human Kinetics), 24.

the desired exercise intensity to inform recommendations for preparticipation medical clearance.

To guide in the decision-making process, a screening algorithm (figure 2.1) has been established by the ACSM to assist in determining if medical clearance is necessary (2, 10).

- Individuals who have been participating in at least moderate physical activity for 30 min on three or more days per week and are asymptomatic without known cardiovascular (i.e., cardiac, peripheral artery, cerebrovascular), metabolic (type 1 or 2 diabetes mellitus), or renal disease can continue their regular exercise program (2, 10).
- Those with known cardiovascular, metabolic, or renal disease who are physically active but asymptomatic and have been cleared to exercise by a physician in the last 12 mo can continue to participate in moderate intensity exercise. If, however, these individuals develop symptoms of cardiovascular, metabolic, or renal disease during resting or exertional symptoms, they should seek medical clearance before continuing their exercise regime (8, 10).
- Inactive individuals who are healthy and asymptomatic may begin light to moderate intensity exercise without seeking medical clearance. If, however, an inactive person has known cardiovascular, metabolic, or renal disease or exhibits signs or symptoms indicative of the presence of these diseases, they should seek medical clearance prior to initiation of an exercise program (10).

In order to apply this new prescreening algorithm, a basic preparticipation health screening questionnaire can be used (form 2.1) (8).

INFORMED CONSENT

Before performing any part of a testing process, you must obtain full **informed consent** from anyone who is going to be tested (3, 4). Doing so documents that the person has been informed about the procedures, benefits, and risks associated with the assessment. At this time, you should also present any available alternative tests (3).

An informed consent form should clearly document the benefits and risks associated with the assessment so that subjects can decide whether the scientific or clinical benefits clearly outweigh the personal risks (3). It is also essential to inform subjects that (a) they are volunteering to participate in the assessment process, (b) they should inform the testers of any problems experienced during the assessment, and (c) they have the right to withdraw from the testing process at any time without any consequences. The accompanying sidebar presents eight items generally included in an informed consent form; for an example, see form 2.2.

Research and Institutional Review

If data are being collected for research purposes or are experimental in nature, all procedures must be approved by an **institutional review board (IRB)** before any testing is undertaken. The IRB is an independent ethics committee that is formally designated to approve, monitor, and review research protocols in order to protect human subjects from physical or psychological harm. This is accomplished by assessing the ethics of the research and its methods. Central to the process is informed consent, which ensures subjects are fully informed and are aware that their participation is voluntary.

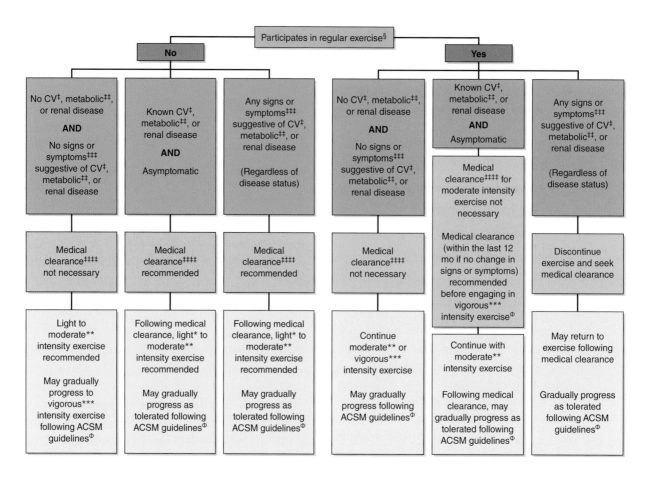

Figure 2.1 Exercise preparticipation health screening algorithm.

§ Exercise participation, performing planned, structured physical activity at least 30 minutes at moderate intensity on at least 3 days per week for at least the last 3 months

★ = Light-intensity exercise, 30% to <40% heart rate reserve or $\dot{V}O_2R$, 2 to <3 METs, 9-11 RPE, an intensity that causes slight increases in heart rate and breathing

★★ = Moderate-intensity exercise, 40% to <60% heart rate reserve or $\dot{V}O_2R$, 3 to <6 METs, 12-13 RPE, an intensity that causes noticeable increases in heart rate and breathing

★★★ = Vigorous-intensity exercise, ≥60% heart rate reserve or $\dot{V}O_2R$, ≥6 METs, ≥14 RPE, an intensity that causes substantial increases in heart rate and breathing

‡ = CV, cardiac disease, peripheral vascular disease, or cerebrovascular disease

‡‡ = Metabolic disease, type 1 and type 2 diabetes mellitus

‡‡‡ = Signs and symptoms, at rest or during exercise; includes pain, discomfort in the chest, neck, jaw, arms, or other areas that may result from ischemia; shortness of breath at rest or with mild exertion; dizziness or syncope; orthopnea or paroxysmal nocturnal dyspnea; ankle edema; palpitations or tachycardia; intermittent claudication; known heart murmur; or unusual fatigue or shortness of breath with usual activities

‡‡‡‡ = Medical clearance, approval from a health care professional to engage in exercise

φ = ACSM guidelines; see *ACSM's Guidelines for Exercise Testing and Prescription, 10th Edition,* 2018

Reprinted, by permission, from D. Riebe et al., 2015, "Updating ACSM's recommendations for exercise preparticipation health screening," *Medicine and Science in Sports and Exercise* 47: 2473-2479.

Items Needed in an Informed Consent Form

- General statement of the testing process and objectives
- Easily understood description of the testing procedures
- Description of all risks associated with the testing process
- Explanation of the benefits associated with the testing procedures
- Statement informing the subject that the testers will answer any questions
- Statement informing the subject that he or she is free to withdraw consent and cease the testing process at any time without any consequences
- Statement informing the subject that he or she is free to refuse to answer any questions or respond to specific items contained in any questionnaires
- General statement of the procedures for maintaining confidentiality of the data obtained from the subject's screening and testing

Adapted from Nieman 2003; American College of Sports Medicine 2009.

PHYSICAL ACTIVITY READINESS QUESTIONNAIRE FOR EVERYONE

The **Physical Activity Readiness Questionnaire for Everyone (PAR-Q+)** contains seven questions designed to identify individuals who need medical clearance before being tested or entering an exercise regime (form 2.3) (6). The PAR-Q+ is widely recognized as a safe prescreening questionnaire for those who plan to undertake low- to moderate-intensity exercise (3, 7). It contains seven yes-or-no questions that are generally easy to understand and answer (3). Additionally, the PAR-Q+ has 10 follow-up questions that are asked based on the response to the initial seven questions. If subjects answer *yes* to any of the initial questions, they should be directed to answer the follow-up questions. If they then answer *yes* to any of the follow-up questions, they should be referred to a physician for a more comprehensive health screening and for medical clearance for any testing procedure or exercise regimen (6).

HEALTH HISTORY QUESTIONNAIRE

During prescreening, you should also use a comprehensive health history questionnaire to establish the subject's medical risk (3, 6). You can use the information gathered from this questionnaire to stratify the subject's risk and determine whether a medical referral is necessary prior to undergoing any testing procedures. This process will also help you determine whether the subject needs medical supervision during any of the testing procedures. A health history questionnaire generally includes 10 items essential to risk assessment (table 2.2) (11).

A sample health history questionnaire is presented in form 2.4 (7). This health history contains all major items recommended for a comprehensive questionnaire. When a subject or client fills out this type of document, make someone available to answer any questions. Once the document has been completed, examine it closely for contraindications to exercise testing. People who have absolute contraindications—such as acute **myocardial infarction** (within 2 d), unstable angina, or uncontrolled

Table 2.2 Components of a Health History Questionnaire

Component	Items of concern (not a comprehensive list)
Any medical diagnosis	Cardiovascular disease, including myocardial infarction Percutaneous coronary artery procedures, including angioplasty Valvular dysfunction and surgeries Symptoms of ischemic coronary syndrome Peripheral vascular disease Hypertension Diabetes Obesity Pulmonary disease Anemia Stroke Cancer
Previous physical examination findings	Heart murmur Abnormal heart sounds Abnormal pulmonary or cardiovascular findings Abnormal blood sugar or heart rate, high blood pressure, high cholesterol, or other lab abnormalities
History of symptoms	Discomfort in the chest, jaw, neck, arms, or back Light-headedness, dizziness, or fainting Unilateral weakness Shortness of breath Palpitation
Recent illness, hospitalization, new medical diagnosis, or surgical procedure	Any recent changes in the subject's medical status
Orthopedic problems	Arthritis Joint swelling Other issues that impede ability to exercise
Medications used and drug allergies	Any medications the subject takes or may be allergic to
Lifestyle habits	Dietary habits, including alcohol use Drug use Tobacco use
Exercise habits	Types, duration, and intensity of exercise currently undertaken
Work history	The subject's current work atmosphere and types of work activities
Family history of disease	Cardiovascular disease Pulmonary disease Metabolic disease Stroke Sudden death

Adapted from Nieman 2003; American College of Sports Medicine 2009.

symptomatic heart failure—should not be tested until the condition is stabilized or resolved (3). For those with relative contraindications—such as left main artery stenosis, severe arterial **hypertension**, or high-degree atrioventricular (AV) block—testing may be performed only after the risk-to-benefit ratio has been carefully evaluated (2). In general, all subjects who present either absolute or relative contraindications should be directed to a physician for a complete examination and to receive clearance for the testing (6). For a more complete list of absolute and relative contraindications to exercise testing, see references 1 and 5.

SIGNS AND SYMPTOMS OF DISEASE AND MEDICAL CLEARANCE

It is important to screen all subjects or clients for signs and symptoms of disease prior to testing (6). This can be done in several ways:

- In the health history questionnaire (form 2.4)
- In a separate checklist you create to address signs and symptoms of disease (form 2.5)
- In the ePARmed-X+ available at http://eparmedx.com/?page_id=24 (6, 7)

Form 2.5 presents a sample checklist for signs and symptoms of disease (6). In this example, subjects with any of the listed conditions or more than two of the listed risk factors would be referred to their physician. The ePARmed-X+ can be used by the physician to evaluate and give medical clearance for subjects who have answered *yes* to any questions on the PAR-Q+ form; that is, the ePARmed-X+ is used either to clear subjects for testing and exercise or to refer them to a medically supervised test or exercise program (6).

CORONARY RISK FACTOR ANALYSIS

In determining whether it is safe to conduct exercise testing with a given subject, it is important to establish the subject's coronary risk profile. Form 2.6 presents a series of items positively associated with **coronary heart disease (CHD)** risk (1, 2). Though these items make up part of the classic risk stratification, the list in form 2.6 is not all inclusive (9). The items should, however, be viewed as clinically relevant thresholds to consider along with other screening items when making decisions about a subject's risk for an adverse coronary event and about the level of medical clearance needed before testing (1, 3).

As a screening tool, form 2.6 can be used to assess the individual's risk by summing the risk factors. If the subject has a high-density lipopro-

tein-C (HDL-C) level equal to or greater than $60 \text{ mg} \cdot \text{dl}^{-1}$, the tester should subtract 1 from the total number of risk factors (2, 7).

LIFESTYLE EVALUATION

Another way to gain insight into an overall risk profile is to develop an understanding of the person's lifestyle (6). A lifestyle questionnaire addresses concerns such as smoking, physical activity, diet, and alcohol consumption. These items can easily be included in the medical or health history questionnaire, as shown in form 2.4 (7). They can also be handled in a separate questionnaire, as shown in form 2.7, which presents a tool called the Fantastic Lifestyle Checklist for assessing current health-related behaviors.

DISEASE RISK STRATIFICATION

Once you have collected the prescreening data, you can establish an overall risk profile to stratify the person's risk and determine the need for a physician's referral or presence during the testing process. A three-tier risk stratification (low, moderate, high) should be used when initially evaluating someone's risk profile (3). For example, taking the data collected in forms 2.4 and 2.5, the tester can stratify the subject's overall risk level using the guidelines presented in table 2.3. This stratification suggests that people who have no more than one risk factor (from form 2.6) are considered **low risk,** while people who have two or more risk factors fall into the **moderate-risk** category. People with known diseases (the diseases listed in table 2.3) are classified as **high risk** and require further medical screening before undergoing exercise testing.

Once the subject's health status has been classified into one of the three tiers, you can make decisions about testing conditions (table 2.4) (3). If subjects are classified as low risk, it is not necessary for them to have a pretest medical examination, and physician supervision of the test is unnecessary. However, if their risk is either moderate or high, then they should undergo a medical screening, and

Table 2.3 Risk Stratification

Low risk	Asymptomatic men and women who have ≤1 cardiovascular disease risk factor from form 2.6
Moderate risk	Asymptomatic men and women who have ≥2 risk factors from form 2.6

High risk	Individuals with
	• known cardiac, peripheral vascular, or cerebrovascular disease; chronic obstructive pulmonary disease (COPD), asthma, interstitial lung disease, or cystic fibrosis; diabetes mellitus (type 1 or type 2), thyroid disorders, or renal or liver disease; or
	• one or more of the following signs or symptoms:
	  - heart murmur;
	  - unexplained fatigue;
	  - dizziness or syncope;
	  - swelling of the ankles;
	  - tachycardia or irregular heartbeat;
	  - unexplained shortness of breath;
	  - intermittent claudication;
	  - breathing discomfort when not in upright position, or interrupted breathing at night; or
	  - pain or discomfort in the jaw, neck, chest, arms, or elsewhere that could be caused by ischemia.

Adapted from American College of Sports Medicine 2010.

Table 2.4 Recommendations for Physician Involvement in Exercise Testing

	Low risk	Moderate risk	High risk
	CURRENT (WITHIN PAST YEAR) MEDICAL EXAMINATION AND EXERCISE TESTING BEFORE PARTICIPATION IN EXERCISE[a]		
Moderate exercise[c]	Not necessary[b]	Not necessary	Recommended
Vigorous exercise[c]	Not necessary	Recommended	Recommended
	PHYSICIAN SUPERVISION OF EXERCISE TESTS		
Submaximal test	Not necessary	Not necessary	Recommended
Maximal test	Not necessary	Recommended	Recommended

[a]When looking at exercise, the classification *Not necessary* suggests that it is not essential to employ medical examination, exercise testing, or physician supervision as part of the preparticipation screening process. However, it could still be done based on the judgment of the exercise physiologist. The *Recommended* classification suggests that prior to beginning an exercise program, the subject should undergo a medical examination and exercise testing conducted under the supervision of a physician.

[b]When applied to exercise testing, the *Not necessary* classification suggests that exercise testing could be done without a pretest medical screening or physician supervision. The *Recommended* classification suggests that the subject should undergo a medical examination prior to undergoing any testing and that the exercise testing should be conducted under the supervision of a physician.

[c]*Moderate exercise* is generally defined as 3 to 6 METs (7), whereas *vigorous exercise* is defined as >6 METs, or >60% of maximal oxygen consumption (1, 2).

Adapted from American College of Sports Medicine 2010.

a physician should be in close proximity and readily available during the test (3). When the risk is high, it is preferable to have the physician present to maximize the safety of the testing procedures.

Another tool for evaluating risk is the Framingham risk equation (4, 12), which can be used to predict multivariate CHD risk in people who do not display overt CHD. These sex-specific equations allow the prediction of overall risk over 10 years (5). Points are allotted based on the following factors:

- Age
- LDL-C (low-density lipoprotein-C) levels or total cholesterol levels

- HDL-C (high-density lipoprotein-C) levels
- BP
- Diabetes
- Smoking

The total points are summed to determine an overall risk stratification. The methodology for using these prediction equations can be found in form 2.8 and form 2.9.

Go to the web study guide to access electronic versions of the forms used in this laboratory.

WWW

Basic Screening Procedures

Equipment

- Exercise health preparticipation health screening questionnaire (form 2.1)
- Informed consent form (form 2.2)
- PAR-Q+ form (form 2.3)
- Health history questionnaire (form 2.4)
- Checklist for signs and symptoms of disease (form 2.5)
- CHD risk factors form (form 2.6)
- Fantastic Lifestyle Checklist (form 2.7)
- Framingham risk equation sheets for males and females (forms 2.8 and 2.9; also available in the web study guide)
- Simulated subject data cards (available in the web study guide)

Pretest Screening

Pair up and administer the seven questionnaires with your partner. One person in the pair serves as the tester, and the other uses the simulated subject cards to act as the subject. The tester should administer the pretesting screening process to simulate what would occur in a laboratory testing scenario. Once the screening is complete, the tester evaluates the subject's overall risk according to the ACSM preparticipation screening guidelines, the classic risk stratification method, and Framingham risk equations. Then reverse roles with your partner and repeat the activity.

When initiating a test, it is important for the tester to follow a step-by-step procedure (3). Many possible procedural steps are available. Here is a basic model.

Step 1: Greet the subject.

Step 2: Explain the process for the pretesting screening. Give an overview of the data that will be collected during the process.

Step 3: Give the subject the informed consent form (form 2.2). Answer any questions while the subject reads, fills out, and signs the document. Form 2.2 is merely an example; a more detailed or specific version may be required for your testing scenario. This example meets the basic criteria for informed consent, but each IRB has its own specific requirements.

Step 4: Provide your subject with the exercise preparticipation health screening questionnaire (form 2.1). Instruct your subject to fill out the questionnaire in complete detail. If possible, ask questions to further explore the presence of or signs of the listed medical conditions.

Step 5: Administer the PAR-Q+ form (form 2.3). Encourage the subject to answer all questions truthfully. Have the subject sign and date the form. Evaluate the form according to the criteria presented on the form.

Step 6: Give the subject a health history questionnaire (form 2.4). Encourage the subject to read the document carefully, to answer all questions, and to ask questions when needed. After the form has been completed, read it over and look for items that may need clarification. Note any contraindications to exercise and testing. (This will aid in the risk stratification process.)

Step 7: Give the subject the checklist for signs and symptoms of disease (form 2.5). Ask the subject to fill out the checklist to the best of his or her knowledge and note any signs or symptoms of disease.

Step 8: Administer the form for CHD risk factors (form 2.6). Ask the subject to answer *yes* to any questions that correspond to his or her health status. After completing the form, add up the points to determine the subject's overall risk.

Step 9: Instruct the subject to fill out the Fantastic Lifestyle Checklist (form 2.7) as truthfully as possible. Evaluate the subject's answers as follows:

a. Total the number of *X*s in each column.

b. Multiply the totals by the numbers indicated on the form.

c. Total the scores to yield a grand score.

Step 10: Evaluate the subject's overall disease risk by comparing the grand total score with the normative data presented in form 2.7. A grand total between 85 and 100 constitutes an excellent score, whereas a grand total between 0 and 34 indicates a need for improvement. Grand total scores between 70 and 84 are classified as very good, scores between 55 and 69 as good, and scores between 35 and 54 as fair.

Step 11: Administer the Framingham Risk Charts. If you are a male use form 2.8. If you are a female use form 2.9.

How to Use the Framingham Risk Charts

If you know the subject's LDL-C and HDL-C levels, you should follow these steps:

Step 1: Based on the subject's age, determine the allotted points (e.g., a 35-year-old woman would be given −4 points) for LDL-C and record the value in step 7 of form 2.8 (males) or 2.9 (females).

Step 2: Based on the subject's LDL-C, determine the allotted points (e.g., if she has an LDL-C of 2.7 mmol·L−1, assign 0 points) and record in step 7 of form 2.8 (males) or 2.9 (females).

Step 3: Based on the subject's HDL-C, determine the allotted points (e.g., if she has an HDL-C of 1.2 mmol·L−1, assign 1 point) and record in step 7 of form 2.8 (males) or 2.9 (females).

Step 4: Assign points based on the subject's BP (e.g., if she has a BP of 135/92 mmHg, assign 2 points) and record in step 7 of form 2.8 (males) or 2.9 (females).

Step 5: Assign points for diabetes (e.g., if she is a diabetic, assign 2 points) and record in step 7 of form 2.8 (males) or 2.9 (females).

Step 6: Assign points for smoking (e.g., if she is not a smoker, assign 0 points) and record in step 7 of form 2.8 (males) or 2.9 (females).

Step 7: Total all the assigned points (for either example, −4 + 0 + 1 + 2 + 2 + 0 = 1) recorded in step 7 of form 2.8 (males) or 2.9 (females).

Step 8: Determine the subject's 10-year CHD risk from the total points (e.g., she has a total score of 1, which means she has a 2% risk of CHD in the next 10 y) and determine comparative risk in step 9 of form 2.8 (males) or 2.9 (females).

If you know the subject's total cholesterol and HDL-C levels, you should follow these steps:

Step 1: Based on the subject's age, determine the allotted points (e.g., a 40-year-old man would be given 1 points) for total cholesterol points and record the value in step 7 of form 2.8 (males) or 2.9 (females).

Step 2: Based on the subject's total cholesterol, determine the allotted points (e.g., if he has a total cholesterol of 6.0 mmol·L−1, assign 1 points) and record the value in step 7 of form 2.8 (males) or 2.9 (females).

Step 3: Based on the subject's HDL-C, determine the allotted points (e.g., if he has an HDL-C of 1.4 mmol·L−1, assign 0 points) and record the value in step 7 of form 2.8 (males) or 2.9 (females).

Step 4: Assign points based on the subject's BP (e.g., if he has a BP of 130/85 mmHg, assign 1 point) and record the value in step 7 of form 2.8 (males) or 2.9 (females).

Step 5: Assign points for diabetes (e.g., if he is not a diabetic, assign 0 points) and record the value in step 7 of form 2.8 (males) or 2.9 (females).

Step 6: Assign points for smoking (e.g., if she is not a smoker, assign 0 points) and record the value in step 7 of form 2.8 (males) or 2.9 (females).

Step 7: Total all the assigned points (for either example, 1 + 1 + 0 + 1 + 0 + 0 = 3) recorded in step 7 of form 2.8 (males) or 2.9 (females).

Step 8: Determine the subject's 10-year CHD risk from the total points (e.g., she has a total score of 3, which means she has a 5% risk of CHD in the next 10 years) and determine comparative risk in step 9 of form 2.8 (males) or 2.9 (females).

Adapted from *ESSA's student manual for health, exercise, and sport assessment*, edited by J. Combes and T. Skinner, Cardiovascular health, J. Combs and A. Williams, pg. 51, copyright 2014, with permission from Elsevier.

Question Set 2.1

1. How does the ACSM preparticipation health screening differ from the classic method?
2. What 10 components should be included in a health history questionnaire?
3. What are the three tiers of an initial risk stratification? What factors determine a subject's risk level?
4. What eight items should be included in an informed consent form?
5. What is a PAR-Q+, and what is it used for?
6. What factors should you concentrate on when evaluating a pretesting screening?
7. Do all clients need medical clearance or supervision during testing? Why or why not?
8. How should you use the information collected in pretesting screening to evaluate a person's risk?
9. How does the initial risk stratification compare with the Framingham risk stratification?

Pretest Results

Equipment

- Clinical case study health history questionnaire for sample client (figure 2.4)
- Checklist for signs and symptoms of disease (form 2.5)
- CHD risk factors form (form 2.6)

Pretest Screening Data

Imagine a new client has come into your health and wellness facility. She has already completed an informed consent form and filled out a comprehensive health history questionnaire. Now it is time to use her answers to make decisions about her overall health and wellness. Form groups of two or three students and go over the client's health history questionnaire (figure 2.4).

When reviewing the results of a pretesting screening, you should perform a comprehensive and systematic analysis of the information collected (4). You can choose from many methods in conducting this evaluation; what follows should serve as a basic model.

Step 1: Examine the client's health history questionnaire for any contraindications to exercise testing, and note them on a copy of the checklist for signs and symptoms of disease (form 2.5). Make sure to comment on any items for which the answer was *yes*. Use this form to make a preliminary decision about the need for physician's clearance for testing.

Step 2: Use the form for CHD risk factors (form 2.6) to evaluate the client's risk in this area. Pay particular attention to the seven major categories of risk factors. Note any risk factors presented by the client, and tabulate the point values.

Step 3: After evaluating the client's health history questionnaire, refer to table 2.3 to determine the initial risk stratification.

Step 4: Once you have determined the client's risk stratification, use table 2.4 to determine whether a current medical examination and exercise testing are needed and whether a physician needs to supervise the test.

Question Set 2.2

1. What general risk factors does this client exhibit? Would these risk factors affect the client's ability to perform physical activity or participate in testing?
2. Does this client present any CHD risk factors? If so, which ones?
3. Does this client exhibit any contraindications to exercise? If so, which ones?
4. Under which risk stratification would this client be classified?
5. Based on the client's health history questionnaire, does she need to be screened by a physician prior to testing?

Figure 2.4 Clinical case study health history questionnaire for sample client.

Contact Information

Jones	*Jamie*	*J*
Last name	First name	Middle initial

9/30/1978	*40*	Sex: ☐ Male ☑ Female
Date of birth (mm/dd/yyyy)	Age	

304-293-6318	*372 Main St*	*Morgantown, WV*	*26505*
Home phone number	Street address	City, state	Zip code

John Jones	*304-293-0647*	*Husband*
Emergency contact	Emergency contact number	Relationship to emergency contact

Matt Lively	*304-293-8828*
Primary physician	Physician's contact number

Background

Please circle the highest grade in school you have completed:

Elementary school:	1	2	3	4	5	6	7	8
High school:	9	10	11	(12)				
College/postgraduate:	13	14	15	16	17	18	19	20+

What is your marital status? ☐ Single ☑ Married ☐ Widowed ☐ Divorced/separated

What is your race or ethnic background?

☑ White, not of Hispanic origin   ☐ American Indian/Alaska Native   ☐ Asian

☐ Black, not of Hispanic origin   ☐ Pacific Islander   ☐ Hispanic

What is your occupation?

☐ Health professional ☐ Disabled, unable to work   ☐ Service   ☐ Manager, educator, professional

☐ Skilled craftsperson ☐ Operator, fabricator, laborer ☑ Homemaker ☐ Technical, sales, support

☐ Retired   ☐ Unemployed   ☐ Student   ☐ Other

Symptoms or Signs Suggestive of Disease

Please check the appropriate box.

Yes	No	
☐	☑	Have you experienced any unusual pain or discomfort in your chest, neck, jaw, arms, or other areas that may be due to heart problems?
☑	☐	Have you experienced unusual fatigue or shortness of breath at rest, during usual activities (e.g., climbing stairs, carrying groceries, brisk walking), or during mild to moderate exercise?
☐	☑	Have you ever had any problems with dizziness or fainting?

(continued)

Figure 2.4 *(continued)*

Yes	No	
☐	☑	When you stand up, do you have difficulty breathing?
☐	☑	Do you have difficulty breathing while sleeping?
☐	☑	Do your ankles swell (ankle edema)?
☑	☐	Have you ever experienced an unusual and rapid heartbeat or fluttering of the heart?
☐	☑	Have you experienced severe pain in your legs while walking?
☐	☑	Have you ever been told by a doctor that you have a heart murmur?

Chronic Disease Risk Factors

Do you know your blood pressure? ☑ = Yes: *130/80* systolic/diastolic   ☐ = No

Do you know your cholesterol levels? ☑ = Yes: *200* total cholesterol; _____ HDL; _____ LDL ☐ = No

Do you know your fasting glucose levels? ☐ = Yes: _____ mg/dl   ☑ = No

Please check the appropriate box.

Yes	No	
☐	☑	Are you a male over age 45 or a female over age 55?
☐	☑	If you are a female, have you experienced premature menopause and are *not* on estrogen replacement therapy?
☑	☐	Has your father or brother had a heart attack or died suddenly from heart disease before age 55?
☐	☑	Has your mother or sister had a heart attack or died suddenly from heart disease before age 65?
☐	☑	Has anyone in your family died before age 40 (excluding accidental death)?
☑	☐	Are you a current cigarette smoker?
☑	☐	Has a doctor told you that you have high blood pressure (more than 130/80 mmHg)?
☑	☐	Are you on medication to control high blood pressure?
☑	☐	Has a doctor ever told you that you have high cholesterol?
☑	☐	Is your serum cholesterol greater than 200 mg/dl?
☑	☐	Do you have diabetes mellitus?
☑	☐	Are you physically inactive or sedentary (i.e., do you perform little physical activity on the job or during leisure time)?
☑	☐	During the past year, have you experienced levels of stress, strain, and pressure that might affect your health?
☑	☐	Do you eat foods that are high in fat and cholesterol (e.g., fatty meats, cheese, fried food, butter, whole milk, eggs) on a daily basis?
☑	☐	Do you tend to avoid foods that are high in fiber (e.g., whole-grain breads and cereals, fresh fruits, vegetables)?
☑	☐	Do you weigh 30 pounds (14 kg) more than you should?
☑	☐	Do you average more than two alcoholic beverages a day?

Medical History

Please check all conditions that you or someone in your family have had or now have.

You   Your family

You	Your family	Condition
☐	☑	Coronary heart disease, heart attack, coronary artery surgery
☐	☑	Angina
☐	☐	Peripheral vascular disease
☐	☐	Phlebitis or emboli
☐	☐	Other heart problems (specify: _____)
☐	☐	Lung cancer
☐	☑	Breast cancer
☐	☐	Prostate cancer
☐	☐	Colorectal cancer (bowel cancer)
☐	☐	Skin cancer
☐	☐	Other cancer (specify: _____)
☐	☐	Stroke
☐	☐	Chronic obstructive pulmonary disease (COPD; e.g., emphysema)
☐	☐	Pneumonia
☐	☑	Asthma
☑	☐	Bronchitis
☑	☑	Diabetes mellitus
☐	☐	Thyroid problems
☐	☐	Kidney disease
☐	☐	Liver disease (cirrhosis of the liver)
☐	☐	Hepatitis (A, B, C, D, or E)
☐	☐	Gallstone or gallbladder disease
☐	☐	Osteoporosis
☐	☑	Arthritis
☐	☐	Gout
☐	☐	Anemia (low iron)
☐	☑	Bone fracture
☐	☐	Major injury to foot, leg, knee, hip, or shoulder
☑	☐	Major injury to back or neck
☐	☐	Stomach or duodenal ulcer
☐	☐	Rectal growth or bleeding

(continued)

Figure 2.4 *(continued)*

You   Your family

❑   ❑   Cataracts

❑   ❑   Glaucoma

❑   ❑   Hearing loss

❑   ❑   Depression

❑   ❑   High anxiety, phobias

❑   ❑   Substance-abuse problems (e.g., alcohol, drugs)

❑   ❑   Eating disorders (anorexia, bulimia)

❑   ❑   Problems with menstruation

❑   ❑   Hysterectomy

❑   ❑   Sleeping problems

❑   ❑   Allergies

❑   ❑   HIV/AIDS

❑   ❑   Any other problems (please be specific and include information on any recent illnesses, hospitalizations, or surgical procedures): _____

_____

_____

Medications

Please check any of the following types of medication that you currently take regularly. Also give the name of the medication.

Medication		Name of medication
❑	Heart medicine	
☑	Blood pressure medicine	*Captopril*
☑	Blood cholesterol medicine	*Vytorin*
❑	Hormones	
☑	Birth control pills	*Seasonale*
❑	Medicine for breathing or lungs	
☑	Insulin	*Humulin N*
❑	Other medicine for diabetes	
❑	Arthritis medicine	
❑	Medicine for depression	
❑	Medicine for anxiety	
❑	Thyroid medicine	
❑	Medicine for ulcers	

❑	Painkiller medicine	_____
❑	Allergy medicine	_____
❑	HIV/AIDS medicine	_____
❑	Hepatitis medicine	_____
❑	Other (please specify)	_____

Physical Fitness, Physical Activity, and Exercise

In general, as compared with other persons your age, rate your physical fitness:

1	2	3	4	5	6	7	8	9	10
❑	❑	❑	☑	❑	❑	❑	❑	❑	❑

Not at all fit   Somewhat fit   Extremely fit

Outside your normal work or daily responsibilities, how often do you engage in exercise that increases your breathing and heart rate at least moderately, and makes you sweat, for at least 20 minutes (e.g., brisk walking, cycling, swimming, aerobic dance, stair-climbing, rowing, basketball, racquetball, vigorous yardwork)?

❑ 5 or more times per week   ❑ 3-4 times per week   ❑ 1-2 times per week
❑ Less than once a week   ☑ Seldom or never

How much hard physical work is required in your job?

❑ A great deal   ❑ A moderate amount   ❑ A little   ☑ None

How long have you exercised or played sports regularly?

☑ Do not exercise regularly   ❑ Less than 1 year   ❑ 1-2 years
❑ 2-5 years   ❑ 5-10 years   ❑ More than 10 years

Diet

On average, how many servings of fruit do you eat per day?
(One serving = 1 medium apple, banana, orange; 1/2 cup chopped, cooked, or canned fruit; 3/4 cup fruit juice.)
☑ None   ❑ 1   ❑ 2   ❑ 3   ❑ 4 or more

On average, how many servings of vegetables do you eat per day?
(One serving = 1/2 cup cooked or chopped raw; 1 cup raw leafy; 3/4 cup vegetable juice.)
❑ None   ☑ 1   ❑ 2   ❑ 3   ❑ 4 or more

On average, how many servings of bread, cereal, rice, or pasta do you eat per day?
❑ None   ❑ 1-3   ☑ 4-6   ❑ 7-9   ❑ 10 or more

When you eat grain and cereal products, which of the following do you emphasize?
❑ Whole grain, high fiber   ❑ Mixture of whole grain and refined   ☑ Refined, low fiber

On average, how many servings of fish, poultry, lean meat, cooked dry beans, peanut butter, or nuts do you eat per day?
(One serving = 2-3 ounces meat; 1/2 cup cooked dry beans; 2 tablespoons peanut butter; 1/3 cup nuts.)
❑ None   ☑ 1   ❑ 2   ❑ 3   ❑ 4 or more

(continued)

Figure 2.4 *(continued)*

On average, how many servings of dairy products do you eat per day?

(One serving = 1 cup milk or yogurt; 1.5 ounces natural cheese; 2 ounces processed cheese.)

❑ None ❑ 1 ☑ 2 ❑ 3 ❑ 4 or more

When you use dairy products, which of the following do you emphasize?

☑ Regular ❑ Low fat ❑ Nonfat

How would you characterize your intake of fat and oil (e.g., regular salad dressing, butter or margarine, mayonnaise, vegetable oil)?

☑ High ❑ Moderate ❑ Low

Body Weight

How tall are you (without shoes)? __5__ feet __5__ inches (1.7 m)

How much do you weigh (with minimal clothing and without shoes)? __190__ pounds (86 kg)

What is the most you have ever weighed? __210__ pounds (95 kg)

Which of the following are you *currently* trying to do?

☑ Lose weight ❑ Gain weight ❑ Stay about the same ❑ Not trying to do anything

Psychological Health

How have you been feeling in general during the past month?

❑ In excellent spirits ❑ In very good spirits ❑ In good spirits

☑ Up and down in spirits a lot ❑ In low spirits ❑ In very low spirits

During the past month, what level of stress would you say that you experienced?

❑ A lot ☑ Moderate ❑ Relatively little ❑ Almost none

In the past year, how much has stress affected your health?

❑ A lot ☑ Some ❑ Hardly or none

On average, how many hours of sleep do you get in a 24-hour period?

❑ Less than 5 hours ☑ 5-6 hours ❑ 7-9 hours ❑ More than 9 hours

Substance Use

Have you smoked at least 100 cigarettes in your life? ☑ Yes ❑ No

How would you describe your cigarette smoking habit?

❑ Never smoked

❑ Used to smoke—how many years since you smoked? _____ years

☑ Currently smoke—how many cigarettes do you smoke on average? __10__ cigarettes/day

How many alcoholic drinks do you consume?

(One drink is a glass of wine, a wine cooler, a bottle or can of beer, a shot glass of liquor, or a mixed drink.)

❑ Never use alcohol ❑ Less than 1 drink per week ☑ 1-6 drinks per week

❑ 1 drink per day ❑ 2-3 drinks per day ❑ More than 3 drinks per day

Occupational Health

Please explain your main job duties: _I am a homemaker. Mostly do things around the house such as cleaning and taking care of the kids._

After a day's work, do you often have pain or stiffness that lasts more than 3 hours?

❏ All the time   ❏ Most of the time   ☑ Some of the time   ❏ Rarely or never

How often does your work entail repetitive pushing and pulling movements, or lifting while bending or twisting, leading to back pain?

❏ All the time   ❏ Most of the time   ❏ Some of the time   ☑ Rarely or never

I hereby state that, to the best of my knowledge, my answers to the preceding questions are complete and correct.

Jane Jones	_Jane Jones_	_10/2/2018_
Printed name of respondent	Signature of respondent	Date (mm/dd/yyyy)

Printed name of parent or guardian	Signature of parent or guardian	Date (mm/dd/yyyy)

James Thomas	_James Thomas_	_10/2/2018_
Printed name of witness	Signature of witness	Date (mm/dd/yyyy)

Adapted, by permission, from D.C. Nieman, 2003, *Exercise testing and prescription: A health-related approach,* 5th ed. (New York, NY: McGraw-Hill), 774. ©The McGraw-Hill Companies.

FORM 2.1

EXERCISE PREPARTICIPATION HEALTH SCREENING QUESTIONNAIRE

Assess your client's health needs by marking all true statements.

Step 1

Symptoms

Does your client experience:

❏ Chest discomfort with exertion

❏ Unreasonable breathlessness

❏ Dizziness, fainting, blackouts

❏ Ankle swelling

❏ Unpleasant awareness of a forceful, rapid, or irregular heart rate

❏ Burning or cramping sensations in your lower legs when walking short distances

If you **did** mark any of these statements under the symptoms, **STOP**. Your client should seek medical clearance before engaging in or resuming exercise. Your client may need to use a facility with a **medically qualified staff**.

If you **did not** mark any symptoms, continue to steps 2 and 3.

Step 2

Current Activity

Has your client performed planned, structured physical activity for at least 30 min at moderate intensity on at least 3 days per week for at least the last 3 months?

❏ Yes

❏ No

Continue to step 3.

Step 3

Medical Conditions

Has your client had or do they currently have:

❏ A heart attack

❏ Heart surgery, cardiac catheterization, or coronary angioplasty

❏ Pacemaker/implantable cardiac defibrillator/rhythm disturbance

❏ Heart valve disease

❏ Heart failure

❏ Heart transplantation

❑ Congestive heart disease

❑ Diabetes

❑ Renal disease

Evaluating Steps 2 and 3

If you **did not mark any of the statements in step 3**, medical clearance is not necessary.

If you marked **"yes"** in step 2 and **marked any of the statements in step 3**, your client may continue to exercise at a light to moderate intensity without medical clearance. Medical clearance is recommended before engaging in vigorous exercise.

If you marked **"no"** in step 2 and **marked any of the statements in step 3**, medical clearance is recommended. Your client may need to use a facility with a **medically qualified staff**.

From G. Haff and C. Dumke, 2019, *Laboratory manual for exercise physiology*, 2nd ed. (Champaign, IL: Human Kinetics). Reprinted, by permission, from M. Magal and D. Riebe, 2016, "New preparticipation health screening recommendations: what exercise professionals need to know," *ACSM Health Fitness Journal* 20(3): 22-27.

FORM 2.2

INFORMED CONSENT FORM

I, _____, have been informed that I will perform a series of tests in order to determine my physical fitness status as well as enhance my understanding of my own health and physical fitness status. I understand that I can voluntarily withdraw from these tests at any time without any penalty. Additionally, I understand that I can ask questions about the tests at any time and will have those questions answered to my satisfaction. In the event of any side effects or injuries related to these tests, I understand that I may contact _____ _____ at any time with my concerns.

Explanation of Tests: I, _____, understand that I will fill out a series of questionnaires, including a health history questionnaire and a PAR-Q+, in order to ensure my safety during the testing process. I understand that if these questionnaires reveal that I am at significant risk of an adverse event, no further tests will be conducted. I understand that if the questionnaires reveal little risk to my safety, I will perform the remainder of the assessments. I will have my blood pressure, body weight, and height assessed with methods typically used in a physician's office. After completing these assessments, I will have my body composition measured by _____, which will evaluate how much fat and fat-free mass my body contains. I will then perform an assessment of my muscular strength and endurance, which will require me to lift weights for a number of repetitions using a _____. After completing this assessment, I will perform a graded exercise test on a _____ in which the workload increases every few minutes until exhaustion or until the test is terminated. I understand that this assessment will give insight into my cardiorespiratory fitness.

Risks and Discomforts: It has been explained to me that during a graded exercise test, certain physical changes can occur, including abnormal blood pressure responses, fainting, irregular heartbeat, and in some instances fatal heart attacks. In order to minimize these risks, the personnel conducting the test are trained to handle such adverse effects. Additionally, they are trained to recognize potential warning signs and will stop any testing if these arise. I understand that the measurement of my height, weight, and body composition are of minimal risk to me. I have also been informed that the assessment of muscular strength and endurance involves some risk of pulling or spraining a muscle and that these risks will be minimized by employing a proper warm-up and by means of technical monitoring by the testing staff. I also understand that after the completion of the testing bouts, I may experience some local muscle soreness that may last for 24 to 48 hours. If muscle soreness does occur, I understand that I can perform a series of stretches that have been demonstrated to me by the testing staff. If these symptoms persist, I will report them to _____ _____.

Benefits from Testing: I understand that the results of these tests will give insight into my overall physical health and wellness. Additionally, I have been informed that this information will reveal any potential health hazards and can be used to better individualize my exercise program.

Inquiries: I understand that if I have any questions, I can ask them of the testing staff. These questions will be answered to my satisfaction by the testing staff.

Confidentiality: I understand that all of my personal health and physical fitness data will be kept confidential.

I have read the information contained in this document and understand it. All questions pertaining to the procedures that I am volunteering to undergo have been answered to my satisfaction. I understand that I am free to decline answering any questions and to withdraw from this testing at any time without penalty. Additionally, I have been informed that all of the information gathered about me and the tests undertaken by me are confidential and will not be disclosed to anyone but me or others in my care or used for exercise prescription without my written permission. Finally, I am aware of all risks associated with this testing and voluntarily give my consent to participate in this testing.

_____   _____
Date   Signature of patient/client

_____   _____
Date   Signature of witness

_____   _____
Date   Signature of supervisor

From G. Haff and C. Dumke, 2019, *Laboratory manual for exercise physiology,* 2nd ed. (Champaign, IL: Human Kinetics). Based on Nieman 2003.

PAR-Q+ FORM

2017 PAR-Q+

The Physical Activity Readiness Questionnaire for Everyone

The health benefits of regular physical activity are clear; more people should engage in physical activity every day of the week. Participating in physical activity is very safe for MOST people. This questionnaire will tell you whether it is necessary for you to seek further advice from your doctor OR a qualified exercise professional before becoming more physically active.

GENERAL HEALTH QUESTIONS

Please read the 7 questions below carefully and answer each one honestly: check YES or NO.	YES	NO
1) Has your doctor ever said that you have a heart condition ☐ OR high blood pressure ☐?	☐	☐
2) Do you feel pain in your chest at rest, during your daily activities of living, **OR** when you do physical activity?	☐	☐
3) Do you lose balance because of dizziness **OR** have you lost consciousness in the last 12 months? Please answer **NO** if your dizziness was associated with over-breathing (including during vigorous exercise).	☐	☐
4) Have you ever been diagnosed with another chronic medical condition (other than heart disease or high blood pressure)? **PLEASE LIST CONDITION(S) HERE:** _____	☐	☐
5) Are you currently taking prescribed medications for a chronic medical condition? **PLEASE LIST CONDITION(S) AND MEDICATIONS HERE:** _____	☐	☐
6) Do you currently have (or have had within the past 12 months) a bone, joint, or soft tissue (muscle, ligament, or tendon) problem that could be made worse by becoming more physically active? Please answer **NO** if you had a problem in the past, but it *does not limit your current ability* to be physically active. **PLEASE LIST CONDITION(S) HERE:** _____	☐	☐
7) Has your doctor ever said that you should only do medically supervised physical activity?	☐	☐

☑ **If you answered NO to all of the questions above, you are cleared for physical activity.**
Go to Page 4 to sign the PARTICIPANT DECLARATION. You do not need to complete Pages 2 and 3.

▶ Start becoming much more physically active – start slowly and build up gradually.

▶ Follow International Physical Activity Guidelines for your age (www.who.int/dietphysicalactivity/en/).

▶ You may take part in a health and fitness appraisal.

▶ If you are over the age of 45 yr and **NOT** accustomed to regular vigorous to maximal effort exercise, consult a qualified exercise professional before engaging in this intensity of exercise.

▶ If you have any further questions, contact a qualified exercise professional.

⬣ **If you answered YES to one or more of the questions above, COMPLETE PAGES 2 AND 3.**

⚠ **Delay becoming more active if:**

✓ You have a temporary illness such as a cold or fever; it is best to wait until you feel better.

✓ You are pregnant - talk to your health care practitioner, your physician, a qualified exercise professional, and/or complete the ePARmed-X+ at **www.eparmedx.com** before becoming more physically active.

✓ Your health changes - answer the questions on Pages 2 and 3 of this document and/or talk to your doctor or a qualified exercise professional before continuing with any physical activity program.

†OSHF
Ontario Society for Health and Fitness

Copyright © 2017 PAR-Q+ Collaboration   1 / 4
01-01-2017

From G. Haff and C. Dumke, 2019, *Laboratory manual for exercise physiology,* 2nd ed. (Champaign, IL: Human Kinetics). Reprinted with permission from the PAR-Q+ Collaboration and the authors of the PAR-Q+ (Dr. Darren Warburton, Dr. Norman Gledhill, Dr. Veronica Jamnik, and Dr. Shannon Bredin).

2017 PAR-Q+

FOLLOW-UP QUESTIONS ABOUT YOUR MEDICAL CONDITION(S)

1.   Do you have Arthritis, Osteoporosis, or Back Problems?

If the above condition(s) is/are present, answer questions 1a-1c   If **NO** ☐ go to question 2

1a.	Do you have difficulty controlling your condition with medications or other physician-prescribed therapies? (Answer **NO** if you are not currently taking medications or other treatments)	YES☐ NO☐
1b.	Do you have joint problems causing pain, a recent fracture or fracture caused by osteoporosis or cancer, displaced vertebra (e.g., spondylolisthesis), and/or spondylolysis/pars defect (a crack in the bony ring on the back of the spinal column)?	YES☐ NO☐
1c.	Have you had steroid injections or taken steroid tablets regularly for more than 3 months?	YES☐ NO☐

2.   Do you currently have Cancer of any kind?

If the above condition(s) is/are present, answer questions 2a-2b   If **NO** ☐ go to question 3

2a.	Does your cancer diagnosis include any of the following types: lung/bronchogenic, multiple myeloma (cancer of plasma cells), head, and/or neck?	YES☐ NO☐
2b.	Are you currently receiving cancer therapy (such as chemotheraphy or radiotherapy)?	YES☐ NO☐

3.   Do you have a Heart or Cardiovascular Condition? *This includes Coronary Artery Disease, Heart Failure, Diagnosed Abnormality of Heart Rhythm*

If the above condition(s) is/are present, answer questions 3a-3d   If **NO** ☐ go to question 4

3a.	Do you have difficulty controlling your condition with medications or other physician-prescribed therapies? (Answer **NO** if you are not currently taking medications or other treatments)	YES☐ NO☐
3b.	Do you have an irregular heart beat that requires medical management? (e.g., atrial fibrillation, premature ventricular contraction)	YES☐ NO☐
3c.	Do you have chronic heart failure?	YES☐ NO☐
3d.	Do you have diagnosed coronary artery (cardiovascular) disease and have not participated in regular physical activity in the last 2 months?	YES☐ NO☐

4.   Do you have High Blood Pressure?

If the above condition(s) is/are present, answer questions 4a-4b   If **NO** ☐ go to question 5

4a.	Do you have difficulty controlling your condition with medications or other physician-prescribed therapies? (Answer **NO** if you are not currently taking medications or other treatments)	YES☐ NO☐
4b.	Do you have a resting blood pressure equal to or greater than 160/90 mmHg with or without medication? (Answer **YES** if you do not know your resting blood pressure)	YES☐ NO☐

5.   Do you have any Metabolic Conditions? *This includes Type 1 Diabetes, Type 2 Diabetes, Pre-Diabetes*

If the above condition(s) is/are present, answer questions 5a-5e   If **NO** ☐ go to question 6

5a.	Do you often have difficulty controlling your blood sugar levels with foods, medications, or other physician-prescribed therapies?	YES☐ NO☐
5b.	Do you often suffer from signs and symptoms of low blood sugar (hypoglycemia) following exercise and/or during activities of daily living? Signs of hypoglycemia may include shakiness, nervousness, unusual irritability, abnormal sweating, dizziness or light-headedness, mental confusion, difficulty speaking, weakness, or sleepiness.	YES☐ NO☐
5c.	Do you have any signs or symptoms of diabetes complications such as heart or vascular disease and/or complications affecting your eyes, kidneys, **OR** the sensation in your toes and feet?	YES☐ NO☐
5d.	Do you have other metabolic conditions (such as current pregnancy-related diabetes, chronic kidney disease, or liver problems)?	YES☐ NO☐
5e.	Are you planning to engage in what for you is unusually high (or vigorous) intensity exercise in the near future?	YES☐ NO☐

OSHF
Ontario Society for Health and Fitness

Copyright © 2017 PAR-Q+ Collaboration   2 / 4
01-01-2017

(continued)

From G. Haff and C. Dumke, 2019, *Laboratory manual for exercise physiology,* 2nd ed. (Champaign, IL: Human Kinetics). Reprinted with permission from the PAR-Q+ Collaboration and the authors of the PAR-Q+ (Dr. Darren Warburton, Dr. Norman Gledhill, Dr. Veronica Jamnik, and Dr. Shannon Bredin).

2017 PAR-Q+

6. **Do you have any Mental Health Problems or Learning Difficulties?** *This includes Alzheimer's, Dementia, Depression, Anxiety Disorder, Eating Disorder, Psychotic Disorder, Intellectual Disability, Down Syndrome*

If the above condition(s) is/are present, answer questions 6a-6b   If **NO** ☐ go to question 7

6a.	Do you have difficulty controlling your condition with medications or other physician-prescribed therapies? (Answer **NO** if you are not currently taking medications or other treatments)	YES☐ NO☐
6b.	Do you have Down Syndrome **AND** back problems affecting nerves or muscles?	YES☐ NO☐

7. **Do you have a Respiratory Disease?** *This includes Chronic Obstructive Pulmonary Disease, Asthma, Pulmonary High Blood Pressure*

If the above condition(s) is/are present, answer questions 7a-7d   If **NO** ☐ go to question 8

7a.	Do you have difficulty controlling your condition with medications or other physician-prescribed therapies? (Answer **NO** if you are not currently taking medications or other treatments)	YES☐ NO☐
7b.	Has your doctor ever said your blood oxygen level is low at rest or during exercise and/or that you require supplemental oxygen therapy?	YES☐ NO☐
7c.	If asthmatic, do you currently have symptoms of chest tightness, wheezing, laboured breathing, consistent cough (more than 2 days/week), or have you used your rescue medication more than twice in the last week?	YES☐ NO☐
7d.	Has your doctor ever said you have high blood pressure in the blood vessels of your lungs?	YES☐ NO☐

8. **Do you have a Spinal Cord Injury?** *This includes Tetraplegia and Paraplegia*

If the above condition(s) is/are present, answer questions 8a-8c   If **NO** ☐ go to question 9

8a.	Do you have difficulty controlling your condition with medications or other physician-prescribed therapies? (Answer **NO** if you are not currently taking medications or other treatments)	YES☐ NO☐
8b.	Do you commonly exhibit low resting blood pressure significant enough to cause dizziness, light-headedness, and/or fainting?	YES☐ NO☐
8c.	Has your physician indicated that you exhibit sudden bouts of high blood pressure (known as Autonomic Dysreflexia)?	YES☐ NO☐

9. **Have you had a Stroke?** *This includes Transient Ischemic Attack (TIA) or Cerebrovascular Event*

If the above condition(s) is/are present, answer questions 9a-9c   If **NO** ☐ go to question 10

9a.	Do you have difficulty controlling your condition with medications or other physician-prescribed therapies? (Answer **NO** if you are not currently taking medications or other treatments)	YES☐ NO☐
9b.	Do you have any impairment in walking or mobility?	YES☐ NO☐
9c.	Have you experienced a stroke or impairment in nerves or muscles in the past 6 months?	YES☐ NO☐

10. **Do you have any other medical condition not listed above or do you have two or more medical conditions?**

If you have other medical conditions, answer questions 10a-10c   If **NO** ☐ read the Page 4 recommendations

10a.	Have you experienced a blackout, fainted, or lost consciousness as a result of a head injury within the last 12 months **OR** have you had a diagnosed concussion within the last 12 months?	YES☐ NO☐
10b.	Do you have a medical condition that is not listed (such as epilepsy, neurological conditions, kidney problems)?	YES☐ NO☐
10c.	Do you currently live with two or more medical conditions?	YES☐ NO☐

PLEASE LIST YOUR MEDICAL CONDITION(S) AND ANY RELATED MEDICATIONS HERE: _____

_____

GO to Page 4 for recommendations about your current medical condition(s) and sign the PARTICIPANT DECLARATION.

OSHF   Copyright © 2017 PAR-Q+ Collaboration   3 / 4
01-01-2017

From G. Haff and C. Dumke, 2019, *Laboratory manual for exercise physiology*, 2nd ed. (Champaign, IL: Human Kinetics). Reprinted with permission from the PAR-Q+ Collaboration and the authors of the PAR-Q+ (Dr. Darren Warburton, Dr. Norman Gledhill, Dr. Veronica Jamnik, and Dr. Shannon Bredin).

2017 PAR-Q+

☑ **If you answered NO to all of the follow-up questions about your medical condition, you are ready to become more physically active - sign the PARTICIPANT DECLARATION below:**

▶ It is advised that you consult a qualified exercise professional to help you develop a safe and effective physical activity plan to meet your health needs.

▶ You are encouraged to start slowly and build up gradually - 20 to 60 minutes of low to moderate intensity exercise, 3-5 days per week including aerobic and muscle strengthening exercises.

▶ As you progress, you should aim to accumulate 150 minutes or more of moderate intensity physical activity per week.

▶ If you are over the age of 45 yr and **NOT** accustomed to regular vigorous to maximal effort exercise, consult a qualified exercise professional before engaging in this intensity of exercise.

⬤ **If you answered YES to one or more of the follow-up questions** about your medical condition:
You should seek further information before becoming more physically active or engaging in a fitness appraisal. You should complete the specially designed online screening and exercise recommendations program - the **ePARmed-X+ at www.eparmedx.com** and/or visit a qualified exercise professional to work through the ePARmed-X+ and for further information.

⚠ **Delay becoming more active if:**

✓ You have a temporary illness such as a cold or fever; it is best to wait until you feel better.

✓ You are pregnant - talk to your health care practitioner, your physician, a qualified exercise professional, and/or complete the ePARmed-X+ **at www.eparmedx.com** before becoming more physically active.

✓ Your health changes - talk to your doctor or qualified exercise professional before continuing with any physical activity program.

⬤ You are encouraged to photocopy the PAR-Q+. You must use the entire questionnaire and NO changes are permitted.
⬤ The authors, the PAR-Q+ Collaboration, partner organizations, and their agents assume no liability for persons who undertake physical activity and/or make use of the PAR-Q+ or ePARmed-X+. If in doubt after completing the questionnaire, consult your doctor prior to physical activity.

PARTICIPANT DECLARATION

⬤ All persons who have completed the PAR-Q+ please read and sign the declaration below.

⬤ If you are less than the legal age required for consent or require the assent of a care provider, your parent, guardian or care provider must also sign this form.

I, the undersigned, have read, understood to my full satisfaction and completed this questionnaire. I acknowledge that this physical activity clearance is valid for a maximum of 12 months from the date it is completed and becomes invalid if my condition changes. I also acknowledge that a Trustee (such as my employer, community/fitness centre, health care provider, or other designate) may retain a copy of this form for their records. In these instances, the Trustee will be required to adhere to local, national, and international guidelines regarding the storage of personal health information ensuring that the Trustee maintains the privacy of the information and does not misuse or wrongfully disclose such information.

NAME _____ DATE _____

SIGNATURE _____ WITNESS _____

SIGNATURE OF PARENT/GUARDIAN/CARE PROVIDER _____

—— **For more information, please contact** ——
www.eparmedx.com
Email: eparmedx@gmail.com

Citation for PAR-Q+
Warburton DER, Jamnik VK, Bredin SSD, and Gledhill N on behalf of the PAR-Q+ Collaboration. The Physical Activity Readiness Questionnaire for Everyone (PAR-Q+) and Electronic Physical Activity Readiness Medical Examination (ePARmed-X+). Health & Fitness Journal of Canada 4(2):3-23, 2011.

Key References
1. Jamnik VK, Warburton DER, Makarski J, McKenzie DC, Shephard RJ, Stone J, and Gledhill N. Enhancing the effectiveness of clearance for physical activity participation; background and overall process. APNM 36(S1):S3-S13, 2011.
2. Warburton DER, Gledhill N, Jamnik VK, Bredin SSD, McKenzie DC, Stone J, Charlesworth S, and Shephard RJ. Evidence-based risk assessment and recommendations for physical activity clearance; Consensus Document. APNM 36(S1):S266-s298, 2011.
3. Chisholm DM, Collis ML, Kulak LL, Davenport W, and Gruber N. Physical activity readiness. British Columbia Medical Journal. 1975;17:375-378.
4. Thomas S, Reading J, and Shephard RJ. Revision of the Physical Activity Readiness Questionnaire (PAR-Q). Canadian Journal of Sport Science 1992;17:4 338-345.

The PAR-Q+ was created using the evidence-based AGREE process (1) by the PAR-Q+ Collaboration chaired by Dr. Darren E. R. Warburton with Dr. Norman Gledhill, Dr. Veronica Jamnik, and Dr. Donald C. McKenzie (2). Production of this document has been made possible through financial contributions from the Public Health Agency of Canada and the BC Ministry of Health Services. The views expressed herein do not necessarily represent the views of the Public Health Agency of Canada or the BC Ministry of Health Services.

✚ **OSHF**
Ontario Society for Health and Fitness

Copyright © 2017 PAR-Q+ Collaboration   4 /4
01-01-2017

From G. Haff and C. Dumke, 2019, *Laboratory manual for exercise physiology*, 2nd ed. (Champaign, IL: Human Kinetics). Reprinted with permission from the PAR-Q+ Collaboration and the authors of the PAR-Q+ (Dr. Darren Warburton, Dr. Norman Gledhill, Dr. Veronica Jamnik, and Dr. Shannon Bredin).

FORM 2.4

HEALTH HISTORY QUESTIONNAIRE

Contact Information

_____ _____ _____
Last name   First name   Middle initial

                                                    Sex: ❏ *Male* ❏ *Female*

_____ _____
Date of birth (mm/dd/yyyy) Age

_____ _____ _____ _____
Home phone number   Street address   City, state   Zip code

_____ _____ _____
Emergency contact   Emergency contact number Relationship to emergency contact

_____ _____
Primary physician   Physician's contact number

Background

Please circle the highest school grade you have completed:

Elementary school:	1	2	3	4	5	6	7	8
High school:	9	11	12					
College/postgraduate:	13	14	15	16	17	18	19	20+

What is your marital status? ❏ Single ❏ Married ❏ Widowed ❏ Divorced or separated

What is your race or ethnic background?

❏ White, not of Hispanic origin ❏ American Indian/Alaska Native ❏ Asian

❏ Black, not of Hispanic origin ❏ Pacific Islander ❏ Hispanic

What is your occupation?

❏ Health professional ❏ Disabled, unable to work ❏ Service ❏ Manager, educator, professional

❏ Skilled craftsperson ❏ Operator, fabricator, laborer ❏ Homemaker ❏ Technical, sales, support

❏ Retired ❏ Unemployed ❏ Student ❏ Other

Symptoms or Signs Suggestive of Disease

Please place a check in the appropriate box:

Yes	No	
❏	❏	Are you a male over age 45 or a female over age 55?
❏	❏	Have you experienced any unusual pain or discomfort in your chest, neck, jaw, arms, or other areas that may be due to heart problems?
❏	❏	Have you experienced unusual fatigue or shortness of breath at rest, during usual activities (e.g., climbing stairs, carrying groceries, walking briskly), or during mild or moderate exercise?

Yes	No	
❏	❏	Have you ever had any problems with dizziness or fainting?
❏	❏	When you stand up, do you have difficulty breathing?
❏	❏	Do you have difficulty breathing while sleeping?
❏	❏	Do your ankles swell (ankle edema)?
❏	❏	Have you ever experienced an unusual or rapid heartbeat or fluttering of the heart?
❏	❏	Have you experienced severe pain in your legs while walking?
❏	❏	Have you ever been told by a doctor that you have a heart murmur?

Chronic Disease Risk Factors

Do you know your blood pressure? ❏ = Yes: _____ / _____ systolic/diastolic ❏ = No

Do you know your cholesterol levels? ❏ = Yes: _____ total cholesterol; _____ HDL; _____ LDL ❏ = No

Do you know your fasting glucose levels? ❏ = Yes: _____ mg/dl ❏ = No

Please place a check in the appropriate box:

Yes	No	
❏	❏	Are you a male over the age of 45 or a female over the age of 55?
❏	❏	If you are a female, have you experienced premature menopause and are *not* on estrogen replacement therapy?
❏	❏	Has your father or brother had a heart attack or died suddenly from heart disease before the age of 55?
❏	❏	Has your mother or sister had a heart attack or died suddenly from heart disease before the age of 65?
❏	❏	Has anyone in your family died before the age of 40 (excluding accidental death)?
❏	❏	Are you a current cigarette smoker?
❏	❏	Has a doctor told you that you have high blood pressure (more than 130/80 mmHG)?
❏	❏	Are you on medication to control high blood pressure?
❏	❏	Has a doctor ever told you that you have high cholesterol?
❏	❏	Is your serum cholesterol greater than 200 mg/dl?
❏	❏	Do you have diabetes mellitus?
❏	❏	Are you physically inactive or sedentary (i.e., do you perform little physical activity on the job or during leisure time)?
❏	❏	During the past year, have you experienced levels of stress, strain, and pressure that might affect your health?
❏	❏	Do you eat foods that are high in fat and cholesterol (e.g., fatty meats, cheese, fried food, butter, whole milk, eggs) on a daily basis?
❏	❏	Do you tend to avoid foods that are high in fiber (e.g., whole-grain breads and cereals, fresh fruits, vegetables)?
❏	❏	Do you weigh 30 pounds (14 kg) more than you should?
❏	❏	Do you average more than two alcoholic beverages a day?

(continued)

Form 2.4 *(continued)*

Medical History

Please check all conditions that you or your family have had or now have.

You   Your family

❏   ❏   Coronary heart disease, heart attack, coronary artery surgery

❏   ❏   Angina

❏   ❏   Peripheral vascular disease

❏   ❏   Phlebitis or emboli

❏   ❏   Other heart problems (specify: _____)

❏   ❏   Lung cancer

❏   ❏   Breast cancer

❏   ❏   Prostate cancer

❏   ❏   Colorectal cancer (bowel cancer)

❏   ❏   Skin cancer

❏   ❏   Other cancer (specify: _____)

❏   ❏   Stroke

❏   ❏   Chronic obstructive pulmonary disease (COPD; e.g., emphysema)

❏   ❏   Pneumonia

❏   ❏   Asthma

❏   ❏   Bronchitis

❏   ❏   Diabetes mellitus

❏   ❏   Thyroid problems

❏   ❏   Kidney disease

❏   ❏   Liver disease (cirrhosis of the liver)

❏   ❏   Hepatitis (A, B, C, D, or E)

❏   ❏   Gallstone/gallbladder disease

❏   ❏   Osteoporosis

❏   ❏   Arthritis

❏   ❏   Gout

❏   ❏   Anemia (low iron)

You   Your family

❏   ❏   Bone fracture

❏   ❏   Major injury to foot, leg, knee, hip, or shoulder

❏   ❏   Major injury to back or neck

❏   ❏   Stomach or duodenal ulcer

❏   ❏   Rectal growth or bleeding

❑	❑	Cataracts
❑	❑	Glaucoma
❑	❑	Hearing loss
❑	❑	Depression
❑	❑	High anxiety, phobia
❑	❑	Substance abuse problems (e.g., alcohol, drugs)
❑	❑	Eating disorders (anorexia, bulimia)
❑	❑	Problems with menstruation
❑	❑	Hysterectomy
❑	❑	Sleeping problems
❑	❑	Allergies
❑	❑	HIV/AIDS
❑	❑	Any other problems (please be specific and include information on any recent illnesses, hospitalizations, or surgical procedures): _____

_____

_____

Medications

Please check any of the following types of medication that you currently take regularly. Also give the name of the medication.

Type of medication   Name of medication

❑ Heart medicine   _____

❑ Blood pressure medicine   _____

❑ Blood cholesterol medicine   _____

❑ Hormones   _____

❑ Birth control pills   _____

❑ Medicine for breathing or lungs _____

❑ Insulin   _____

❑ Other medicine for diabetes   _____

❑ Arthritis medicine   _____

❑ Medicine for depression   _____

❑ Medicine for anxiety   _____

❑ Thyroid medicine   _____

❑ Medicine for ulcers   _____

(continued)

Form 2.4 *(continued)*

❏ Painkiller medicine   _____

❏ Allergy medicine   _____

❏ HIV/AIDS medicine   _____

❏ Hepatitis medicine   _____

❏ Other (please specify)   _____

Physical Fitness, Physical Activity, and Exercise

In general, as compared with other persons your age, rate your physical fitness:

1	2	3	4	5	6	7	8	9	10
❏	❏	❏	❏	❏	❏	❏	❏	❏	❏

Not at all fit   Somewhat fit   Extremely fit

Outside your normal work or daily responsibilities, how often do you engage in exercise that at least moderately increases your breathing and heart rate, and makes you sweat, for at least 20 minutes (e.g., brisk walking, cycling, swimming, aerobic dance, stair-climbing, rowing, basketball, racquetball, vigorous yardwork)?

❏ 5 or more times per week   ❏ 3-4 times per week   ❏ 1-2 times per week

❏ Less than once a week   ❏ Seldom or never

How much hard physical work is required in your job?

❏ A great deal   ❏ A moderate amount   ❏ A little   ❏ None

How long have you exercised or played sports regularly?

❏ I do not exercise regularly   ❏ Less than 1 year   ❏ 1-2 years

❏ 2-5 years   ❏ 5-10 years   ❏ More than 10 years

Diet

On average, how many servings of fruit do you eat per day?

(One serving = 1 medium apple, banana, orange; 1/2 cup chopped, cooked, or canned fruit; 3/4 cup fruit juice.)

❏ None   ❏ 1   ❏ 2   ❏ 3   ❏ 4 or more

On average, how many servings of vegetables do you eat per day?

(One serving = 1/2 cup cooked or chopped raw; 1 cup raw leafy; 3/4 cup vegetable juice.)

❏ None   ❏ 1   ❏ 2   ❏ 3   ❏ 4 or more

On average, how many servings of bread, cereal, rice, or pasta do you eat per day?

❏ None   ❏ 1-3   ❏ 4-6   ❏ 7-9   ❏ 10 or more

When you eat grain and cereal products, which of the following do you emphasize?

❏ Whole grain, high fiber   ❏ Mixture of whole grain and refined   ❏ Refined, low fiber

On average, how many servings of fish, poultry, lean meat, cooked dry beans, peanut butter, or nuts do you eat per day?

(One serving = 2-3 ounces meat; 1/2 cup cooked dry beans; 2 tablespoons peanut butter; 1/3 cup nuts.)

❏ None   ❏ 1   ❏ 2   ❏ 3   ❏ 4 or more

On average, how many servings of dairy products do you eat per day?

(One serving = 1 cup milk or yogurt; 1.5 ounces natural cheese; 2 ounces processed cheese.)

❑ Non ❑ 1 ❑ 2 ❑ 3 ❑ 4 or more

When you use dairy products, which of the following do you emphasize?

❑ Regular ❑ Low fat ❑ Nonfat

How would you characterize your intake of fat and oil (e.g., regular salad dressing, butter or margarine, mayonnaise, vegetable oil)?

❑ High ❑ Moderate ❑ Low

Body Data

How tall are you (without shoes)? _____ feet _____ inches

How much do you weigh (with minimal clothing and without shoes)? _____ pounds

What is the most you have ever weighed? _____ pounds

Which of the following are you *currently* trying to do?

❑ Lose weight ❑ Gain weight ❑ Stay about the same ❑ Not trying to do anything

Psychological Health

How have you been feeling in general during the past month?

❑ In excellent spirits ❑ In very good spirits ❑ In good spirits
❑ Up and down in spirits a lot ❑ In low spirits ❑ In very low spirits

During the past month, what level of stress would you say that you experienced?

❑ A lot ❑ Moderate ❑ Relatively little ❑ Almost none

In the past year, how much has stress affected your health?

❑ A lot ❑ Some ❑ Hardly or none

On average, how many hours of sleep do you get in a 24-hour period?

❑ Less than 5 hours ❑ 5-6 hours ❑ 7-9 hours ❑ More than 9 hours

Substance Use

Have you smoked at least 100 cigarettes in your entire life? ❑ Yes ❑ No

How would you describe your cigarette smoking habit?

❑ Never smoked

❑ Used to smoke—how many years has it been since you smoked? _____ years

❑ Currently smoke—how many cigarettes do you smoke on average? _____ cigarettes/day

How many alcoholic drinks do you consume?

(One drink is a glass of wine, a wine cooler, a bottle or can of beer, a shot glass of liquor, or a mixed drink.)

❑ None ❑ Less than 1 drink per week ❑ 1-6 drinks per week
❑ 1 drink per day ❑ 2-3 drinks per day ❑ More than 3 drinks per day

(continued)

Form 2.4 *(continued)*

Occupational Health

Please explain your main job duties: _____

_____

_____

After a day's work, do you often have pain or stiffness that lasts more than 3 hours?

❑ All the time   ❑ Most of the time   ❑ Some of the time   ❑ Rarely or never

How often does your work entail repetitive pushing and pulling movements, or lifting while bending or twisting, leading to back pain?

❑ All the time   ❑ Most of the time   ❑ Some of the time   ❑ Rarely or never

I hereby state that, to the best of my knowledge, my answers to the preceding questions are complete and correct.

_____   _____   _____

Printed name of respondent   Signature of respondent   Date (mm/dd/yy)

_____   _____   _____

Printed name of parent of guardian   Signature of parent or guardian   Date (mm/dd/yy)

_____   _____   _____

Printed name of witness   Signature of witness   Date (mm/dd/yy)

From G. Haff and C. Dumke, 2019, *Laboratory manual for exercise physiology*, 2nd ed. (Champaign, IL: Human Kinetics). Adapted, by permission, from D.C. Nieman, 2003, *Exercise testing and prescription: A health-related approach*, 5th ed. (New York: McGraw-Hill), 774. © The McGraw-Hill Companies.

FORM 2.5

CHECKLIST FOR SIGNS AND SYMPTOMS OF DISEASE

Instructions: Ask if the individual has any of the following conditions and risk factors. If so, refer the individual to a physician to obtain a signed medical clearance prior to any exercise testing or participation.

_____   _____   _____

Last name   First name   Middle initial

Cardiovascular

Yes	No	Condition	Comments
❑	❑	Bone fracture	_____
❑	❑	Hypertension	_____
❑	❑	Hypercholesterolemia	_____
❑	❑	Heart murmur	_____
❑	❑	Myocardial infarction	_____
❑	❑	Fainting/dizziness	_____
❑	❑	Claudication	_____
❑	❑	Chest pain	_____
❑	❑	Palpitation	_____
❑	❑	Ischemia	_____
❑	❑	Tachycardia	_____
❑	❑	Ankle edema	_____
❑	❑	Stroke	_____

Pulmonary

Yes	No	Condition	Comments
❑	❑	Asthma	_____
❑	❑	Bronchitis	_____
❑	❑	Emphysema	_____
❑	❑	Nocturnal dyspnea	_____
❑	❑	Coughing up blood	_____
❑	❑	Exercise-induced asthma	_____
❑	❑	Breathlessness during or after mild exertion	_____

(continued)

Form 2.5 *(continued)*

Metabolic

Yes	No	Condition	Comments
❑	❑	Diabetes	_____
❑	❑	Obesity	_____
❑	❑	Glucose intolerance	_____
❑	❑	McArdle syndrome	_____
❑	❑	Hypoglycemia	_____
❑	❑	Thyroid disease	_____
❑	❑	Cirrhosis	_____

Musculoskeletal

Yes	No	Condition	Comments
❑	❑	Osteoporosis	_____
❑	❑	Osteoarthritis	_____
❑	❑	Low back pain	_____
❑	❑	Prosthesis	_____
❑	❑	Muscular atrophy	_____
❑	❑	Swollen joints	_____
❑	❑	Orthopedic pain	_____
❑	❑	Artificial joint	_____

Risk Factors

An individual who has two or more of these factors should consult a physician for clearance to participate in exercise testing.

❑ ❑ Male older than 45 _____

❑ ❑ Female older than 55 or has had a hysterectomy or is postmenopausal _____

❑ ❑ Smoking or quit smoking within past 6 months _____

❑ ❑ Blood pressure >130/80 mmHg _____

❑ ❑ Doesn't know blood pressure _____

❑ ❑ Blood cholesterol >200 mg/dl _____

❑ ❑ Doesn't know blood cholesterol _____

❑ ❑ Close relative had a heart attack before age 55 (father or brother) or age 65 (mother or sister)

❑ ❑ Physically inactive (does not engage in >30 min of physical activity more than 4 days per week)

❑ ❑ Overweight by more than 20 lb (9 kg) _____

From G. Haff and C. Dumke, 2019, *Laboratory manual for exercise physiology*, 2nd ed. (Champaign, IL: Human Kinetics). Adapted, by permission, from V. Heyward and A.L. Gibson, 2014, *Advanced fitness assessment and exercise prescription*, 7th ed. (Champaign, IL: Human Kinetics), 368-369.

RISK FACTORS FOR CORONARY HEART DISEASE

Risk factors	Criteria	Yes	No	Points
POSITIVE RISK FACTORS				
age	Men ≥45 y; Women ≥55 y	❑	❑	
Family history	Myocardial infarction, coronary revascularization, or sudden death before age 55 for father or first-degree male relative (brother or son) or before age 65 for mother or first-degree female relative (sister or daughter)	❑	❑	
Cigarette smoking	Current cigarette smoking, smoking cessation within previous 6 months, or exposure to environmental tobacco smoke	❑	❑	
Hypertension	SBP ≥140 mmHg or DBP ≥90 mmHg measured on two separate occasions or use of antihypertensive medication	❑	❑	
Dyslipidemia	Total cholesterol ≥200 mg · dl^{-1}, HDL-C <40 mg · dl^{-1}, LDL-C ≥130 mg · dl^{-1}, or on lipid-lowering medication	❑	❑	
Impaired fasting glucose	Fasting glucose ≥100 mg · dl^{-1} but <126 mg · dl^{-1}, or 2 h OGTT (oral glucose tolerance test) value ≥140 mg · dl^{-1} but <200 mg · dl^{-1} as measured on two separate occasions	❑	❑	
Obesity	BMI ≥30 kg · m^{-2} or waist circumference >102 cm (40 in.) for men or >88 cm (35 in.) for women	❑	❑	
Physical inactivity	Not participating in a regular exercise program or not meeting minimum physical activity recommendations from the U.S. Surgeon General's report (accumulating 30 min or more of moderate [40%-60% $\dot{V}O_2R$] physical activity on most days of the week for at least 3 months)	❑	❑	

TOTAL POSITIVE RISK FACTOR POINTS =

	NEGATIVE RISK FACTORS			
High HDL-C	Serum HDL-C ≥60 mg · dl^{-1}	❑	❑	

TOTAL NEGATIVE RISK FACTOR POINTS =

TOTAL RISK FACTOR POINTS =

Add 1 point if the individual has the positive risk factor. If HDL-C is elevated, subtract 1 point from the sum of the positive risk factors.

From G. Haff and C. Dumke, 2019, *Laboratory manual for exercise physiology*, 2nd ed. (Champaign, IL: Human Kinetics). Adapted from American College of Sports Medicine 2010.

FORM 2.7

FANTASTIC LIFESTYLE CHECKLIST

Instructions: Unless otherwise specified, place an *X* beside the box that best describes your behavior or situation in the past month. Explanations of questions and scoring are provided at the end of the form.

Family and friends	I have someone to talk to about things that are important to me.	Almost never	Seldom	Some of the time	Fairly often	Almost always
	I give and receive affection.	Almost never	Seldom	Some of the time	Fairly often	Almost always
Physical activity	I am vigorously active for at least 30 min per day (e.g., running, cycling).	Less than 1 time per week	1-2 times per week	3 times per week	4 times per week	5 or more times per week
	I am moderately active (e.g., gardening, climbing stairs, walking, doing housework).	Less than 1 time per week	1-2 times per week	3 times per week	4 times per week	5 or more times per week
Nutrition	I eat a balanced diet (see explanation).	Almost never	Seldom	Some of the time	Fairly often	Almost always
	I often eat excess sugar, salt, animal fat, or junk food.	Four of these	Three of these	Two of these	One of these	None of these
	I am within _____ kg of my healthy weight.	Not within 8 kg (20 lb)	8 kg (20 lb)	6 kg (15 lb)	4 kg (10 lb)	2 kg (5 lb)
Tobacco and toxics	I smoke tobacco.	More than 10 times per week	1-10 times per week	None in the past 6 months	None in the past year	None in the past 5 years
	I use drugs such as marijuana or cocaine.	Sometimes	———			Never
	I overuse prescribed or over-the-counter drugs.	Almost daily	Fairly often	Only occasionally	Almost never	Never
	I drink caffeine-containing coffee, tea, or cola.	More than 10 per day	7-10 per day	3-6 per day	1-2 per day	never
Alcohol	My average alcohol intake per week is _____ (see explanation).	More than 20 drinks	13-20 drinks	11-12 drinks	8-10 drinks	0-7 drinks
	I drink more than four drinks on an occasion.	Almost daily	Fairly often	Only occasionally	Almost never	Never
	I drive after drinking.	Sometimes	———			Never

Sleep, seat belt use, stress, and safe sex	I sleep well and feel rested.	Almost never		Seldom		Some of the time		Fairly often	Almost always
	I use seat belts.	Never		Seldom		Some of the time		Most of the time	Always
	I am able to cope with the stresses in my life.	Almost never		Seldom		Some of the time		Fairly often	Almost always
	I relax and enjoy leisure time.	Almost never		Seldom		Some of the time		Fairly often	Almost always
	I practice safe sex (see explanation).	Almost never		Seldom		Some of the time		Fairly often	Always
Type of behavior	I seem to be in a hurry.	Almost always		Fairly often		Some of the time		Seldom	Almost never
	I feel angry or hostile	Almost always		Fairly often		Some of the time		Seldom	Almost never
Insight	I am a positive or optimistic thinker.	Almost never		Seldom		Some of the time		Fairly often	Almost always
	I feel tense or uptight.	Almost always		Fairly often		Some of the time		Seldom	Almost never
	I feel sad or depressed.	Almost always		Fairly often		Some of the time		Seldom	Almost never
Career	I am satisfied with my job or role.	Almost never		Seldom		Some of the time		Fairly often	Almost always

	TEST SCORING								
Step 1	Total the *X*s in each column.	____		____		____		____	____
Step 2	Multiply the totals by the numbers indicated (write your answer in the box).	× 0		× 1		× 2		× 3	× 4
Step 3	Add your scores across the bottom for the grand total.			____	+	____	+	____ +	____
	Grand total								=

(continued)

Form 2.7 *(continued)*

Balanced Diet

According to Canada's Food Guide (for people 4 y and older), different people need different amounts of food. The amount of food you need every day from the four food groups and other foods depends on your age, body size, activity level, whether you are male or female, and whether you are pregnant or breastfeeding. That's why the Food Guide gives a lower and higher number of servings for each food group. For example, young children can choose the lower number of servings, while male teenagers can select the higher number. Most other people can choose servings somewhere in between.

Grain products	Vegetables and fruit	Milk products	Meat and alternatives	Other foods
Choose whole-grain and enriched products more often.	Choose dark-green and orange vegetables more often.	Choose lower-fat milk products more often.	Choose leaner meats, poultry, and fish, as well as dried peas, beans, and lentils, more often.	Taste and enjoyment can also come from other foods and beverages that are not part of the four food groups. Some of these are higher in fat or calories, so use these foods in moderation.
RECOMMENDED NUMBER OF SERVINGS PER DAY				
5-12	5-10	Children aged 4-9: 2-3 Youth aged 10-16: 3-4 Adults: 2-4 Pregnant and breast-feeding women: 3-4	2-3	

Alcohol Intake

1 drink equals		Canadian	Metric	U.S.
1 bottle of beer	5% alcohol	12 oz	340.8 ml	10 oz
1 glass of wine	12% alcohol	5 oz	142 ml	4.5 oz
1 shot of spirits	40% alcohol	1.5 oz	42.6 ml	1.25 oz

Safe Sex

Refers to the use of methods to prevent infection or conception.

What Does the Score Mean?

85-100	70-84	55-69	35-54	0-34
Excellent	Very good	Good	Fair	Needs improvement

A low total score does not mean that you have failed. There is always the chance to change your lifestyle. Look at the areas where you scored a 0 or 1 and decide which areas you want to work on first.

Tips

1. Don't try to change all areas at once. This will be too overwhelming.

2. Writing down your proposed changes and your overall goal will help you succeed.

3. Make changes in small steps toward the overall goal.

4. Enlist the help of a friend to make similar changes or to support you in your attempts.

5. Congratulate yourself for achieving each step. Give yourself appropriate rewards.

6. Ask your personal trainer, coach, family physician, nurse, or health department for more information on any of these areas.

From G. Haff and C. Dumke, 2019, *Laboratory manual for exercise physiology,* 2nd ed. (Champaign, IL: Human Kinetics). Adapted from D. Wilson, 1998, *Fantastic lifestyle assessment.*

FORM 2.8

FRAMINGHAM RISK EQUATION FOR MALES

Step 1

Age		
Years	LDL Pts	Chol Pts
30-34	-1	[-1]
35-39	0	[0]
40-44	1	[1]
45-49	2	[2]
50-54	3	[3]
55-59	4	[4]
60-64	5	[5]
65-69	6	[6]
70-74	7	[7]

Step 2

LDL - C		
(mg/dl)	(mmol/L)	LDL Pts
<100	<2.59	-3
100-129	2.60-3.36	0
130-159	3.37-4.14	0
160-190	4.15-4.92	1
≥190	≥4.92	2

Cholesterol		
(mg/dl)	(mmol/L)	Chol Pts
<160	<4.14	[-3]
160-199	4.15-5.17	[0]
200-239	5.18-6.21	[1]
240-279	6.22-7.24	[2]
≥280	≥7.25	[3]

Step 3

HDL - C			
(mg/dl)	(mmol/L)	LDL Pts	Chol Pts
<35	<0.90	2	[2]
35-44	0.91-1.16	1	[1]
45-49	1.17-1.29	0	[0]
50-59	1.30-1.55	0	[0]
≥60	≥1.56	-1	[-2]

Step 4

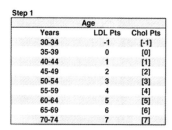

Blood Pressure					
Systolic	Diastolic (mm Hg)				
(mm Hg)	<80	80-84	85-89	90-99	≥100
<120	0 [0] pts				
120-129		0 [0] pts			
130-139			1 [1] pts		
140-159				2 [2] pts	
≥160					3 [3] pts

Note: When systolic and diastolic pressures provide different estimates for point scores, use the higher number

Step 5

Diabetes		
	LDL Pts	Chol Pts
No	0	[0]
Yes	2	[2]

Step 6

Smoker		
	LDL Pts	Chol Pts
No	0	[0]
Yes	2	[2]

(sum from steps 1-6)

Step 7

Adding up the points	
Age	_____
LDL-C or Chol	_____
HDL - C	_____
Blood Pressure	_____
Diabetes	_____
Smoker	_____
Point total	_____

Key	
Color	Relative Risk
green	Very low
white	Low
yellow	Moderate
rose	High
red	Very high

(determine CHD risk from point total)

Step 8

CHD Risk			
LDL Pts Total	10 Yr CHD Risk	Chol Pts Total	10 Yr CHD Risk
<-3	1%		
-2	2%		
-1	2%	[<-1]	[2%]
0	3%	[0]	[3%]
1	4%	[1]	[3%]
2	4%	[2]	[4%]
3	6%	[3]	[5%]
4	7%	[4]	[7%]
5	9%	[5]	[8%]
6	11%	[6]	[10%]
7	14%	[7]	[13%]
8	18%	[8]	[16%]
9	22%	[9]	[20%]
10	27%	[10]	[25%]
11	33%	[11]	[31%]
12	40%	[12]	[37%]
13	47%	[13]	[45%]
≥14	≥56%	[≥14]	[≥53%]

(compare to average person your age)

Step 9

Comparative Risk			
Age (years)	Average 10 Yr CHD Risk	Average 10 Yr Hard* CHD Risk	Low** 10 Yr CHD Risk
30-34	3%	1%	2%
35-39	5%	4%	3%
40-44	7%	4%	4%
45-49	11%	8%	4%
50-54	14%	10%	6%
55-59	16%	13%	7%
60-64	21%	20%	9%
65-69	25%	22%	11%
70-74	30%	25%	14%

* Hard CHD events exclude angina pectoris

** Low risk was calculated for a person the same age, optimal blood pressure, LDL-C 100-129 mg/dL or cholesterol 160-199 mg/dl, HDL-C 45 mg/dL for men or 55 mg/dL for women, non-smoker, no diabetes

Risk estimates were derived from the experience of the Framingham Heart Study, a predominantly Caucasian population in Massachusetts, USA

FORM 2.9

FRAMINGHAM RISK EQUATION FOR FEMALES

Step 1

Age		
Years	LDL Pts	Chol Pts
30-34	-9	[-9]
35-39	-4	[-4]
40-44	0	[0]
45-49	3	[3]
50-54	6	[6]
55-59	7	[7]
60-64	8	[8]
65-69	8	[8]
70-74	8	[8]

Step 2

LDL - C		
(mg/dl)	(mmol/L)	LDL Pts
<100	<2.59	-2
100-129	2.60-3.36	0
130-159	3.37-4.14	0
160-190	4.15-4.92	2
≥190	≥4.92	2

Cholesterol		
(mg/dl)	(mmol/L)	Chol Pts
<160	<4.14	[-2]
160-199	4.15-5.17	[0]
200-239	5.18-6.21	[1]
240-279	6.22-7.24	[1]
≥280	≥7.25	[3]

Step 3

HDL - C			
(mg/dl)	(mmol/L)	LDL Pts	Chol Pts
<35	<0.90	5	[5]
35-44	0.91-1.16	2	[2]
45-49	1.17-1.29	1	[1]
50-59	1.30-1.55	0	[0]
≥60	≥1.56	-2	[-3]

Step 4

Blood Pressure					
Systolic	Diastolic (mm Hg)				
(mm Hg)	<80	80-84	85-89	90-99	≥100
<120	-3 [-3] pts				
120-129		0 [0] pts			
130-139			0 [0] pts		
140-159				2 [2] pts	
≥160					3 [3] pts

+ Note: When systolic and diastolic pressures provide different estimates for point scores. use the higher number

Step 5

Diabetes		
	LDL Pts	Chol Pts
No	0	[0]
Yes	4	[4]

Step 6

Smoker		
	LDL Pts	Chol Pts
No	0	[0]
Yes	2	[2]

Step 7

(sum from steps 1-6)

Adding up the points	
Age	_____
LDL-C or Chol	_____
HDL - C	_____
Blood Pressure	_____
Diabetes	_____
Smoker	_____
Point total	_____

Key	
Color	Relative Risk
green	Very low
white	Low
yellow	Moderate
rose	High
red	Very high

Step 8

(determine CHD risk from point total)

CHD Risk			
LDL Pts Total	10 Yr CHD Risk	Chol Pts Total	10 Yr CHD Risk
≤-2	1%	[≤-2]	[1%]
-1	2%	[-1]	[2%]
0	2%	[0]	[2%]
1	2%	[1]	[2%]
2	3%	[2]	[3%]
3	3%	[3]	[3%]
4	4%	[4]	[4%]
5	5%	[5]	[4%]
6	6%	[6]	[5%]
7	7%	[7]	[6%]
8	8%	[8]	[7%]
9	9%	[9]	[8%]
10	11%	[10]	[10%]
11	13%	[11]	[11%]
12	15%	[12]	[13%]
13	17%	[13]	[15%]
14	20%	[14]	[18%]
15	24%	[15]	[20%]
16	27%	[16]	[24%]
≥17	≥32%	[≥17]	[≥27%]

Step 9

(compare to average person your age)

Comparative Risk			
Age (years)	Average 10 Yr CHD Risk	Average 10 Yr Hard* CHD Risk	Low** 10 Yr CHD Risk
30-34	<1%	<1%	<1%
35-39	<1%	<1%	1%
40-44	2%	1%	2%
45-49	5%	2%	3%
50-54	8%	3%	5%
55-59	12%	7%	7%
60-64	12%	8%	8%
65-69	13%	8%	8%
70-74	14%	11%	8%

* Hard CHD events exclude angina pectoris

** Low risk was calculated for a person the same age, optimal blood pressure, LDL-C 100-129 mg/dL or cholesterol 160-199 mg/dl, HDL-C 45 mg/dL for men or 55 mg/dL for women, non-smoker, no diabetes

Risk estimates were derived from the experience of the Framingham Heart Study. a predominantly Caucasian population in Massachusetts. USA

Flexibility Testing

Objectives

- Define *flexibility* and the factors that affect it.
- Differentiate between direct and indirect methods of assessing flexibility.
- Describe and conduct the six most common variants of the sit-and-reach test.
- Learn the shoulder elevation and back-scratch methods for evaluating shoulder flexibility.
- Compare individual and group results from tests performed in the laboratory activity with those represented in the normative literature.

DEFINITIONS

flexibility—Range of motion (ROM) in a joint or related series of joints (3, 35).

goniometer—Device, similar to a protractor and containing two arms, used to measure joint angles (16).

inclinometer—Gravity-based goniometer used to measure range of motion (ROM) in a joint (16).

percentile rank—Method for comparing how a subject performs with how other subjects perform; typically between 1% and 99% (19).

range of motion (ROM)—Degree of movement that occurs in a joint (26).

Flexibility is generally defined as the **range of motion (ROM)** in a joint or related series of joints (35, 37) or as the ability to move a muscle or group of muscles through a ROM (15). Overall flexibility can be affected by the structure of a joint (2, 5), the age and sex of the individual (26, 41), the elasticity and plasticity of the connective tissue (15, 42), and the individual's activity level (26).

Ball-and-socket joints (e.g., shoulder, hip) enable the greatest ROM and offer the capacity to move in all anatomical planes. The second most flexible are the ellipsoidal joints, such as the wrist (an oval-shaped condyle that fits into an elliptical cavity); they have less ROM and move primarily in the sagittal and frontal planes (26). The smallest ROM is found in the hinge joints (e.g., knee, elbow), which move primarily in the sagittal plane. The ROM of any joint is affected by the articulating surfaces of the joint and by the soft tissues surrounding it, such as tendons, ligaments, fascial sheaths, joint capsules, and skin (12, 16, 26). Flexibility involves both elasticity, which is the ability to return to a resting length, and plasticity, which is the ability to change the length of the soft tissues.

Generally, younger people are more flexible than their older counterparts (26, 27), and women tend to be more flexible than men (26). Sex-specific differences in flexibility are most likely related to anatomical and structural differences, as well as differences in the physical activities typically undertaken by members of each sex (26). Regardless of sex, flexibility declines by 20% to 30% between the ages of 30 and 70 (10). This reduction may be attributed to a drop in physical activity as one ages (24); as a whole, people who are more active tend to be more flexible regardless of age or sex (26). This relationship highlights the importance of appropriately designed training programs and the potential for people to increase their flexibility and improve their overall fitness.

DIRECT AND INDIRECT ROM ASSESSMENT

As one of the five major components of fitness (4, 5), flexibility is commonly tested as part of health-related test batteries. Flexibility is joint specific, and no single test can be used to evaluate total-body flexibility (5); however, flexibility testing can be conducted on any joint of the body thanks to a variety of direct and indirect methods (16, 18).

The most common method for directly assessing joint ROM is to use a **goniometer** (figure 3.1), a device similar to a protractor that is used to measure joint angles (18). Goniometers have two arms coupled with a protractor for measuring the degrees of joint displacements (39). They are simple to use (39) and are considered highly reliable when standardized procedures are employed (8, 29). Goniometer size varies according to the joint being measured (29, 39).

Despite its usefulness, the goniometer may not be the best tool for directly assessing spinal movement or complex movements such as supination, pronation, inversion, and eversion. For these movements, an **inclinometer** (figure 3.2) is considered more accurate (39). An inclinometer uses a universal center of gravity as a consistent starting point and a weighted needle and protractor to determine ROM (29, 33). Positioning and securing the device to the subject can be difficult, but both electronic and mechanical inclinometers are reliable methods for assessing ROM (29).

When direct methods for evaluating ROM are unavailable, you can assess flexibility using indirect methods (31). These methods usually report results in inches or centimeters (18), and they have been determined to be reliable. The most common

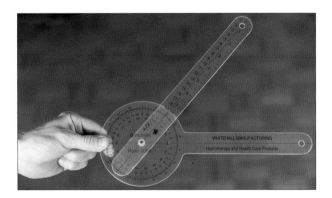

Figure 3.1 Goniometer.

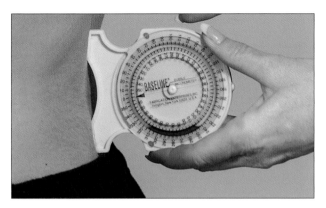

Figure 3.2 Inclinometer.

indirect methods for assessing flexibility are the sit-and-reach test (36), the shoulder elevation test (1), and the back-scratch test (16).

BODY AREAS

The body areas typically assessed for flexibility in a health fitness testing battery are the back, hamstrings, and shoulders. The flexibility of these areas can be assessed with a variety of methods.

Back and Hamstring Flexibility Tests

Poor flexibility of the low back extensors and the hamstrings has been associated with low back pain (28, 42) and poor performance in sporting activities requiring flexibility in these areas (24, 35). For these reasons, assessment of low back and hamstring flexibility is common in most exercise physiology laboratories and health fitness settings (7). This assessment can be achieved by means of direct or indirect methods.

Goniometer testing is a direct measure of hamstring flexibility when used in conjunction with a straight-leg raise (13). This assessment procedure is considered the criterion, or gold standard, for assessing hamstring flexibility. Even though goniometer-based testing has a subjective end-point criterion (2), it is a highly reliable method for assessing hamstring flexibility (8).

Regardless of the joint being assessed, the methodology for using the goniometer is easy to implement (18, 39). The goniometer is placed with the center of the protractor fixed to the axis of rotation while one arm is lined up with the proximal articulating segment and the second arm corresponds to the distal segment (figure 3.3) (39). The movement of the second arm allows quantification of the ROM for the distal segment.

The most commonly used indirect test of hamstring and low back flexibility is the sit-and-reach test (figure 3.4) (5, 19, 31). However, this test is generally considered to be a poor indicator of low back function and is more commonly accepted as an indicator of hamstring flexibility (20, 21, 23, 24, 31, 32, 34). Therefore, this assessment should be considered primarily as an indicator of hamstring (i.e., semitendinosus, semimembranosus, and biceps femoris) flexibility and secondarily as a measurement of low back (erector spinae), buttocks (gluteus maximus and gluteus medius), and calf (gastrocnemius) flexibility (2).

The sit-and-reach test is considered by many to be a field-based assessment (31) but may involve characteristics typically seen in laboratory testing (2). When performed with standardized methods, the sit-and-reach test has produced consistently reliable assessments ranging from .70 to .98 depending upon the population examined (20, 21, 31, 38). From a validity standpoint, the sit-and-reach test appears to produce a valid assessment only of hamstring flexibility ($r = .70-.76$, $p < .05$) (31).

The sit-and-reach test can be used in several forms, including the traditional method first presented by the American Alliance of Health, Physical Education, Recreation and Dance (AAHPERD) (2, 3, 13). Modifications include the YMCA (2, 5, 35), Canadian (2, 22), wall (2, 17), V-sit (2, 11), backsaver (20, 21), and chair (30) sit-and-reach approaches.

Shoulder Flexibility Tests

Shoulder flexibility can affect the ability to perform activities of daily living, such as combing one's hair, dressing, and reaching for the seat belt in a car (16). In sport, deficient shoulder flexibility can limit performance and increase the risk of injury (40). As a result, assessment of shoulder flexibility is common in both sport and fitness testing. Two of the most common tests are the shoulder elevation test (1, 19) and the back-scratch test (16, 36).

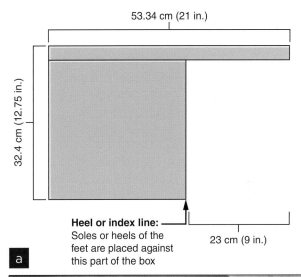

Figure 3.4 Sit-and-reach box: *(a)* dimensions and *(b)* sample sit-and-reach box.

Go to the web study guide to access electronic versions of individual and group data sheets, the question sets from each laboratory activity, and case studies for each laboratory activity.

Figure 3.3 Goniometer placement for measuring ROM.

Traditional, Wall, V-Sit, and Chair Sit-and-Reach Test Comparisons

Equipment

- Physician's scale or equivalent electronic scale
- Stadiometer
- Sit-and-reach box
- Yardstick or meterstick
- Chair
- Individual data sheet
- Group data sheet
- Microsoft Excel or equivalent spreadsheet program

Warm-Up

Regardless of the sit-and-reach test method, all subjects should perform a structured warm-up beforehand to increase the reliability and validity of the test. The warm-up should include a 5 min general warm-up and a 5 min dynamic stretching warm-up (39).

The 5 min general warm-up can involve activities such as jogging, cycling, skipping, jumping rope, or doing calisthenics (26). The activity should be undertaken with the goal of increasing HR, blood flow, muscle temperature, respiration rate, and perspiration while decreasing the viscosity of joint fluids (12, 26). Taken collectively, these responses prepare the subject for the more specific warm-up activities that are performed next.

After completing the general warm-up, the subject should perform dynamic warm-up movements that work through the ROMs required for the testing activity (26). Dynamic warm-up activities appropriate for the sit-and-reach test include walking knee lifts, leg swings, trunk rotations, lunges, standing stiff-leg deadlifts, and the inchworm (25). See table 3.1 for a sample 10 min warm-up that includes 5 min of general warm-up and 5 min of dynamic stretching. After completing the warm-up, the subject can begin the testing process using the following protocol.

Table 3.1 Warm-Up for Sit-and-Reach Testing

Warm-up	Activity	Total time
General*	Cycling	5 min
Specific	Walking lunge Walking lunge with dynamic rotation High-knee walk Leg swing Trunk rotation Squat	5 min

*The general warm-up activity can be selected based on the equipment available.

Traditional Sit-and-Reach Test

The traditional sit-and-reach test requires a sit-and-reach box with the index or heel line marked at 23 cm (9.1 in.) (see figure 3.4).

Step 1: Place the sit-and-reach box against a wall or object that prevents it from slipping during the testing process.

Step 2: Gather basic data (e.g., age, height, weight) for the individual data sheet as described in laboratory activity 1.1. Have the subject put his or her shoes back on for the warm-up.

Step 3: Have the subject perform a 10 min warm-up (see sample warm-up in table 3.1).

Step 4: Have the subject remove his or her shoes and sit on the floor with the heels and soles of the feet against the index or heel line (23 cm mark), with the legs fully extended, and with the medial sides of the feet about 20 cm (8 in.) apart (figure 3.5a).

Step 5: Place your hands across the subject's knees to ensure full leg extension is maintained.

Step 6: Instruct the subject to extend the arms with one hand on top of the other and palms facing down (figure 3.5b).

Step 7: Have the subject bend forward and reach with both hands along the measuring scale on top of the sit-and-reach box. This position should be held for 1 to 2 s. If the subject's knees bend or the fingertips become uneven, the test does not count, and the process should be repeated. Record the resulting measure on the individual data sheet.

Step 8: Perform steps 6 and 7 three more times and consider the fourth trial as the maximal stretch.

Step 9: Record the results of the fourth trial in the appropriate location on the individual data sheet.

Step 10: Compare the results of the fourth trial with the **percentile ranks** or normative data presented in table 3.2, and note the subject's ranking in the appropriate spot on the individual data sheet.

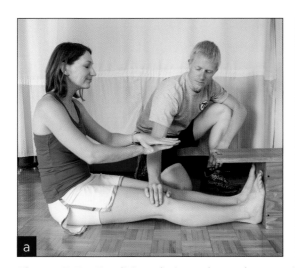

Figure 3.5 Traditional sit-and-reach test: (a) start and (b) extended position.

Table 3.2 Percentile Ranks and Normative Data for the Traditional Sit-and-Reach Test (cm)

AGE (y)	20-29		30-39		40-49		50-59		60-69	
Sex	**M**	**F**	**M**	**F**	**M**	**F**	**M**	**F**	**M**	**F**
90	39	40	37	39	34	37	35	37	32	34
80	35	37	34	36	31	33	29	34	27	31
70	33	35	31	34	27	32	26	32	23	28
60	30	33	29	32	25	30	24	29	21	27
50	28	31	26	30	22	28	22	27	19	25
40	26	29	24	28	20	26	19	26	15	23
30	23	26	21	25	17	23	15	23	13	21
20	20	23	18	22	13	21	12	20	11	20
10	15	19	14	18	9	16	9	16	8	15

(Percentile rank — left axis label)

May be used for the traditional, backsaver, and V-sit tests. If values fall between the percentile ranks, interpolate the data. For example, if a male subject is 21 years old and achieves a result of 37 cm, then his percentile rank is 85.

Source: *Canadian Physical Activity, Fitness & Lifestyle Approach: CSEP-Health & Fitness Program's Appraisal and Counselling Strategy*, 3rd edition, © 2003. Reprinted with permission from the Canadian Society for Exercise Physiology.

Wall Sit-and-Reach Test

The wall sit-and-reach test is a modification in which the heel or index line is set according to the individual (17); specifically, the subject places the hips, back, and head against a wall when establishing the zero point. This step allows you to correct for the proportion of leg length to trunk length.

Step 1: Begin this lab activity by gathering basic data (e.g., age, height, weight) for the individual data sheet as described in laboratory activity 1.1. Have the subject put his or her shoes back on prior to initiating the warm-up step.

Step 2: Have the subject perform a 10 min warm-up (table 3.1).

Step 3: Instruct the subject to remove his or her shoes and sit on the floor with the hips, back, and head against a wall.

Step 4: Have the subject extend the legs with the feet roughly 20 to 30 cm (8-12 in.) apart.

Step 5: Position the sit-and-reach box against the subject's heels. To prevent the box from slipping, brace it with your feet or a suitable object.

Step 6: Instruct the subject to place one hand on top of the other with the palms facing down.

Step 7: Instruct the subject to reach forward as far as possible while keeping the hips, back, and head in contact with the wall. It is okay if the shoulder moves forward. Determine how far the person's fingertips reach, and record this measurement to the nearest 1.25 cm (0.5 in.) (figure 3.6a). This is called the *index line* or *zero position*.

Step 8: Having established the zero position, instruct the subject to reach forward three times during the same movement along the device while making sure to keep the palms against the measuring device (figure 3.6b). The subject should hold the third movement for 2 s while you measure and record the distance on the individual data sheet.

Step 9: Subtract the measure recorded in step 7 from the value determined in step 8. Record this value on the individual data sheet.

Step 10: Repeat steps 8 and 9.

Step 11: Record the best of the two trials in the appropriate location on the individual and group data sheets.

Step 12: Interpret the results of the test by comparing them with the normative data and percentile ranks presented in table 3.3. Note the percentile rank in the appropriate box on the individual data sheet.

Figure 3.6 Wall sit-and-reach test: *(a)* start and *(b)* extended position.

Table 3.3 Percentile Ranks and Normative Data for the Wall Sit-and-Reach Test

AGE (y)		<35			36-49			>50			
SEX		M		F		M		F		M		F
Units	cm	in.	cm	in.	cm	in.	cm	in.	cm	in.	cm	in.
90	45	17.7	45	17.7	41	16.1	44	17.3	38	15.0	38	15.0
80	43	17.0	42	16.5	37	14.6	41	16.1	34	13.4	36	14.2
70	40	15.7	41	16.1	35	13.8	39	15.4	31	12.2	35	13.8
60	38	15.0	40	15.7	34	13.4	37	14.6	29	11.4	31	12.2
50	37	14.6	38	15.0	32	12.6	34	13.4	26	10.2	28	11.0
40	34	13.4	37	14.6	29	11.4	33	13.0	25	9.8	26	10.2
30	33	13.0	35	13.8	27	10.6	31	12.2	24	9.4	23	9.1
20	29	11.4	32	12.6	25	9.8	28	11.0	22	8.7	21	8.3
10	23	9.1	26	10.2	21	8.3	25	9.8	20	7.9	19	7.5

If values fall between the percentile ranks, interpolate the data. The reference or heel line is set at 0.
Adapted from Adams 1998.

V-Sit Sit-and-Reach Test

The V-sit method is another modification of the traditional sit-and-reach test (2, 11). The V-sit does not require a sit-and-reach box, but it does require the heels to be separated by 30 cm (about 12 in.) to create the V.

Step 1: Begin this lab activity by gathering basic data (e.g., age, height, weight) for the individual data sheet as described in laboratory activity 1.1. Have the subject put on his or her shoes.

Step 2: Instruct the subject to perform a 10 min warm-up (see table 3.1).

Step 3: Have the subject sit on the floor without shoes on and fully extend the legs with the feet separated by 30 cm.

Step 4: Place a meterstick between the subject's legs so that the 23 cm (9.1 in.) mark aligns with the subject's heels. To prevent the meterstick from moving, tape it to the floor.

Step 5: Hold the subject's knees to ensure the legs do not bend during the test.

Step 6: Have the subject place one hand on top of the other with the palms facing down and the fingertips aligned (figure 3.7a).

Step 7: Instruct the subject to lean forward and reach with both hands along the meterstick until fully extended (figure 3.7b). This position should be held for 1 to 2 s. Record the distance achieved on the individual data sheet.

Step 8: Perform steps 5 through 7 three more times and consider the fourth trial as the maximal stretch. Record the results of the fourth trial in the appropriate part of the individual and group data sheets.

Step 9: Calculate the average of the four trials, and record this value in the appropriate location on the individual data sheet.

Step 10: Compare the average achieved during the tests with the normative data and percentile ranks used for the traditional sit-and-reach test (see table 3.2), and note the percentile rank in the appropriate location on the individual data sheet.

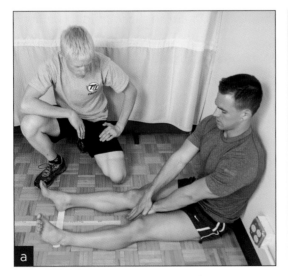 

Figure 3.7 V-sit sit-and-reach test: (a) start and (b) extended position.

Chair Sit-and-Reach Test

The chair sit-and-reach test is a modified sit-and-reach test that is designed so that the participant does not have to get up and down from the floor (29). This test requires a chair and meterstick (or ruler).

Step 1: Begin this lab activity by gathering basic data (e.g., age, height, weight) for the individual data sheet as described in laboratory activity 1.1. Have the subject put on his or her shoes.

Step 2: Instruct the subject to perform a 10 min warm-up (see table 3.1).

Step 3: Place the chair firmly against the wall to prevent it from tipping over during the testing.

Step 4: Instruct the subject to sit on the chair so that the crease at the angle of the trunk and the anterior thigh (i.e., inguinal fold) is parallel with the edge of the chair.

Step 5: Instruct the subject to bend one leg while the foot is flat on the floor. Then the subject extends the opposite leg as straight as possible in front of the hip, with the heel of the foot placed securely on the floor. Instruct the subject to dorsiflex the foot so that it is at an approximate 90° angle.

Step 6: Hold the meterstick (or ruler) parallel to the extended leg, with one end resting on the toes and the other supported by your hand (figure 3.8a).

Step 7: Instruct the subject to exhale and drop the head between the arms while reaching forward during the test.

Step 8: Instruct the subject to slowly reach with both hands along the meterstick toward the toes of the foot on the extended leg. The subject should reach as far as possible and hold the maximally extended position only momentarily (figure 3.8b). It is important that the hands stay parallel to the floor and that the subject does not reach farther with one hand.

Step 9: During the test, make sure the knee stays extended and does not bend (i.e., remains at 180° throughout the test). If the subject either hyperextends or bends the leg, the test should be considered invalid and steps 6 through 8 should be repeated.

Step 10: Record the distance reached to the nearest 0.1 cm on the individual data sheet.

Step 11: Repeat steps 6 through 10 two more times.

Step 12: Compare the best trial with the percentile ranks in table 3.4 (note this test is for people aged >60, so only compare with the 60-year-old column; no normative data exist for younger populations).

Step 13: Repeat steps 4 through 12 with the opposite leg.

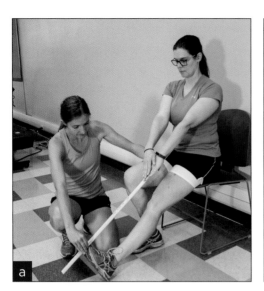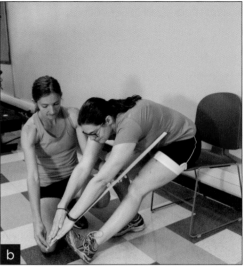

Figure 3.8 Chair sit-and-reach test: *(a)* start and *(b)* extended position.

Table 3.4 Percentile Ranks and Normative Data for the Chair Sit-and-Reach Test (cm)

AGE (y)		60-64		65-69		70-74		75-79		80+	
Rating	Percentile	Male	Female	Male	Female	Male	Female	Male	Female	Male	Female
Excellent	90	20.4	20.8	16.0	18.8	15.0	17.6	9.9	15.0	7.0	12.1
Good	70	10.7	12.7	7.0	10.9	6.0	10.0	2.1	6.3	1.0	5.9
Average	50	6.5	6.0	2.0	6.0	0.0	4.0	–3.0	3.8	–4.0	0.0
Poor	30	2.0	3.0	–5.4	1.0	–6.0	0.0	–11.0	0.7	–13.0	–2.0
	10	–2.0	–5.4	–16.0	–4.6	–18.0	–9.0	–19.7	–7.3	–28.0	–13.2

Adapted from *ESSA's student manual for health, exercise and sport assessment*, edited by J. Coombes and T. Skinner, Flexibility, S. Lark, T. Brancato, and T. Skinner, pg. 182, copyright 2014, with permission from Elsevier.

Question Set 3.1

1. What major factors can affect flexibility?
2. What do sit-and-reach tests tell us about back and hamstring flexibility?
3. Compare the results of your sit-and-reach tests with the normative data. What do the results indicate?
4. Compare the class sit-and-reach data as a whole with the normative data. What do these results indicate?
5. Create a bar graph that presents means for the class data collected for the traditional, wall, V-sit, and chair tests.
6. Use a paired *t*-test to compare the results of the sit-and-reach tests for the five most flexible men with the results for the five most flexible women. How do these findings compare with what you would expect to find?
7. Fill in the table in the individual data sheet comparing the methods used in this lab. Indicate the best application for each test, and describe the pros and cons of each test.

LABORATORY ACTIVITY 3.1

INDIVIDUAL DATA SHEET

Name or ID number: _____ Date: _____

Tester: _____ Time: _____

Sex: M / F (circle one) Age: _____ y Height: _____ in. _____ m

Weight: _____ lb _____ kg Temperature: _____ °F _____ °C

Barometric pressure: _____ mmHg Relative humidity: _____ %

Sit-and-reach test results		Trial 1	Trial 2	Trial 3	Trial 4	Best	4th trial	Average	Index/heel line (cm)
Traditional						____		____	23
Wall				____	____		____	____	
V-sit						____	____		
Chair	Right leg				____				____
	Left leg				____				____

Individual results ranking		Results (cm)	Ranking/category/percentile
Traditional			Percentile =
Wall sit			Percentile =
V-sit			Percentile =
Chair	Right leg		Percentile =
	Left leg		Percentile =

YMCA, Backsaver, and Goniometer Test Comparisons

Equipment

- Physician's scale or equivalent electronic scale
- Stadiometer
- Goniometer
- Sit-and-reach box
- Yardstick or meterstick
- Tape
- Individual data sheet
- Group data sheet
- Microsoft Excel or equivalent spreadsheet program

Warm-Up

Regardless of the sit-and-reach test method, all subjects should perform a structured warm-up beforehand to increase the reliability and validity of the test. The warm-up session should include a 5 min general warm-up and a 5 min dynamic stretching warm-up (39).

The 5 min general warm-up can involve activities such as jogging, cycling, skipping, jumping rope, or doing calisthenics (26). The activity should be undertaken with the goal of increasing HR, blood flow, muscle temperature, respiration rate, and perspiration while decreasing viscosity of joint fluids (12, 26). Taken collectively, these responses prepare the subject for the more specific warm-up activities that are performed next.

After completing the general warm-up, the subject should perform dynamic warm-up movements that work through the ROMs required for the testing activity (26). Dynamic warm-up activities appropriate for the sit-and-reach test might include walking knee lifts, leg swings, trunk rotations, lunges, standing stiff-leg deadlifts, and the inchworm (25). See table 3.1 for a sample 10 min warm-up that includes 5 min of general warm-up and 5 min of dynamic stretching. After completing the warm-up, the subject can begin the testing process.

YMCA Sit-and-Reach Test

The YMCA sit-and-reach test is similar to the traditional test except that no sit-and-reach box is used (5, 14). In this test, use a yardstick or meterstick placed on the floor so that the zero point is directed toward the subject. As a result of not using the sit-and-reach box, the YMCA test generally produces about a 2.5 cm (1 in.) difference in result when compared with the traditional sit-and-reach test (2).

Step 1:   Place a yardstick or meterstick on the floor so that the zero end faces the subject. Place a piece of tape so that it intersects with the stick at 38 cm (15 in.).

Step 2:   Gather basic data (e.g., age, height, weight) for the individual data sheet as described in laboratory activity 1.1. Have the subject put on his or her shoes.

Step 3:   Have the subject perform a 10 min warm-up (see table 3.1).

Step 4:   Instruct the subject to remove his or her shoes and sit on the floor with the stick between the legs. The legs should be separated by 25 to 30 cm (10-12 in.) and extended at right angles to the taped line. The heels should touch the taped line (figure 3.9a).

Step 5:   Place your hands across the subject's knees to ensure full leg extension is maintained.

Step 6:   Instruct the subject to extend the arms with one hand on top of the other and with the palms facing down (figure 3.9b).

Step 7:   Have the subject bend forward and reach with both hands along the stick while keeping them parallel to one another. This position should be held for 1 to 2 s. Measure the distance reached and record it on the individual data sheet. If the subject's knees bend or the fingertips become uneven, the test does not count, and it should be repeated.

Step 8:   Perform steps 5 through 7 three more times.

Step 9:   Record the best result from the four trials on the individual data sheet, and compare this value with the percentile ranks or normative data presented in table 3.2. Note the results in the appropriate spot on the individual data sheet.

Figure 3.9 YMCA sit-and-reach test: (a) start and (b) extended position.

Backsaver Sit-and-Reach Test

The backsaver sit-and-reach test is another modification of the traditional sit-and-reach test. It reduces the posterior compression typically seen when bending forward with both legs extended (2, 9, 11). Instead, the backsaver test requires the subject to bend forward with one leg extended and the other bent (2, 11, 21). Even so, some subjects report discomfort, in this case in the hip joint of the bent leg (20, 21). Other than the altered leg positions and the fact that each leg is measured separately, the backsaver test is performed in a manner similar to that of the traditional sit-and-reach test.

Step 1: Place the sit-and-reach box against a wall or an object that will prevent it from slipping during the testing process.

Step 2: Gather basic data (e.g., age, height, weight) for the individual data sheet as described in laboratory activity 1.1. Instruct the subject to put on his or her shoes.

Step 3: Have the subject perform a 10 min warm-up (see table 3.1).

Step 4: Instruct the subject to sit on the floor without shoes on and with the right leg extended so that the sole of the foot is placed against the sit-and-reach box at the index or heel line (at the 23 cm or 9.1 in. mark). The left leg should be bent at approximately 90°, with the sole of the left foot placed flat on the floor (21) about 5 to 7.5 cm (2-3 in.) from the sit-and-reach box (2).

Step 5: Have the subject extend the arms with one hand on top of the other and with the palms facing down.

Step 6: For the test, the subject should bend forward while reaching with both hands along the measuring scale on top of the sit-and-reach box. This position should be held for 1 to 2 s. The subject may have to move the bent leg to the side while extending forward to allow movement through the ROM. If the extended knee bends or the fingertips become uneven, the test does not count, and it should be repeated. Record the resulting measurement on the individual data sheet.

Step 7: Perform steps 4 through 6 three more times with the right leg.

Step 8: Compare the results of the fourth trial with the percentile ranks or normative data presented in table 3.2, and note the results in the appropriate spot on the individual data sheet.

Step 9: Have the subject change positions so that the left leg is extended and the sole of the left foot is placed against the sit-and-reach box at the index or heel line (at the 23 cm or 9.1 in. mark). The right leg should be bent at approximately 90°, with the sole of the right foot placed flat on the floor (21) about 5 to 7.5 cm from the sit-and-reach box (2). See figure 3.10a.

Step 10: Have the subject extend the arms with one hand on top of the other and with the palms facing down. See figure 3.10b.

Step 11: Instruct the subject to bend forward while reaching with both hands along the measuring scale on top of the sit-and-reach box. This position should be held for 1 to 2 s. The subject may have to move the bent leg to the side while extending forward to allow movement through the ROM. If the subject's extended knee bends or the fingertips become uneven, the test does not count, and it should be repeated. Record the resulting measurement on the individual sit data sheet.

Step 12: Perform steps 9 through 11 three more times with the left leg.

Step 13: Compare the results of the fourth trial with the percentile ranks or normative data presented in table 3.2, and note the results in the appropriate spot on the individual data sheet.

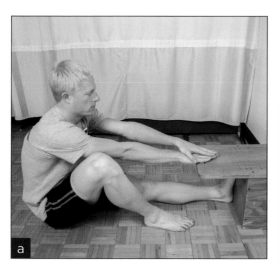 

Figure 3.10 Backsaver sit-and-reach test: *(a)* start and *(b)* extended position.

Goniometer Test of Hamstring Flexibility

The goniometer test of hamstring flexibility is often considered the criterion measure or gold standard for evaluating hamstring flexibility (21). It evaluates hamstring flexibility during a passive straight-leg raise and is performed separately on each leg (2, 21). The criterion score achieved for this test is registered as the maximal angle in degrees achieved when maximum hip flexion occurs (21). Although the test can be performed by one tester, you may prefer to use two testers—one to lift the subject's leg while the other uses the goniometer to measure hip flexion (2).

Step 1: Direct the subject through a 10 min warm-up (see table 3.1).

Step 2: Gather basic data (e.g., age, height, weight) for the individual data sheet as described in laboratory activity 1.1.

Step 3: Have the subject lie on his or her back on a table or on the floor.

Step 4: Align the goniometer with the axis of the subject's right hip.

Step 5: Align the stationary arm of the goniometer with the subject's trunk, and position the moveable arm in line with the right femur.

Step 6: Once the stationary and moveable arms are aligned, move the right leg toward hip flexion while making sure the knee of the moving leg remains straight. The end point of flexion is determined when you feel tightness. Once this position is achieved, hold it for 1 to 2 s while measuring the angle achieved to the nearest degree.

Step 7: Return the right leg to a resting position, and record the measurement on the individual data sheet.

Step 8: Align the goniometer with the axis of the subject's left hip (figure 3.11a).

Step 9: Align the stationary arm of the goniometer with the subject's trunk, and position the moveable arm in line with the left femur.

Step 10: Once the stationary and moveable arms are aligned, move the left leg toward hip flexion while making sure the knee of the moving leg remains straight. The end point of flexion is determined when you feel tightness. Once this position is achieved, hold it for 1 to 2 s while measuring the angle achieved to the nearest degree (figure 3.11b).

Step 11: Return the left leg to a resting position, and record the measurement on the individual data sheet.

Step 12: Perform steps 3 through 11 three more times.

Step 13: Calculate the average of the four trials for both the right and left legs, and record this information in the appropriate location on the individual and group data sheets. When interpreting the degree of flexion achieved, a smaller angle indicates better flexibility.

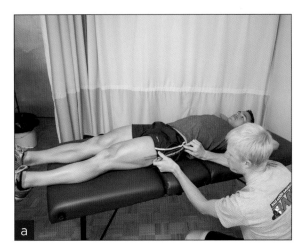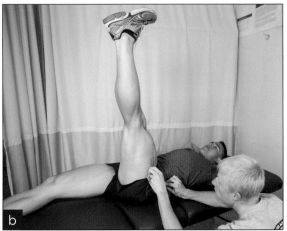

Figure 3.11 Goniometer hamstring test: *(a)* start and *(b)* with full hip flexion.

Question Set 3.2

1. How does the YMCA sit-and-reach test differ from the backsaver version?

2. Compare the results of your sit-and-reach test with the normative data. What do these results indicate?

3. Compare the class data for the sit-and-reach test as a whole with the normative data. What do these results indicate?

4. Create a bar graph that presents means for your data and for the class data.

5. Use a paired *t*-test to compare the results of the sit-and-reach tests for the five most flexible men with the results for the five most flexible women. How do these findings compare with what you would expect to find?

6. How do the percentile ranks compare between the three tests performed in this laboratory activity?

7. Fill in the table in the individual data sheet comparing the methods used in the present lab. Indicate the best application for each test, and describe the pros and cons of each test.

LABORATORY ACTIVITY 3.2

INDIVIDUAL DATA SHEET

Name or ID number: _____ Date: _____

Tester: _____ Time: _____

Sex: M / F (circle one) Age: _____ y Height: _____in. _____m

Weight: _____lb _____kg Temperature: _____°F _____°C

Barometric pressure: _____ mmHg Relative humidity: _____%

Sit-and-reach test results		Trial 1	Trial 2	Trial 3	Trial 4	Best	4th trial	Index/heel line (cm)
YMCA							_____	38
Backsaver	Right leg					_____		23
	Left leg					_____		23
Goniometer test results		**Trial 1**	**Trial 2**	**Trial 3**	**Trial 4**	**Average**	**Degrees**	
Goniometer	Right leg							
	Left leg							

Test		Results (cm)	Ranking/category/percentile
YMCA			Category =
Backsaver	Right leg		Percentile =
	Left leg		Percentile =
Test		**Degrees**	
Goniometer	Right leg		
	Left leg		

Canadian, Traditional, and Backsaver Sit-and-Reach Test Comparisons

Equipment

- Physician's scale or equivalent electronic scale
- Stadiometer
- Sit-and-reach box
- Yardstick or meterstick
- Individual data sheet
- Group data sheet
- Microsoft Excel or equivalent spreadsheet program

Canadian Sit-and-Reach Test

In comparison with the traditional sit-and-reach test, the Canadian variant differs in several ways (2, 11). First, it involves a standardized warm-up protocol that includes static stretching. Second, the subject's feet are separated by 5 cm (2 in.)—significantly less than the 23 cm (9.1 in.) required by the traditional sit-and-reach test. Finally, the height of the sit-and-reach box is adjusted so that the meterstick or measuring device is positioned at the height of the toes, and the heel or index line is set at 26 cm (10.2 in.). Otherwise, the procedure is similar to that of the traditional test.

Step 1:   Set up the testing area by securing the sit-and-reach box against a wall or an object that prevents it from slipping.

Step 2:   Gather basic data (e.g., age, height, weight) for the individual data sheet as described in laboratory activity 1.1. Have the subject put on his or her shoes.

Step 3:   Have the subject perform a 10 min warm-up (see table 3.1).

Step 4:   Instruct the subject to remove his or her shoes and to perform a modified hurdle stretch with one leg extended and the opposite leg bent so that the sole of that foot is positioned against the extended leg. The stretch is held for 20 s and is repeated twice for each leg.

Step 5:   Adjust the sit-and-reach box so that it is positioned at the height of the toes and so that the heel is positioned against the heel or index line, which is at 26 cm (10.2 in.) (figure 3.12a). If you do not have an adjustable sit-and-reach box, place a meterstick at the subject's toe level with the heels crossing the meterstick at the 26 cm mark.

Step 6:   Instruct the subject to place one hand on top of the other with the palms facing downward. Have the subject slowly bend forward while running both hands along the meterstick (figure 3.12b). It is important that the subject keeps his or her head down and hands placed evenly with one on top of the other. Once the subject can no longer extend, the position is held for 2 s. If the knees bend or the fingertips become uneven, the test does not count, and it should be repeated.

Step 7: Read the distance achieved to the nearest centimeter, and record it on the individual data sheet.

Step 8: Perform steps 4 through 7 three more times, and note the best of the four trials on the laboratory data sheet.

Step 9: Interpret the results of the test by comparing them with the normative data and percentile ranks presented in table 3.5. Note the subject's percentile ranking on the individual data sheet.

Figure 3.12 Modified hurdle stretch for the Canadian sit-and-reach test: *(a)* start and *(b)* extended position.

Table 3.5 Percentile Ranks and Normative Data for the Canadian Sit-and-Reach Test (cm)

AGE (y)		15-19		20-29		30-39		40-49		50-59		60-69	
Classification	Percentile	M	F	M	F	M	F	M	F	M	F	M	F
High	81-100	38	42	39	40	37	40	34	37	34	38	32	34
Above average	61-80	34	38	34	37	33	36	29	34	28	33	25	31
Average	41-60	29	34	30	33	28	32	24	30	24	30	20	27
Below average	21-40	24	29	25	28	23	27	18	25	16	25	15	23
Low	1-20	24	29	25	28	23	27	18	25	16	25	15	23

Percentile rank

The index for the Canadian sit-and-reach test is 26 cm (10.2 in.). If using these norms with a traditional sit-and-reach test, subtract 3 cm (1.2 in.) from the norms presented here. M = male and f = female.

Source: *The Canadian Physical Activity, Fitness & Lifestyle Approach: CSEP-Health & Fitness Program's Health-Related Appraisal and Counselling Strategy,* 3rd Edition © 2003. Reprinted with permission of the Canadian Society for Exercise Physiology.

Traditional Sit-and-Reach Test

The traditional sit-and-reach test requires a sit-and-reach box where the index or heel line is marked at 23 cm (9.1 in.) (see figure 3.4).

Step 1: Place the sit-and-reach box against a wall or object that prevents it from slipping during the testing process.

Step 2: Gather basic data (e.g., age, height, weight) for the individual data sheet as described in laboratory activity 1.1. Have the subject put on his or her shoes.

Step 3: Have the subject perform a 10 min warm-up (see table 3.1).

Step 4: Instruct the subject to remove his or her shoes and have them sit on the floor with the heels and soles of the feet against the index or heel line (23 cm or 9.1 in. mark), with the legs fully extended and feet about 20 cm (8 in.) apart.

Step 5: Place your hands across the subject's knees to ensure full leg extension is maintained (see figure 3.5a).

Step 6: Instruct the subject to extend the arms with one hand on top of the other and palms facing down (see figure 3.5b).

Step 7: Have the subject bend forward and reach with both hands along the measuring scale on top of the sit-and-reach box. This position should be held for 1 to 2 s. If the subject's knees bend or the fingertips become uneven, the test does not count, and the process should be repeated. Record the resulting measure on the individual data sheet.

Step 8: Perform steps 6 and 7 three more times.

Step 9: Record the results of the fourth trial in the appropriate location on the individual data sheet.

Step 10: Compare the results of the fourth trial with the percentile ranks or normative data presented in table 3.2, and note the subject's ranking in the appropriate spot on the individual data sheet.

Backsaver Sit-and-Reach Test

The backsaver sit-and-reach test is a modification of the traditional sit-and-reach test. It reduces the posterior compression typically seen when bending forward with both legs extended (2, 6, 35). Instead, the backsaver test requires the subject to bend forward with one leg extended and the other bent (2, 11, 21). Even so, some subjects report discomfort, in this case in the hip joint of the bent leg (20, 21). Other than the altered leg positions and the fact that each leg is measured separately, the backsaver test is performed in a manner similar to that of the traditional sit-and-reach test.

Step 1: Place the sit-and-reach box against a wall or an object that will prevent it from slipping during the testing process.

Step 2: Gather basic data (e.g., age, height, weight) for the individual data sheet as described in laboratory activity 1.1. Have the subject put on his or her shoes.

Step 3: Have the subject perform a 10 min warm-up (see table 3.1).

Step 4: Instruct the subject to remove his or her shoes and sit on the floor with the right leg extended so that the sole of the foot is placed against the sit-and-reach box at the index or heel line (23 cm or 9.1 in. mark). The subject's left leg should be bent at approximately 90°, with the sole of the left foot placed flat on the floor (21) about 5 to 7.5 cm (2-3 in.) from the sit-and-reach box (2).

Step 5: Have the subject extend the arms with one hand on top of the other and with the palms facing down.

Step 6: For the test, the subject should bend forward while moving both hands along the measuring scale on top of the sit-and-reach box. This position should be held for 1 or 2 s. The subject may have to move the bent leg to the side while extending forward to allow movement through the ROM. If the extended knee bends or the fingertips become uneven, the test does not count, and it should be repeated. Record the resulting measurement on the individual data sheet.

Step 7: Perform steps 4 through 6 three more times with the right leg.

Step 8: Compare the results of the fourth trial with the percentile ranks or normative data presented in table 3.2, and note the results in the appropriate spot on the individual data sheet.

Step 9: Have the subject change positions so that the left leg is extended and the sole of the left foot is placed against the sit-and-reach box at the index or heel line (23 cm or 9.1 in. mark). The right leg should be bent to approximately 90°, with the sole of the right foot placed flat on the floor (21) about 5 to 7.5 cm (2-3 in.) from the sit-and-reach box (2). See figure 3.10*a*.

Step 10: Have the subject extend the arms with one hand on top of the other and with the palms facing down. See figure 3.10*b*.

Step 11: Instruct the subject to bend forward while moving both hands along the measuring scale on top of the sit-and-reach box. This position should be held for 1 to 2 s. The subject may have to move the bent leg to the side while extending forward to allow movement through the ROM. If the subject's extended knee bends or the fingertips become uneven, the test does not count, and it should be repeated. Record the resulting measurement on the individual data sheet.

Step 12: Perform steps 9 through 11 three more times with the right leg.

Step 13: Compare the results of the fourth trial with the percentile ranks or normative data presented in table 3.2, and note the results in the appropriate spot on the individual data sheet.

Question Set 3.3

1. How does the Canadian sit-and-reach test differ from the traditional version? How does it differ from the various other tests that could be used when evaluating lower-body flexibility?

2. How do the percentile ranks for the Canadian sit-and-reach test differ from those for the backsaver variant?

3. Compare the results of your sit-and-reach test with the normative data. What do these results indicate?

4. Compare the class data as a whole for the sit-and-reach test with the normative data. What do these results indicate?

5. Create a bar graph of the means and standard deviations for the five most flexible men and the five least flexible men.

6. Create a bar graph of the means and standard deviations for the five most flexible women and the five least flexible women.

7. Fill in the table in the individual data sheet comparing the methods used in this lab. Indicate the best application for each test, and describe the pros and cons of each test.

LABORATORY ACTIVITY 3.3

INDIVIDUAL DATA SHEET

Name or ID number: _____ Date: _____

Tester: _____ Time: _____

Sex: M / F (circle one) Age: _____ y Height: _____ in. _____ m

Weight: _____ lb _____ kg Temperature: _____ °F _____ °C

Barometric pressure: _____ mmHg Relative humidity: _____ %

Sit-and-reach test results		Trial 1	Trial 2	Trial 3	Trial 4	Best	4th trial	Index/heel line (cm)
Canadian							_____	26
Traditional						_____		23
Backsaver	Right leg					_____		23
	Left leg					_____		23

Individual results ranking		Results (cm)	Ranking/category/percentile
Canadian			Percentile =
Traditional			Percentile =
Backsaver	Right leg		Percentile =
	Left leg		Percentile =

Shoulder Flexibility Test Comparisons

Equipment

- Mat
- 18 in. (46 cm) ruler
- Two yardsticks or metersticks
- Individual data sheet
- Group data sheet

Warm-Up

Have the subject perform a structured warm-up beforehand to increase the reliability and validity of the test. The warm-up session should include a 5 min general warm-up and a 5 min dynamic stretching warm-up (38).

The 5 min general warm-up can involve activities such as jogging, cycling, skipping, jumping rope, or doing calisthenics (26). The activity should be undertaken with the goal of increasing HR, blood flow, muscle temperature, respiration rate, and perspiration while decreasing viscosity of joint fluids (12, 26). Taken collectively, these responses prepare the subject for the more specific warm-up activities that are performed next.

After completing the general warm-up, the subject should perform dynamic warm-up movements that work through the ROMs required for the testing activity (26). Dynamic warm-up activities appropriate for the shoulder flexibility test might include arm swings, arm circles, shoulder rotations, or other activities (25). After completing the warm-up, the subject can begin the testing process.

Shoulder Elevation Test

This is a commonly used test of shoulder and chest flexibility (1).

Step 1: Set up the testing area. Place a mat on the floor in an area with enough room to perform testing.

Step 2: Gather basic data (e.g., age, height, weight) for the individual data sheet as described in laboratory activity 1.1. Have the subject put on his or her shoes.

Step 3: Instruct the subject to perform a 10 min warm-up that includes dynamic activities such as arm swings and shoulder rotations.

Step 4: Have the subject grasp a meterstick in front of the body with both hands using a pronated grip (knuckles facing forward). Make sure the subject keeps the arms relaxed in this position.

Step 5: Determine arm length by measuring the distance from the acromion process to the top of the meterstick being held by the subject (figure 3.13a). Measure to the nearest centimeter. Record the result on the individual data sheet for laboratory activity 3.4.

Step 6: Have the subject assume a prone position (lying facedown) on the mat so that the chin touches the floor and the arms are extended overhead while holding the meterstick with the same grip used in step 4.

Step 7: Have the subject slowly raise the meterstick as high as possible while maintaining chin contact with the floor and keeping the elbows extended. Encourage the subject to hold the highest possible position for 1 or 2 s.

Step 8: Measure the distance between the floor and the bottom of the meterstick when the subject achieves the highest position possible (figure 3.13*b*). Record this measure in the individual data sheet.

Step 9: Perform steps 4 through 8 twice more.

Step 10: After completing three testing trials, calculate the shoulder elevation score with the following formula (1, 16), and record it on the individual data sheet:

$$\text{Shoulder elevation score} = \frac{\text{shoulder elevation (cm)} \times 100}{\text{arm length (cm)}}$$

Step 11: Compare the shoulder elevation score with the normative data and ratings presented in table 3.6, and record the result on the individual data sheet.

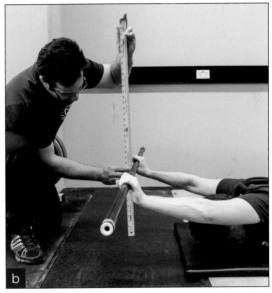

Figure 3.13 Shoulder extension test: *(a)* determining arm length and *(b)* measuring shoulder elevation.

Courtesy of Greg Haff

Table 3.6 Percentile Ranks and Norms for Shoulder Elevation

	Classification	Percentile	MALE in.	MALE cm	FEMALE in.	FEMALE cm
Percentile rank	Well above average	90	106-123	269-312	105-123	267-312
	Above average	70	88-105	224-267	86-104	218-264
	Average	50	70-87	178-221	68-85	173-216
	Below average	30	53-69	135-175	50-67	127-170
	Well below average	10	35-52	89-132	31-49	79-124

Adapted from Acevedo 2001; Adapted from B.L. Johnson and J. K. Nelson, 1986, *Practical measurement for evaluation in physical education*, 4th ed. (Minneapolis, MN: Lea & Febiger).

Back-Scratch Test

The back-scratch test is a simple shoulder flexibility test that requires minimal equipment.

Step 1: Have the subject perform a 10 min warm-up that includes dynamic activities such as arm swings and shoulder rotations.

Step 2: Gather basic data (e.g., age, height, weight) for the individual data sheet as described in laboratory activity 1.1. Have the subject put on his or her shoes.

Step 3: After the warm-up, have the subject raise the right arm, bent at the elbow, and reach across the back as far as possible while also placing the left arm down and behind the back. The subject should attempt to cross the fingers of the left and right hands behind the back (figure 3.14*a*).

Step 4: Measure the distance of the finger overlap or gap to the nearest 0.5 in. (1.3 cm) (figure 3.14*b*). An overlap is considered a positive score, whereas a gap between the hands is considered a negative score. If the fingers merely touch, the score is 0. Record the test result in the appropriate location on the individual data sheet.

Step 5: Perform steps 3 and 4 twice more.

Step 6: Have the subject raise the left arm, bent at the elbow, and reach across the back as far as possible while also placing the right arm down and behind the back. The subject should attempt to cross the fingers of the left and right hands behind the back.

Step 7: Measure the distance of the finger overlap or gap to the nearest 0.5 in. An overlap is considered a positive score, whereas a gap between the hands is considered a negative score. If the fingers merely touch, the score is 0. Record the test result in the appropriate location on the individual data sheet.

Step 8: Perform steps 6 and 7 twice more.

Step 9: Calculate the average for the raised right arm and raised left arm, and record the results on the individual and group data sheets. Compare the average results for each arm with the normative data presented in table 3.7 and record the result on the individual data sheet.

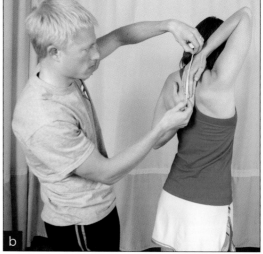

Figure 3.14 Back-scratch test: *(a)* arm position and *(b)* measurement.

Table 3.7 Back-Scratch Test Normative Data for College-Aged Students

Classification	MALE		FEMALE	
	in.	cm	in.	cm
Excellent	≥5	≥12.7	≥5	≥12.7
Above average	2.0-4.8	5.1-12.2	2.0-4.75	5.1-12.1
Average	0.0-1.8	0.0-4.6	0.0-1.75	0.0-4.6
Below average	−1.0--0.25	−2.5--0.6	−1.0--0.25	−2.5--0.6
Poor	<−1	<−2.5	<−1	<−2.5

Adapted from Nieman 2003.

Question Set 3.4

1. What are the differences between the back-scratch and shoulder elevation tests?

2. Compare the results of your shoulder elevation and back-scratch tests with the normative data. What do these results indicate?

3. Compare the class data as a whole for the shoulder elevation and back-scratch tests with the normative data. What do these results indicate?

4. Create bar graphs of the means and standard deviations for the back-scratch and shoulder elevation tests—one for each sex and one with both sexes combined.

5. Use a paired t-test to compare the results of the shoulder elevation and back-scratch tests for the five most flexible men with the results for the five most flexible women. How do these findings compare with what you would expect to find?

LABORATORY ACTIVITY 3.4

INDIVIDUAL DATA SHEET

Name or ID number: _____ Date: _____

Tester: _____ Time: _____

Sex: M / F (circle one) Age: _____ y Height: _____ in. _____ m

Weight: _____ lb _____ kg Temperature: _____ °F _____ °C

Barometric pressure: _____ mmHg Relative humidity: _____ %

Shoulder Extension Test

Arm length test: _____ in. _____ cm

Trial	Shoulder elevation score (cm)	[Shoulder elevation (cm) × 100] / arm length	Flexibility classification
1		(_____ × 100) /_____	
2		(_____ × 100) /_____	
3		(_____ × 100) /_____	
Best		(_____ × 100) /_____	

Overall percentile rank: _____

Back-Scratch Test

	Trial 1 (cm)	Trial 2 (cm)	Trial 3 (cm)	Average (cm)
Right arm				
Left arm				
Average				

Ranking: _____

Blood Pressure Measurements

Objectives

- Understand the concept of systemic blood pressure (BP) fluctuations during the cardiac cycle.
- Define the terms *cardiac cycle, systolic blood pressure*, and *diastolic blood pressure*.
- Develop the skill of taking BP during rest and exercise.
- Examine BP responses to changes in body position and aerobic and isometric exercise.
- Understand risk stratifications of stages of hypertension and contraindications to exercise testing.

DEFINITIONS

blood pressure (BP)—Pressure in the arterial peripheral circulation during the cardiac cycle.

cardiac cycle—Mechanical and electrical events occurring during one heartbeat; includes both systole and diastole.

cardiac output (CO or $\dot{Q}$)—Amount of blood produced by the heart every minute ($L \cdot min^{-1}$).

diastole—Phase of the cardiac cycle during ventricular relaxation.

diastolic blood pressure (DBP)—Pressure (mmHg) in the peripheral large arteries during diastole.

heart rate (HR)—Number of heart beats per minute ($beats \cdot min^{-1}$).

Korotkoff phases—Change in sounds heard through a stethoscope as pressure is released in a sphygmomanometer as blood flow transitions from zero flow, to turbulent, to laminar flow.

laminar blood flow—Type of blood flow not resulting in an audible sound through a stethoscope; defined by streamline flow or flow in parallel layers without turbulence.

orthostatic hypotension—Reduction in BP as the result of a change in body position.

stroke volume (SV)—Amount of blood per heartbeat ($ml \cdot beat^{-1}$).

systole—Phase of cardiac cycle during ventricular contraction.

systolic blood pressure (SBP)—Pressure (mmHg) in the peripheral large arteries during systole.

turbulent blood flow—Blood flow resulting in an audible sound through a stethoscope; defined by turbulent flow due to variations of pressure and velocity.

vascular resistance—Resistance to blood flow as determined by Poiseuille's law, according to which resistance equals the length of the vessel multiplied by the blood viscosity divided by vessel radius to the fourth power.

vasovagal syncope—Fainting due to low BP perfusion of the brain as the result of increased parasympathetic tone (vagal nerve) or withdrawal of sympathetic tone stimulated by a trigger.

venous return—Return of blood to the heart from the venous side of the circulation. Venous circulation has less pressure and thus is suspect to external influences such as body position, muscular contractions, and thoracic pressure.

Blood pressure (BP) is one of the basic measures in human physiology, and it provides a good measure of the work of the heart. It is considered so important that every trip to the doctor's office includes a BP measurement. It can be defined as the pressure exerted against the walls of the various vessels of the arterial side of the circulatory system by the heart as it pumps blood to the body. BP depends on two factors: the volume of blood delivered to a vessel and the amount of resistance exerted against the blood being delivered (vascular resistance or total peripheral resistance). **Cardiac output (CO or Q̇)** is determined by **heart rate (HR)** and **stroke volume (SV)**: CO = HR × SV. **Vascular resistance** during exercise is determined largely by the diameter of the vessels; a small decrease in the radius of the vessels results in large increases in vascular resistance. A change in either vascular resistance or CO results in changes in BP.

BP is usually recorded as systolic pressure over diastolic pressure. **Arterial systolic blood pressure (SBP)** is the maximum pressure attained during peak ventricular ejection or **systole** during the **cardiac cycle**. **Diastolic blood pressure (DBP)** is the minimum arterial pressure within the cardiac cycle during ventricular relaxation or **diastole**. These are typically reported as SBP over DBP, or SBP/DBP in mmHg. Average normal resting BP is <120 mmHg over <80 mmHg. Table 4.1 and table 4.2 present resting BP norms and their classifications. In November 2017, the classifications were changed to define hypertension as ≥130/80 mmHg, instead of the previous definition of ≥140/90 mmHg (1). This small change in defining hypertension results in 46% of the general population being classified as hypertensive, instead of 32% by the previous definition. This amounts to more than 25 million more people classified as hypertensive in the United States alone. The new classification does not necessarily mean that earlier pharmaceutical intervention is needed; instead, it allows doctors and patients to earlier address this critical risk

Table 4.1 Percentile Norms for BP in Active Women and Men

| | RESTING BP (SBP/DBP; mmHg) | | | | | |
| | WOMEN | | | MEN | | |
	20-49	50-59	60+	20-49	50-59	60+
Very low (>80%)	<104/<70	<110/<70	<120/<75	<111/<75	<116/<78	<120/<76
Low (60%-80%)	106-112/70-75	110-120/70-79	120-128/75-80	112-120/74-80	116-122/78-80	120-130/76-80
Average (40%-60%)	110-120/72-80	120-130/79-82	128-136/80	120-126/80-84	122-130/80-86	130-140/80-84
High (20%-40%)	118-130/78-82	130-140/82-90	136-142/80-88	127-138/84-90	130-140/86-90	140-150/84-90
Very high (<20%)	<130/>82	>140/>90	>142/>88	>138/>90	>140/>90	>150/>90

SBP = systolic blood pressure; DBP = diastolic blood pressure.

Adapted from M.L. Pollock, J.H. Wilmore, and S.M. Fox, 1978, *Health and fitness through physical activity* (New York, NY: John Wiley and Sons).

Table 4.2 Classification of BP for Adults 18 Years or Older*

Category	SBP (mmHg)**	DBP (mmHg)
Normal	<120	<80
High	120-129	<80
Stage 1 hypertension	130-139	80-89
Stage 2 hypertension	≥140	≥90

*This classification applies to individuals not taking antihypertensive medication and not acutely ill. It is based on the average of two or more readings on two or more occasions.

**When systolic and diastolic pressures fall into different categories, use the higher category for classification.

Data from Whelton et al. 2017.

factor by lifestyle changes. Hypertension (table 4.2) may contribute as a risk factor toward cardiovascular disease. It may also prove to be a contraindication to beginning or continuing an exercise test in a clinical situation (2). This lends importance to the accuracy of BP measurements by exercise professionals.

BP depends on many factors, including body position, hydration, sex, muscle actions, stress, diet, and disease (e.g., atherosclerosis). This lab will demonstrate the response of BP to changes in body position, which are often called *orthostatic changes*. When a person is supine, the heart works less hard to deliver blood to the periphery, especially the brain, because it does not have to work against gravitational forces. In addition, HR is typically lower due to the benefit of enhanced **venous return** and thus increased SV. When a person moves from supine to standing, venous return is reduced due to the effect of gravity, and thus BP drops. This common occurrence is often referred to as **orthostatic hypotension**. Many people have experienced dizziness or light-headedness upon standing up; if severe, this effect can result in fainting, or **vasovagal syncope**, which is more common in people with low BP. Baroreceptors in the carotid arch sense this lowering of BP and stimulate the cardiovascular brain centers to adjust BP by increasing venous return and strength of heart contraction. As a result of these processes, BP is often lower (by 10-20 mmHg) immediately after standing than while supine, whereas steady-state standing BP is higher than supine BP.

Again, many factors can influence BP besides body position and exercise. Even the mere act of measuring BP may cause it to rise due to stress experienced by the subject or patient. This occurrence, often referred to as *white coat syndrome*, can be exacerbated in a hospital or other medical environment. Thus, perceived stress can artificially elevate BP measurements. As a result, providing a relaxing environment—along with other techniques, such as ensuring standardized body position, allowing a few minutes of rest prior to measurement, and having the subject avoid crossed legs or undue muscular actions during measurement—can be essential for accurate measurement of BP.

BLOOD PRESSURE RESPONSES TO EXERCISE

In the transition from rest to exercise, SBP rises rapidly and then levels off once steady state is attained (3, 6). Typical systolic pressures during submaximal exercise range from 140 to 160 mmHg. Graded dynamic exercise normally produces a progressive increase in systolic pressure, and values can reach as high as 250 mmHg during maximal exercise.

Figure 4.1 shows a typical SBP and DBP responses to exercise of increasing intensity. The rise in systolic pressure reflects the increased force of contraction of the ventricles in order to increase CO from sympathetic stimulation. Diastolic pressure may show either no change or a slight increase or decrease during graded exercise due to a redistribution of blood flow to the capillary beds in the large exercising muscle groups (3, 6). If large muscle groups are involved, a small (10-20 mmHg) decrease in diastolic pressure may be observed near maximal effort due to the large amount of vessel dilation.

Following heavy exercise, SBP drops rapidly. This rapid drop (hypotension) results from the pooling of blood in vessels that were dilated during the heavy exercise. The pooling reduces venous return and thus decreases CO. If the reduction in CO is severe enough, the person may experience light-headedness due to insufficient blood flow to the brain. For this reason,

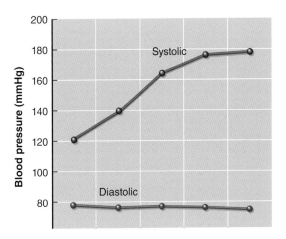

Figure 4.1 Systolic and diastolic pressure changes in response to exercise of increasing intensity.

a cool-down period is recommended following intense exercise in order to maintain HR and SV and allow the cardiovascular system to readjust gradually.

Pathological BP responses to exercise in those at risk for cardiovascular disease include a lack of increase in SBP with increase in intensity, severe increases or decreases in DBP, syncope, and pallor or other signs of poor perfusion (2). Because of this, general indications for stopping an exercise test include a drop in SBP ≥10 mmHg with an increase in intensity or if SBP decreases below the value at rest in the same position, an increase of >250 mmHg in SBP or >115 mmHg in DBP, a lack of an increase in HR with an increase in intensity, and poor signs of perfusion or dyspnea (2). BP responds differently during isometric exercise or even slow concentric actions. Sustained muscle actions that occlude blood flow can increase resistance tremendously and can result in extremely high SBP and DBP. Thus, even though muscle is contracting and in need of blood flow, continuous muscular contraction surrounding the blood vessels can significantly increase resistance and decrease blood flow. Indeed, it has been demonstrated that isometric contractions of the forearm at 50% of max reduce oxygen delivery to downstream vascular beds by 80% after only 20 s (4). In an effort to overcome this occlusion, the heart increases BP.

ACCURATE BLOOD PRESSURE CHECKS

Measuring BP is a basic skill in exercise physiology because it provides a good indication of the work of the heart. As a result, this skill is critical for properly prescribing exercise for at-risk individuals. Indirect measurement of BP is done with a stethoscope and a sphygmomanometer. The indirect measurement of BP monitors the sounds of blood flow in the brachial artery, or **Korotkoff phases**, that are audible through the stethoscope (for more information about the Korotkoff phases, see the accompanying highlight box). **Laminar blood flow** makes little or no sound in the arteries, whereas **turbulent blood flow** due to occlusion from the BP cuff can make a variety of sounds. It is important to rely on these sounds or phases, and not the movement of mercury or needle on the sphygmomanometer. Blood is capable of pushing against the cuff without actually getting through the aperture created by the cuff.

Although BP can differ slightly in the left and right arms, the difference is minimal in

Korotkoff Phases

When pressure in the cuff is greater than SBP, blood flow is occluded in the distal brachial artery. As pressure in the cuff is slowly released, blood is allowed to flow through the widening aperture. The amount of turbulent flow creates differing sounds, which are described by the Korotkoff phases:

- Phase I—Systolic pressure is indicated by the first tapping sound that becomes audible—read the dial or mercury column to the nearest mmHg.
- Phase II—A murmur or swishing sound is heard.
- Phase III—Sound increases in intensity.
- Phase IV—A distinct, abrupt muffling of sound is noted, though it may not be heard at rest. This is diastolic pressure during exercise.
- Phase V—At rest, diastolic pressure is marked by the disappearance of the sound; record this pressure to the nearest mmHg. During exercise, you may hear sounds all the way down to 0 mmHg. In this case, record phase IV for diastolic pressure.

You can listen to the Korotkoff sounds at www.thinklabs.com/heart-sounds.

healthy adults; for the purposes of skill acquisition in this lab, use the most convenient arm. Note, however, that in clinical assessment both arms should be measured, and the highest recording should be used as the BP measurement (5). When doing serial measurements, the same arm should be used. The arm should be as free of clothing as possible because clothing can muffle sounds during auscultation; however, simply rolling up or bunching a sleeve too tightly can mimic a BP cuff and thus partially occlude blood flow. Cuffs of varying sizes are available, including for children, adults, and large adults. Index lines on the cuff should indicate whether the cuff is appropriate for a given person. The bladder should cover about 80% of the arm to enable it to occlude blood flow effectively when inflated.

BP measurement can be affected by the testing situation and by preparation. Ideally, subjects should refrain from using stimulants, such as caffeine, prior to testing. They should wear loose clothing, be euhydrated, and avoid strenuous exercise for several hours before the test. While being tested, they should not cross their legs, and they should avoid any form of isometric muscle action, such as pressing down on their legs, dangling their feet off the ground, or sitting erect with the back unsupported. In addition, the environment should be free of stimuli such as loud music or noises or unnecessary activity (5).

Go to the web study guide to access electronic versions of individual and group data sheets, the question sets from each laboratory activity, and case studies for each laboratory activity.

WWW

Effects of Body Position on BP

Equipment

- Sphygmomanometer
- Stethoscope
- Stopwatch
- Individual data sheet
- Group data sheet

Begin this lab activity by gathering basic data (e.g., age, height, weight) for the individual data sheet as described in laboratory activity 1.1.

Resting BP

Step 1:   Have the subject sit in a chair for ~3 min to attain steady state at rest.

Step 2:   Palpate for brachial pulse and measure HR (see figure 4.2a and "Measurement of Heart Rate" in lab 7 and "Procedures for Palpating HR" in lab 8).

Step 3:   The subject should be comfortably seated with the arm slightly bent, the palm facing up, and the forearm supported nearly horizontal at the level of the heart.

Step 4:   Place the BP cuff on the subject's arm.

Step 5:   Place the cuff with the lower margin about 1 in. (2.5 cm) above the antecubital space. Ensure the bladder covers the brachial artery. During the measurement, the arm should be supported at the level of the heart (see figure 4.2b).

Step 6:   Palpate to find the brachial artery in the antecubital space and place the flat side of the stethoscope bell lightly over the brachial artery (figure 4.2c). Too much pressure on the bell can affect blood flow in the artery and cause sounds independent of the cuff pressure. Too little pressure can reduce the ability to hear Korotkoff phases. Be sure the bell is flat against the skin and you have the earpieces in correctly (pointed forward).

Step 7:   Holding the stethoscope in place, close the air-release screw (clockwise) and inflate the BP cuff to 200 mmHg or more by pumping the air bulb. The pressure occludes the flow to the brachial artery, causing the pulse sound to stop. At rest, you may need only to inflate to 180 mmHg to be above SBP (figure 4.2c).

Step 8:   Release pressure slowly and smoothly at a rate of 2 or 3 mmHg · s^{-1} by releasing the air-release screw (counterclockwise).

Step 9:   As pressure is released, the Korotkoff sounds become audible. The appearance of a faint tapping sound marks the SBP; at this point, note the mmHg reading. This sound will increase in intensity before fading.

Step 10: When the subject is at rest, such as sitting or standing, the DBP is marked by the disappearance of sound in the brachial artery, which indicates laminar blood flow. During exercise, the DBP is marked by an abrupt muffling of the pulse sound. Note the mmHg at the appropriate Korotkoff phase.

Step 11: Once systolic and diastolic measurements are obtained, release all pressure out of the cuff.

Step 12: Record measurements (HR, SBP, and DBP) on the individual data sheet.

Step 13: Repeat steps 1 through 12 with the same subject for trial 2 data collection, allowing at least 2 to 3 min between measurements. Each lab member should act as both the subject and the BP data collector. Members should fill out the individual data sheet for their own BP measurements.

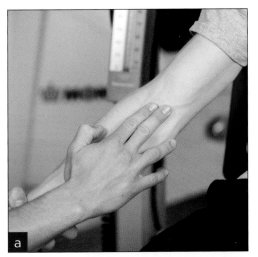
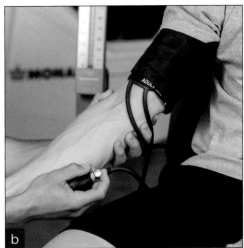
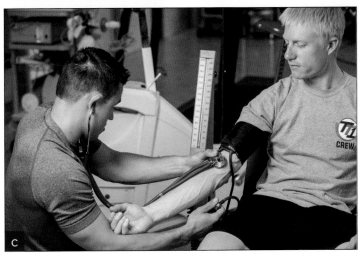

Figure 4.2 Measuring resting BP: *(a)* palpation of the brachial artery, *(b)* placement of the BP cuff, and *(c)* measuring resting BP.

Supine and Standing BP

Step 1: Have the subject quietly maintain a supine position for ~3 min with the BP cuff already attached.

Step 2: After ~3 min, measure and record the supine HR and BP as described previously in the procedure for measuring resting BP. While taking the BP, have the subject keep the cuffed arm still; lifting it will affect the measurement. The subject should let the cuffed arm rest comfortably next to the body with the palm up to allow access to the antecubital space.

Step 3: Pump the cuff to 200 mmHg just before having the subject stand up.

Step 4: Have the subject stand up, and immediately measure HR and BP.

Step 5: Have the subject remain standing for at least 3 min. Then repeat the HR and BP measurements.

Step 6: Repeat steps 1 through 5 to collect data for trial 2. Record the data in the appropriate location on the individual and group data sheets.

Question Set 4.1

1. Prepare a data table reporting the subject's name, height, and weight, as well as the sitting, supine, and standing HR and BP collected during the laboratory session.

2. What changes did you observe in HR and BP with the three body positions? What accounts for these changes?

3. What is happening physiologically when a person stands up quickly and gets light-headed?

LABORATORY ACTIVITY 4.1

INDIVIDUAL DATA SHEET

Name or ID number: _____ Date: _____

Tester: _____ Time: _____

Sex: M / F (circle one) Age: _____ y Height: _____ in. _____ m

Weight: _____ lb _____ kg Temperature: _____ °F _____ °C

Barometric pressure: _____ mmHg Relative humidity: _____ %

| Trial | HR (beats · min⁻¹) | BP (mmHg) | | | | | | | |
| | | SITTING POSITION | | SUPINE POSITION | | INITIAL STANDING | | >2 MIN STANDING | |
		Systolic	Diastolic	Systolic	Diastolic	Systolic	Diastolic	Systolic	Diastolic
1									
2									
Mean									
SD									

Effects of Dynamic Exercise on BP

Equipment

- Sphygmomanometer
- Cycle ergometer
- Stethoscope
- Stopwatch
- Individual data sheet

Techniques for HR and BP Measurement During Exercise

A moving subject complicates the measurement of HR and BP because of the added background noise. Here are some tricks to help you perform HR palpation and BP auscultation during exercise on a cycle ergometer:

- Avoid looking at the pedals while taking HR—the cadence can throw off your counting.
- While counting the HR for 15 s, you might close your eyes to concentrate.
- Control the noise in the room as much as you can—keep your labmates' talking to a minimum!
- While taking BP, be conscious of the stethoscope and sphygmomanometer tubing. Any bumping or rubbing against knees, arms, or the ergometer will add noise to your auscultation.
- Control and support the subject's arm to avoid motion and undue muscle actions.
- Instruct the subject to refrain from gripping the handlebars tightly; engaging in this isometric action can increase BP.
- Initial cuff pressure needs to increase as intensity increases to ensure cuff pressure exceeds SBP. At low levels of exercise, 200 mmHg is sufficient, but at higher intensities you may have to go to 250 mmHg or more before releasing pressure.
- Because the subject will have a higher HR during exercise, you can afford to release air pressure faster (at a rate of 5 or 6 $mmHg \cdot s^{-1}$).
- Continue monitoring the subject's adherence to the prescribed cadence, and be aware of the subject's subjective symptoms (e.g., pallor, dyspnea).

Because students are generally new to taking HR and BP during exercise, do not worry about sticking to the 3 min stage time during these lab activities. If you miss an HR or a BP, just try again. It is more important to practice the skill.

Submaximal Exercise BP

Step 1: Gather basic data (e.g., age, height, weight) for the individual data sheet as described in laboratory activity 1.1.

Step 2: Fit the subject on the cycle ergometer comfortably with a slight knee angle (5°-15°) when the leg is fully extended (figure 4.3*a*).

Step 3: Before the test, read the subject the rating of perceived exertion (RPE) instructions (see lab 10 for a full discussion of RPE):

During the test, we want you to pay close attention to how hard you feel the exercise work rate is. This feeling should reflect your total amount of exertion and fatigue, combining all sensations and feelings of physical stress, effort, and fatigue. Don't concern yourself with any one factor, such as leg pain, shortness of breath, or exercise intensity, but try to concentrate on your total, inner feeling of exertion. Try not to underestimate or overestimate your feelings of exertion; be as accurate as you can.

Step 4: Allow the subject to ask questions prior to being connected to the BP equipment.

Step 5: Place the BP cuff on the subject's convenient arm and measure resting HR and BP while the subject remains seated on the cycle ergometer. (Remind the subject not to place undue pressure on the pedals or grip the handlebars too tightly, because this will elevate BP.) Record resting measurements on the individual data sheet. Calculate age-predicted HRmax and 85% of age-predicted HRmax.

Step 6: Have the subject begin pedaling at 70 revolutions per min (rev·min^{-1}). If necessary, use a metronome to maintain constant cadence, which can affect power on the bike and thus intensity. (BP is a function of exercise intensity.) Add 0.5 kg (1.1 lb) of resistance to the ergometer for stage 1.

Step 7: Allow 3 min of pedaling at 70 rev · min^{-1} to attain steady state. Then measure HR, BP, and RPE. (Use the steps in lab 4.1 for BP measurement.)

Step 8: Increase the ergometer resistance to 1 kg (2.2 lb) for stage 2. Continue having the subject pedal at 70 rev · min^{-1}.

Step 9: Allow 3 min to attain steady state, and then measure HR, BP, and RPE.

Step 10: Increase the ergometer resistance to 1.5 kg (3.3 lb) for stage 3. Continue having the subject pedal at 70 rev · min^{-1}. Do not exceed 85% of age-predicted HRmax.

Step 11: Allow 3 min to attain steady state, and then measure HR, BP, and RPE.

Step 12: Decrease the ergometer resistance to 0.5 kg in stage 4 for recovery. Continue having the subject pedal at 70 rev · min^{-1}.

Step 13: Allow 3 min to attain steady state, and then measure HR, BP, and RPE.

Step 14: Perform steps 1 through 13 on a second test subject, and record the data on the individual data sheet (subject B section).

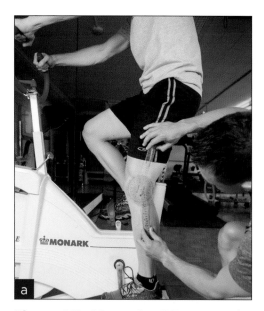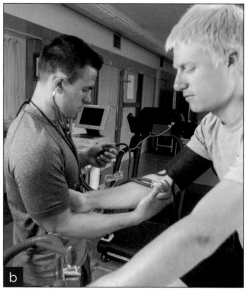

Figure 4.3 Measuring BP on a cycle ergometer: *(a)* proper positioning of the subject with a slight knee angle (5°-15°) with the leg fully extended and *(b)* technique for measuring BP.

Question Set 4.2

1. Prepare a data table reporting the subject's name, height, and weight, as well as the HR, BP, and RPE data collected during the laboratory session.
2. How do SBP and DBP respond to exercise of increasing intensity?
3. What are the indications for stopping a test that relate to BP response during an exercise test?

LABORATORY ACTIVITY 4.2

INDIVIDUAL DATA SHEET

Date: _____ Time: _____

Tester: _____

Temperature: _____°F _____°C Barometric pressure: _____mmHg

Relative humidity: _____ %

Subject A

Name or ID number: _____

Sex: M / F (circle one)   Age: _____ y

Height: _____ in. _____ m   Weight: _____lb _____kg

Resting BP: _____ / _____ mmHg   Resting HR: _____ beats·min⁻¹

Age-predicted HRmax: _____ beats·min⁻¹   85% age-predicted HRmax: _____ beats·min⁻¹

	Time (min)	Workload (kg)	HR (beats · min⁻¹)	BP (mmHg)	RPE
Stage 1	3	0.5			
Stage 2	6	1.0			
Stage 3	9	1.5			
Stage 4	12	0.5			

Subject B

Name or ID number: _____

Sex: M / F (circle one)   Age: _____ y

Height: _____ in. _____ m   Weight: _____lb _____kg

Resting BP: _____ / _____ mmHg   Resting HR: _____ beats·min⁻¹

Age-predicted HRmax: _____ beats·min⁻¹   85% age-predicted HRmax: _____ beats·min⁻¹

	Time (min)	Workload (kg)	HR (beats · min⁻¹)	BP (mmHg)	RPE
Stage 1	3	0.5			
Stage 2	6	1.0			
Stage 3	9	1.5			
Stage 4	12	0.5			

Effects of Isometric Contractions on BP

Equipment

- Leg extension machine (if unavailable, use a wall)
- Hand dynamometer
- Sphygmomanometer
- Stethoscope
- Stopwatch
- Individual data sheet
- Group data sheet

Begin this lab activity by gathering basic data (e.g., age, height, weight) for the individual data sheet as described in laboratory activity 1.1. Two subjects (A and B) will complete this lab activity to fill out the individual data sheet.

Upper-Body Isometric Exercise BP

Step 1:   Place the BP cuff on the subject's arm.

Step 2:   Have the subject sit quietly for ~3 min.

Step 3:   After ~3 min, measure and record the resting HR (palpated) and BP.

Step 4:   Instruct the subject to perform a maximum voluntary contraction (MVC) on the handgrip dynamometer with the dominant hand (figure 4.4; also see lab 12 for handgrip instructions). Record handgrip MVC on the individual data sheet.

Step 5:   Calculate and record 30% of MVC in kg.

Step 6:   After ~2 min of rest, have the subject hold the handgrip at 30% of MVC for 1 min.

Step 7:   Toward the end of the 1 min isometric contraction, measure HR, BP, and RPE while the subject continues to hold 30% MVC. Measure BP in the nonexercising, nondominant arm. It is important that the nonexercising arm is as relaxed as possible during the BP measurement.

Step 8:   Record the data in the appropriate locations on the individual data sheet for subject A.

Step 9:   Repeat steps 1 through 8 for subject B.

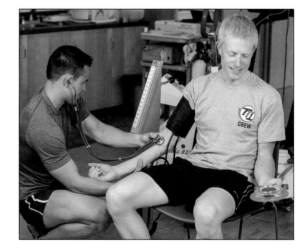

Figure 4.4   BP measurement during handgrip exercise.

Lower-Body Isometric Exercise BP

This part of the lab involves lower-body isometric exercise, which can be done with a leg extension machine or against a wall (figure 4.5 and figure 4.6).

Step 1:   If using the leg extension machine, have the subject select a weight on the machine that is approximately 50% of 1RM, and record on the individual data sheet.

Step 2:   Have the subject hold the weight at full extension for as long as possible or for 3 min, whichever comes first. When close to muscle failure, have the subject indicate that it is time to take HR, BP, and RPE.

Step 3:   While the subject continues to hold the 50% of 1RM, measure HR, BP, and RPE. Be sure the subject relaxes the arm as much as possible while you take the BP measurement.

Step 4:   Record the data in the appropriate locations on the individual data sheet for subject A.

Step 5:   Repeat steps 1 through 4 for subject B.

If a leg extension machine is not available, the subject can perform a wall sit, which is sometimes referred to as *waiting for the bus* (figure 4.6).

Step 1:   Have the subject do a wall sit for as long as possible or for 3 min, whichever comes first. When close to muscle failure, have the subject indicate that it is time to take HR, BP, and RPE.

Step 2:   While the subject continues to hold the wall sit, measure HR, BP, and RPE. Be sure the subject relaxes the arm while you take the BP measurement.

Step 3:   Record the data in the appropriate locations on the individual data sheet for subject A.

Step 4:   Repeat steps 1 through 3 for subject B.

Figure 4.5 BP measurement during isometric leg extension.

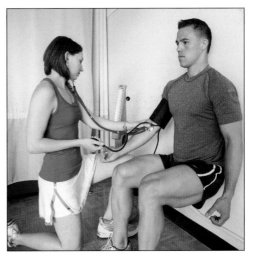

Figure 4.6 BP measurement during wall sit.

Question Set 4.3

1. Complete a data table reporting the subject's name, height, and weight, as well as the HR, BP, and RPE data collected during the laboratory session.

2. How does isometric exercise affect SBP and DBP? Why?

3. How does the BP response during arm exercise compare with the response during leg exercise? What explains this?

4. What does this lab illustrate about exercise prescription in hypertensive clients?

5. What do peak BPs during isometric actions depend on?

LABORATORY ACTIVITY 4.3

INDIVIDUAL DATA SHEET

Date: _____ Time: _____

Tester: _____

Temperature: _____ °F _____ °C Barometric pressure: _____ mmHg

Relative humidity: _____ %

Subject A

Name or ID number: _____

Sex: M / F (circle one)   Age: _____ y

Height: _____ in. _____ m Weight: _____ lb _____ kg

Resting BP: _____ / _____ mmHg Resting HR: _____ beats·min⁻¹

Age-predicted HRmax: _____ beats·min⁻¹ 85% age-predicted HRmax: _____ beats·min⁻¹

Handgrip MVC: _____ kg 30% handgrip MVC: _____ kg

Leg extension 50% 1RM: _____ kg

	Time (min)	Workload (kg)	HR (beats·min⁻¹)	BP (mmHg)	RPE
Resting		0			
Handgrip					
Leg extension					
Wall sit					

Subject B

Name or ID number: _____

Sex: M / F (circle one)   Age: _____ y

Height: _____ in. _____ m   Weight: _____lb _____kg

Resting BP: _____ / _____ mmHg   Resting HR: _____ beats·min^{-1}

Age-predicted HRmax: _____ beats·min^{-1} 85% age-predicted HRmax: _____ beats·min^{-1}

Handgrip MVC: _____ kg   30% handgrip MVC: _____ kg

Leg extension 50% 1RM: _____ kg

	Time (min)	Workload (kg)	HR (beats·min^{-1})	BP (mmHg)	RPE
Resting		0			
Handgrip					
Leg extension					
Wall sit					

Resting Metabolic Rate Determinations

Objectives

- Understand the indirect calorimetry measurement of energy metabolism.
- Understand the respiratory parameters indicating metabolic status and energy utilization ($\dot{V}O_2$, $\dot{V}CO_2$, RER).
- Determine RMR by measuring oxygen uptake.
- Compare the determination of RMR via oxygen uptake with the determination of RMR via estimation equations.

DEFINITIONS

adenosine triphosphate (ATP)—Molecule with energy harnessed in phosphate bonds used to perform mechanical or metabolic work.

aerobic metabolism—Production of ATP in the presence of oxygen-utilizing oxidative phosphorylation.

anaerobic metabolism—Production of ATP without the need for oxygen.

basal metabolic rate (BMR)—Minimal level of energy required to sustain the body's vital functions in the waking state; measured via indirect calorimetry in a supine position after 8 h of sleep and a 12 h fast with the subject having spent the night in the laboratory to avoid movement prior to measurement.

bioenergetics—Metabolic processes that result in stored energy being converted into a useable form of energy (ATP).

carbon dioxide production ($\dot{V}CO_2$)—Amount of CO_2 produced by the body in the conversion of carbon-containing macronutrients to usable energy.

indirect calorimetry—Measurement of energy expenditure from oxygen consumption ($\dot{V}O_2$) and carbon dioxide production ($\dot{V}CO_2$).

metabolic cart—Automated machine that measures expired CO_2 and O_2 and ventilation to calculate $\dot{V}O_2$ and $\dot{V}CO_2$.

metabolic equivalent (MET)—Amount of energy expended at or above rest. Rest is assumed to be 3.5 ml · kg^{-1} · min^{-1}, therefore calculated by dividing relative $\dot{V}O_2$ by 3.5 ml · kg^{-1} · min^{-1}.

oxygen consumption ($\dot{V}O_2$)—Amount of oxygen taken up and utilized by a person in a given time.

respiratory exchange ratio (RER)—Ratio of carbon dioxide production ($\dot{V}CO_2$) to oxygen consumption ($\dot{V}O_2$).

resting metabolic rate (RMR)—Energy expenditure at rest in a fasted, well-rested state in a supine position but, unlike BMR, not requiring the subject to stay overnight in the laboratory.

steady state—Point where aerobic metabolism contributes 100% of the energy requirement at a given absolute intensity.

total energy expenditure (TEE)—The sum of the major energy-consuming processes that contribute to a person being in energy balance. The three components of TEE are thermic effect of food, physical activity, and RMR.

The conversion of food into the usable form of energy for metabolism—**adenosine triphosphate (ATP)**—is called **bioenergetics**. ATP is a high-energy phosphate used for the energy required during muscular contractions.

$$Food + O_2 \rightarrow CO_2 + H_2O + ATP + heat$$

When this conversion is completed in the presence of oxygen (O_2), it is referred to as **aerobic metabolism**. Humans are also capable of **anaerobic metabolism**—energy production in the absence of oxygen, which is discussed further in laboratory 13. As exercise physiologists, we measure **oxygen consumption ($\dot{V}O_2$)** and **carbon dioxide production ($\dot{V}CO_2$)** by using **indirect calorimetry** to estimate energy expenditure. As a result, these measurements are often called *respiratory gas exchange*. This process can be conducted while subjects are at rest or during exercise. We must assume that these measurements include little or no anaerobic contribution to energy expenditure, which typically holds true during steady-state exercise at submaximal intensities. **Steady state** indicates that parameters of aerobic metabolism are in equilibrium with energy expenditure. To ensure steady-state measurements during rest and exercise, physiologists typically require several minutes to be completed at a given submaximal intensity.

Measurement of $\dot{V}O_2$ and $\dot{V}CO_2$ requires special equipment such as a **metabolic cart** (see appendix D). It uses the knowledge that our atmosphere contains 20.93% oxygen, 0.03% carbon dioxide, and 79.04% nitrogen and that the concentrations of these gases change in our expired breath. The amount of oxygen decreases as we consume oxygen, whereas the amount of carbon dioxide increases and nitrogen remains unchanged in our expired breath; for a more detailed explanation of these calculations and of the Haldane transformation, see Kenney (10) and appendix C. The consumption of oxygen and the production of carbon dioxide thus become functions of the fractions of these gases in the air and the expired breath, as well as the volume of air per breath. Therefore, metabolic carts measure only three critical components—the percentages of oxygen and carbon dioxide in the expired breath and the

volume of that breath. As we know from physics, the volume of a gas is affected by both temperature and pressure, which means the local barometric pressure and ambient temperature are important for these measurements. Many metabolic carts can measure expired gases, ventilation, and temperature and pressure for every breath, whereas others use a small mixing chamber to average the numbers over a discrete period of time (about 15 s). Part of this lab requires knowledge of how to use the metabolic equipment at your institution (see appendix D); your lab instructor will guide you through its use.

AEROBIC METABOLISM AND RESPIRATORY EXCHANGE RATIO

One important measurement that can be accomplished with knowledge of $\dot{V}O_2$ and $\dot{V}CO_2$ is the type of fuel being used. On this point, the respiratory quotient (RQ) or **respiratory exchange ratio (RER)** provides information regarding the macronutrient being oxidized at any given time during steady state. RQ and RER differ only in the place where they are measured—that is, at the cellular level and in the expired gases, respectively. For the purposes of this lab, we will refer to RER.

$$RQ = \frac{\dot{V}CO_2}{\dot{V}O_2} \text{ (at the cellular level)}$$

$$RER = \frac{\dot{V}CO_2}{\dot{V}O_2} \text{ (measured in expired gases)}$$

These ratios measure the quantity of CO_2 produced in relation to the quantity of O_2 consumed. Because of inherent differences in the chemical composition of carbohydrate, fat, and protein, each requires a different amount of oxygen to completely oxidize the molecules to the end products of bioenergetics—ATP, CO_2, and H_2O. Thus, because the caloric equivalent for oxygen differs somewhat depending on the nutrient oxidized, one must know the RER and the amount of oxygen consumed in order to precisely estimate the body's energy expenditure in kcal · min^{-1} (see table 5.1). The number of kilocalories produced for each liter of oxygen consumed (kcal · L^{-1}) depends on the substrate being utilized or RER.

The RER is 1.0 for carbohydrate, 0.70 for fat, and 0.82 for protein. The protein contribution to energy production is low (generally less than 5%). To gain a true understanding of the contribution of protein to energy production, urine must be collected because urea is the metabolic end point for the amino groups of amino acids that have been oxidized. However, due to the complexity of analysis and the inconvenience of collecting urine, as well as the small contribution of protein to energy expenditure, RER is generally considered to be nonprotein RER. Therefore, at any given time, a mixture of carbohydrate and fat is being oxidized by the mitochondria for energy needs, and the kilocalories produced per liter of oxygen consumed varies accordingly. Table 5.1 reflects the percentage of fat and carbohydrate contributing to energy production at a given RER.

Normally, people who consume a diet of mixed carbohydrate and fat have an RER value of about 0.85 at rest, which means that for every 1 L of oxygen consumed, approximately 4.86 kcal are produced. In measuring **resting metabolic rate (RMR)**, the researcher must have a measure of oxygen consumption and an RER value for each minute of the measurement period in order to determine the kilocalories produced per minute (see the accompanying highlight box for a sample calculation).

Many exercise physiology students never realize *why* carbohydrate has an RER of 1.0 and fat has an RER of 0.7. Here are the equations for the overall oxidation of carbohydrate and fat:

Carbohydrate

$$C_6H_{12}O_6 + 6\ O_2 \rightarrow 6\ CO_2 + 6\ H_2O + 38\ ATP$$

$$RER = \frac{\dot{V}CO_2}{\dot{V}O_2} = \frac{6\ CO_2}{6\ O_2} = 1.0$$

Fat

$$C_{16}H_{32}O_2 + 23\ O_2 \rightarrow 16\ CO_2 + 16\ H_2O + 129\ ATP$$

$$RER = \frac{\dot{V}CO_2}{\dot{V}O_2} = \frac{16\ CO_2}{23\ O_2} = 0.7$$

The 6 carbons contained in glucose require 6 molecules of oxygen for complete aerobic combustion through oxidative phosphorylation in the mitochondrion. The production of 6 carbon dioxides yields a ratio of 1.0. Conversely, 23 oxygen molecules are required in the example of a 16-carbon fatty acid, which results in a ratio of 0.7. Knowing that an RER value of 0.7 implies 100% of fat burned, we can calculate percentages of carbohydrate and fat burned from RER that are not listed in table 5.1.

$$\%\ \text{fat burned} = \frac{1 - RER}{1 - 0.7}$$

Table 5.1 Caloric Equivalence of RER and %kcal From Carbohydrate and Fat

RER	ENERGY kcal · L^{-1}	%KCAL Carbohydrate	Fat
0.71	4.69	0	100
0.75	4.74	16	84
0.80	4.80	33	67
0.85	4.86	51	49
0.90	4.92	68	32
0.95	4.99	84	16
1.00	5.05	100	0

Reprinted, by permission, from W.L. Kenney, J.H. Wilmore, and D.L. Costill, 2015, *Physiology of sport and exercise*, 6th ed. (Champaign, IL: Human Kinetics), 124.

Predicting Energy Expenditure Using $\dot{V}O_2$ and RER

For this example, our test subject Greg, who weighs 200 lb (90.9 kg), has walked up a 3% grade at 3 mph (4.8 km · h^{-1}). Using indirect calorimetry, you have measured his $\dot{V}O_2$ at 15.2 ml · kg^{-1} · min^{-1} and his RER at 0.92.

$$200\ \text{lb} \times \frac{1\ \text{kg}}{2.2\ \text{lb}} = 90.9\ \text{kg}$$

$$15.2\ \frac{\text{ml}}{\text{kg} \cdot \text{min}} \times 90.9\ \text{kg} \times \frac{1\ \text{L}}{1{,}000\ \text{ml}} = 1.38\ \frac{\text{L}}{\text{min}}$$

$$1.38\ \frac{\text{L}}{\text{min}} \times 4.95\ \frac{\text{kcal}}{\text{L}} = 6.83\ \frac{\text{kcal}}{\text{min}}$$

If Greg walked at this intensity for 20 min, he would burn 137 kcal. Often, RER is not known, in which case these values are rounded to 5.0 kcal · L^{-1} of O_2 consumed. This is often the case when using equipment found in health clubs or clinics to predict energy expenditure.

Percent carbohydrate is then calculated simply by subtracting percent fat burned from 100 (we are assuming nonprotein RER).

Although RQ and RER are measured by the same parameters ($\dot{V}O_2$ and $\dot{V}CO_2$), these ratios often yield different results based on the location of the measurement. Under resting or submaximal steady-state exercise conditions, the measurement of RER by expired gases reflects the actual gas exchange at the cellular level. An RQ of 1.0 reflects 100% utilization of carbohydrate and is the highest value possible. However, at maximal exhaustive exercise, when anaerobic energy production increases, these ratios may vary considerably. Because more CO_2 is expired due to the high levels of lactic acid associated with maximal exercise, the RER will be greater than the RQ under these conditions. RER is often recorded above 1.15 during a $\dot{V}O_2$max test. In fact, an RER of above 1.1 is a criterion for a valid $\dot{V}O_2$max test. (These concepts are elaborated further in labs 10 and 11.)

TOTAL ENERGY EXPENDITURE AND RMR

The minimum level of energy required to sustain the body's vital functions in the waking state is the **basal metabolic rate (BMR)**, which is usually expressed in kcal · d^{-1}. True BMR is difficult to obtain because the subject must sleep for at least 8 h and fast for at least 12 h. This typically requires staying overnight in a metabolic laboratory. As a result, most researchers measure the RMR, which is considered synonymous with resting energy expenditure (REE).

RMR still requires resting and fasting, but it does not necessitate staying overnight in the lab. The subject should

- abstain from strenuous exercise for at least 48 h prior to measurement,
- abstain from eating for 12 to 14 h prior to measurement,
- lie or recline quietly in a dark, quiet environment (while awake) for 30 min prior to the measurement and during the O_2 sampling, and
- have a normal body temperature (absence of a fever).

The test is conducted by measuring the oxygen consumed during a 10 to 15 min sampling period to ensure steady state at rest. Values for oxygen consumption during the RMR test usually range between 200 and 300 ml · min^{-1} (0.2-0.3 L · min^{-1} or 0.8-1.43 kcal · min^{-1}), depending on the person's size, age, and a few other factors. Expressed as relative oxygen consumption, values should approximate 3.5 ml · kg^{-1} · min^{-1}, or 1 **metabolic equivalent (MET)**.

The assessment of RMR can be important because of its contribution to the total energy expended every day. RMR is one of three components that contribute to an individual's **total energy expenditure (TEE)**. For a majority of people, approximately 67% of TEE in a 24 h period is a result of RMR. The other two components are thermic effect of food (TEF), sometimes called *dietary-induced thermogenesis (DIT)*, and energy expenditure from physical activity (PA). TEF and DIT both refer to the energy needed to digest energy intake through diet (TEI, or total energy intake). This amount typically ranges only from 100 to 200 kcal · d^{-1} (7.5% of TEI), so it is often ignored. PA, however, can be a large component of TEE. It is the energy expended above rest during any purposeful activity—whether simply walking across campus or playing a vigorous game of basketball. Thus, PA can vary tremendously between individuals and can range from fewer than 300 kcal in sedentary people to more than 3,000 kcal in people who are very active.

$$\text{TEE (kcal · d}^{-1}) = \text{RMR} + \text{TEF} + \text{PA}$$

Interest in RMR typically pertains to weight loss or weight gain. Practical use of RMR includes establishing a baseline of energy expenditure for constructing a sound program of weight control or weight management. Because RMR is only one of two main components of TEE, TEE often needs to be estimated from predicted physical activity levels (PALs). The Institute of Medicine (IOM), the health arm of the National Academy of Sciences, developed four PALs as a means of predicting the PA component of TEE (1). PAL is the ratio of TEE divided by RMR. See table 5.2.

If the PAL is 1.0, then the subject's TEE is essentially the same as the RMR. Conversely, for a very active female with a PAL of 1.95, TEE is 1.95 times RMR. Thus, if her RMR is estimated to be 1,650 kcal, then her TEE would be 1.95 × 1,650 = 3,218 kcal · d^{-1}.

Another option for predicting the PA component of TEE is to use the category descriptor (male or female) to determine the PAC (physical activity coefficient; see table 5.2) in the following equations for estimating TEE, with the results expressed in kcal · d^{-1}.

Males 19+ Years Old

$$\text{TEE} = 662 - 9.53 \times \text{age in y} + [\text{PAC} \times (15.91 \times \text{BW in kg} + 539.6 \times \text{HT in m})]$$

Females 19+ Years Old

$$\text{TEE} = 354 - 6.91 \times \text{age in y} + [\text{PAC} \times (9.361 \times \text{BW in kg} + 726 \times \text{HT in m})]$$

Estimations of TEE are more accurate if the caloric expenditure of physical activity and RMR are determined more precisely; however, using the PAL or PAC may provide a starting point for programs involving weight loss or gain.

RMR varies from person to person, which is why the previously mentioned pretest instructions are essential for obtaining a precise measurement. The individual must be fasted and rested because digesting a meal or recovering

Table 5.2 Physical Activity Levels

Category	Physical activity level (PAL)	Physical activity coefficient (PAC) for males/females
Sedentary	>1.0-<1.4	1.00/1.00
Low active	≥1.4-<1.6	1.11/1.12
Active	≥1.6-<1.9	1.25/1.27
Very active	≥1.9-<2.5	1.48/1.45

Adapted from the Food and Nutrition Board, 2005, *Dietary reference intakes for energy, carbohydrate, fiber, fat, fatty acids, cholesterol, protein, and amino acids* (Macronutrients), 161.

Sample Calculations for TEE

Our test subject is a 35-year-old female who weighs 154 lb (70 kg), stands 5 ft 6 in. (1.68 m) tall, and considers herself active. The following calculations illustrate that she would have a TEE of 2,648 kcal · d^{-1}.

$$TEE = 354 - 6.91 \times 35 + [1.27 \times (9.361 \times 70 + 726 \times 1.68)]$$

$$= 354 - 241.85 + [1.27 \times (655.27 + 1{,}219.68)]$$

$$= 354 - 241.85 + 2051.8$$

$$= 2{,}648 \ kcal \cdot d^{-1}$$

from a recent bout of exercise can affect RMR. Even if a subject is compliant with the pretest instructions, individual differences can occur.

RMR can be affected by factors including the following:

- Males typically have a higher RMR than females.
- RMR increases with body weight.
- A person with a large body surface area will have a higher RMR; a tall person will have a higher RMR than a short person.
- In general, a younger person has a higher RMR. Children and adolescents require large amounts of energy for growth.
- From age 20 to age 70, RMR can decrease by 10% to 20%. The decline has been estimated at about 2% per decade. Some of this decline may be due to loss in lean body mass (LBM) (i.e., sarcopenia).
- As body temperature increases, so does RMR. A high body temperature can raise RMR by 30% to 40%.
- RMR is increased by an increase in LBM and decreased by a loss of LBM. However, large increases in LBM need to occur in order to see even small increases in RMR. At rest, muscle consumes little energy. However, energy conservation, as experienced during a low-calorie diet, is capable of decreasing RMR.
- Fasting (negative energy balance) decreases RMR by 10% to 25%, depending on the extent of the fast or diet.
- Endurance training may decrease RMR due to decreases in HR, BP, and respiration associ-

ated with a training regimen—the body may become more efficient in daily energy needs. Others have found that being fit may increase RMR. Thus, research in this area is equivocal.

- An increase in ambient temperature causes RMR to decrease, whereas cold weather adaptation increases RMR (often referred to as *cold-induced thermogenesis*).
- Stress increases RMR by mobilizing the sympathetic nervous system. Similarly, circulating catecholamines (epinephrine and norepinephrine) from the adrenal glands can increase RMR.
- The amount of thyroxine produced by the thyroid gland is an important hormonal control of RMR.

For further reading on factors that influence RMR, see the following material from this laboratory's reference list: Carpenter et al. (2), Compher et al. (3), Haugen et al. (7), Heshka et al. (8), Jorgensen et al. (9), Lemmer et al. (11), Owen et al. (13), Schmidt et al. (14), Sparti et al. (15), and Speakman and Selman (16).

Because of the current understanding of RMR from direct measurement using a metabolic cart, RMR can also be estimated with reasonable validity. The following lab activities involve both estimating RMR and measuring it directly.

Go to the web study guide to access electronic versions of individual and group data sheets, the question sets from each laboratory activity, and case studies for each laboratory activity.

WWW

Predicting RMR

Equipment

- Platform scale (or digital or other scale)
- Stadiometer (wall or freestanding) or physician's scale with attached anthropometer
- RMR prediction equations
- Calculator
- Individual sheet
- Group data sheet

Because the direct measurement of oxygen uptake is impractical in many situations, various formulas and nomograms have been developed to estimate RMR. Of course, certain information must be known: height (HT), weight (WT), age, and sex. In general, equations that use more factors known to affect RMR are considered to be more accurate. Some prediction equations are given in table 5.3. These and more prediction equations were favorably compared to measured RMR by indirect calorimetry (4, 6). Specifically, the Harris-Benedict (5) and World Health Organization (WHO) (16) equations estimate RMR within ±10% (4, 6). However, other equations may be best used for certain populations based on sex, BMI, age, or ethnicity (6).

Table 5.3 Prediction Equations for Estimating RMR

Number	Male	Female	Source
1	66.473 + 13.7516(BW) + 5.0033(HT) – 6.755(age)	655.0955 + 9.5634(BW) + 1.8496(HT) – 4.6756(age)	(5)
2	879 + 10.2(BW)	795 + 7.18(BW)	(12, female only)
3	BSA × 38 kcal · h^{-1} × 24 h	BSA × 35 kcal · h^{-1} × 24 h	
4	BW × 24.2 kcal · kg^{-1}	BW × 22.0 kcal· kg^{-1}	
5	BW × 1.0 × 24 h	BW × 0.9 × 24 h	
6	15.3 × BW + 679	14.7 × BW + 496	(17)
7	9.99(BW) + 6.25(HT) – 4.92(age) + 5.0	9.99(BW) + 6.25(HT) – 4.92(age) – 161	(11)

All answers are given in kilocalories per day. HT = height in centimeters; BW = body weight in kilograms; age = age in years; BSA = body surface area. BSA in m^2 is the square root of BW × HT / 3,600.

RMR Calculations

Step 1: Gather basic data (e.g., age, height, weight) for the individual data sheet as described in laboratory activity 1.1.

Step 2: Determine your own RMR by using the prediction equations (show all calculations). Express RMR in kcal per minute, per hour, and per day. Complete your own individual data sheet for each equation.

Step 3: Record these values in the group data sheet.

Step 4: Compute the mean value for males and for females in the class, and then compare your values with the class average.

Question Set 5.1

1. Which of the equations do you believe best predicts your RMR? Why?
2. What difference did you find between males and females in the class? What explains this difference?
3. Calculate your predicted TEE using the previous equations (from the PAL and PAC in table 5.2). What assumptions are made with these equations?

LABORATORY ACTIVITY 5.1

INDIVIDUAL DATA SHEET

Name or ID number: _____ Date: _____

Tester: _____ Time: _____

Sex: M / F (circle one) Age: _____ y Height: _____ in. _____ m

Weight: _____ lb _____ kg Temperature: _____ °F _____ °C

Barometric pressure: _____ mmHg Relative humidity: _____ %

Body surface area (BSA): _____ m²

		kcal · min⁻¹	kcal · h⁻¹	kcal · d⁻¹
RMR prediction equations	1			
	2			
	3			
	4			
	5			
	6			
	7			

Measuring RMR

Equipment

- Quiet room with a place for the subject to lie down
- Platform scale (or digital or other scale)
- Stadiometer (wall or freestanding) or physician's scale with attached anthropometer
- Metabolic cart, including gas flow meter and O_2 and CO_2 analyzers
- Hans Rudolph valve, mouthpiece, and noseclips
- RMR prediction equations provided in laboratory activity 5.1
- Calculator
- Individual data sheet

Metabolic Cart RMR Test

Step 1: After two students volunteer to be measured for RMR, other students split and assist in measuring one of the two subjects. Report data for both subjects A and B on the individual data sheet.

Step 2: If possible, the test should be done in the early morning, and the subjects should be fasted for >8 h.

Step 3: If the fast is not possible, then subjects should abstain from eating for at least 3 h prior to testing and avoid strenuous exercise for 24 h beforehand.

Step 4: Gather basic data (e.g., age, height, weight) for the individual data sheet as described in laboratory activity 1.1.

Step 5: Have the subject lie quietly for 15 min.

Step 6: During this time, calculate the subject's predicted RMR with one of the equations from table 5.3, and fill out the individual data sheet using the prediction equations.

Step 7: Outfit the subject with the breathing apparatus (i.e., Hans Rudolph value and mouthpiece) and noseclip from the calibrated metabolic cart for the 10 min measurement period. Do not disturb the subject while collecting data.

Step 8: Record the $\dot{V}O_2$ in L · min^{-1} and RER at the end of each minute of the 10 min measurement period on the individual data sheet.

Step 9: Complete the calculations for the remainder of the table. Include kcal · min^{-1}, kcal · h^{-1}, and kcal · d^{-1}, as well as percent carbohydrate and percent fat, using the kcal · L^{-1} of O_2 provided in table 5.1.

Question Set 5.2

1. Compare the values measured from the metabolic cart with the values predicted from the prediction equations. Which equation do you think gives the best prediction? Why?

2. From your knowledge of the subject's adherence to the typical RMR pretest instructions, which of the pretest instructions do you feel most had an effect on the measurement?

3. How much of the RMR is from carbohydrate? From fat? What does this information suggest about the prevailing fuel of choice for our bodies at rest?

LABORATORY ACTIVITY 5.2

INDIVIDUAL DATA SHEET

Date: _____ Time: _____

Tester: _____

Temperature: _____°F _____°C   Barometric pressure: _____mmHg

Relative humidity: _____%

Subject A

Name or ID number: _____

Sex: M / F (circle one)   Age: _____y

Height: _____in. _____m

Weight: _____lb _____kg

BSA: _____m²

		kcal · min⁻¹	kcal · h⁻¹	kcal · d⁻¹
RMR prediction equations	1			
	2			
	3			
	4			
	5			
	6			
	7			

Subject B

Name or ID number: _____

Sex: M / F (circle one)   Age: _____ y

Height: _____ in. _____ m

Weight: _____ lb _____ kg

BSA: _____ m²

		kcal · min⁻¹	kcal · h⁻¹	kcal · d⁻¹
RMR prediction equations	1			
	2			
	3			
	4			
	5			
	6			
	7			

Measuring RMR Using Indirect Calorimetry: Subject A

Min	$\dot{V}O_2$ (L · min⁻¹)	RER	% carbohydrate	% fat	kcal · min⁻¹	kcal · h⁻¹	kcal · d⁻¹
1							
2							
3							
4							
5							
6							
7							
8							
9							
10							
Mean							

Measuring RMR Using Indirect Calorimetry: Subject B

Min	$\dot{V}O_2 (L \cdot min^{-1})$	RER	% carbohydrate	% fat	kcal · min⁻¹	kcal · h⁻¹	kcal · d⁻¹
1							
2							
3							
4							
5							
6							
7							
8							
9							
10							
Mean							

Laboratory 6

Oxygen Deficit and EPOC Evaluations

Objectives

- Demonstrate how data are collected for the measurement of O_2 deficit and EPOC (excess post-oxygen consumption).
- Draw and label an O_2 deficit and EPOC graph.
- Calculate the O_2 deficit and EPOC from laboratory-collected data.
- Discuss biochemical concepts related to O_2 deficit and EPOC.

DEFINITIONS

anaerobic glycolysis—Metabolic pathway that can provide energy quickly; does not require oxygen and produces lactate.

excess post-oxygen consumption (EPOC)—Amount of oxygen consumption above rest after cessation of exercise (often used synonymously with *oxygen debt*).

lactate accumulation—A greater reliance on fast-glycolysis that results in the buildup of lactic acid.

oxidative phosphorylation—Production of ATP using oxygen as the final electron acceptor in the electron transport chain.

oxygen debt—Amount of oxygen consumption above rest after cessation of exercise (often used synonymously with *excess post-oxygen consumption*).

oxygen deficit—Lag in oxygen consumption in the transition from rest to submaximal exercise.

phosphocreatine (PCr)—High-energy phosphate that can donate a phosphate group to adenosine diphosphate (ADP) to quickly resynthesize ATP; does not require oxygen but does not have the storage capacity to support ATP provision for longer than about 10 s.

Now that we have a general understanding of oxygen consumption and energy expenditure from laboratory 5, we can apply these concepts to the transition from rest to exercise. Oxygen consumption implies that energy (ATP) is being provided through the oxidative phosphorylation pathway in the mitochondria. Oxidative phosphorylation has no trouble meeting the energy need for resting energy metabolism. When the body transitions to exercise, however, energy requirements increase, and this new demand

for ATP is not met immediately by oxidative phosphorylation.

TRANSITION FROM REST TO EXERCISE

When a person makes the transition from rest to a submaximal exercise intensity, ATP demand immediately increases to the level required for the work output; however, there is a delay in the measured oxygen uptake responses. During this time, early in exercise, some of the ATP is provided by the immediate, short-term anaerobic sources, **phosphocreatine (PCr)** and **anaerobic glycolysis**. By about 3 min into the activity, oxygen consumption plateaus to reach steady state, which indicates that the ATP requirement is now being met aerobically by **oxidative phosphorylation**. The volume of oxygen missing in those first minutes of work is known as the **oxygen deficit**, as shown in figure 6.1.

During exercise of submaximal intensity (<75%-80% of maximal), the energy requirement of the task can be met by means of oxygen consumption ($\dot{V}O_2$) or aerobic metabolism through oxidative phosphorylation. Steady-state $\dot{V}O_2$ then represents the energy expenditure of the submaximal intensity (see figures 6.1 and 6.2a). When exercise stops, oxygen uptake does not immediately fall to the resting value; instead, it decreases over several minutes to hours. The volume of oxygen consumed above the resting value during recovery from exercise is known as the **oxygen debt** or **excess post-oxygen consumption (EPOC)**, as shown in figure 6.1.

Let's say a subject straddles the treadmill in preparation for exercise. The energy demand for standing on the treadmill is relatively low—for most people, it approximates 3.5 ml · kg⁻¹ · min⁻¹, or 1 MET. When the subject steps on the treadmill and begins moving at a moderate pace, the energy requirement or ATP demand increases immediately in a square wave fashion. Oxidative phosphorylation, however, does not immediately meet this new energy requirement, and, as a result, the body relies on anaerobic ATP pathways to meet the demand until the aerobic system can be stimulated to the new level of ATP production. The magnitude or amount of the deficit depends on a number of factors, including intensity of exercise, genetics, and fitness level. For example, a trained person incurs less oxygen deficit than an untrained person in reaching the same oxygen consumption (figure 6.3). The trained person relies less on anaerobic pathways and thus causes less **lactate accumulation** and PCr depletion, both of which can contribute to fatigue in certain situations and also contribute to the magnitude of the deficit and EPOC.

OXYGEN UPTAKE DURING EXERCISE AND RECOVERY

The steep initial increase in oxygen consumption and the subsequent slower increase to steady state are often referred to as the *fast* and *slow* portions of oxygen uptake kinetics. For further

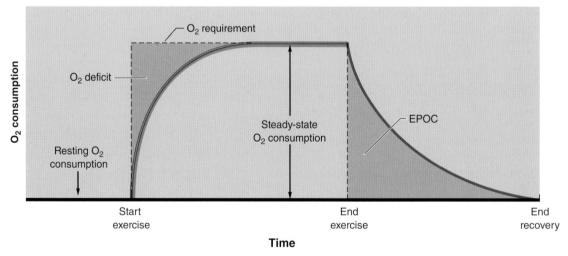

Figure 6.1 Oxygen consumption during rest-to-exercise transition, steady state, and recovery.

Reprinted, by permission, from J.H. Wilmore, D.L. Costill, and W.L. Kenney, 2015, *Physiology of sport and exercise*, 6th ed. (Champaign, IL: Human Kinetics), 129.

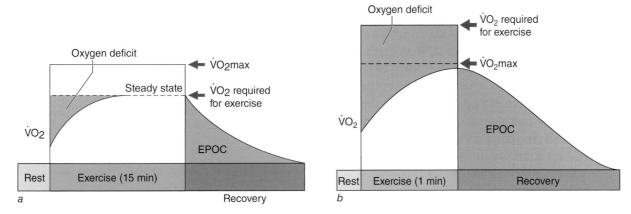

Figure 6.2 Oxygen consumption and requirements during *(a)* low-intensity, steady-state exercise and *(b)* high-intensity, non-steady-state exercise.

Adapted, by permission, from NSCA, 2016, Bioenergetics of exercise and training, by T.J. Herda and J.T. Cramer. In *Essentials of strength training and conditioning*, 4th ed., edited by G.G. Haff and N.T. Triplett (Champaign, IL: Human Kinetics), 57, 58.

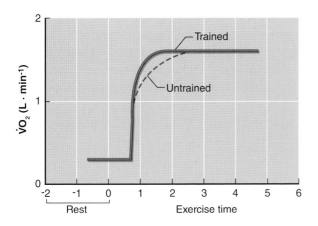

Figure 6.3 The relationship between exercise time and oxygen uptake, illustrating the time to reach a set level of $\dot{V}O_2$max in a trained and an untrained man.

reading in this active area of research, see the following items in the reference list for this laboratory: 3, 5, 6, 7, 8, 9, and 10.

During steady-state oxygen consumption (figure 6.2*a*), the energy needed to support submaximal exercise is provided by the aerobic system or by oxidative phosphorylation. Therefore, the $\dot{V}O_2$ represents the total energy expenditure of that exercise. Once exercise ceases, oxygen uptake (oxygen debt or EPOC) falls quickly in the first 2 to 4 min, then more slowly, toward the preexercise value. The fast component is believed to be due to rapid reestablishment of the PCr and oxygen stores. The slow component was once believed to be related to the oxygen needed to clear lactic acid from the blood by the liver, where it was converted back into glucose (gluconeogenesis through the Cori cycle). Now, however, it is thought that only about 20% of recovery oxygen uptake is used for this process and that the vast majority relates to other processes. It is also clear that oxygen uptake during recovery can remain elevated for a wide variety of reasons unrelated to restoration of the high-energy phosphate stores—for example, HR, breathing rate, body temperature, muscle inflammation, hormone levels, sodium-potassium pump activity, and the reestablishment of ion gradients (2, 4).

The magnitude of the oxygen debt (or EPOC) is therefore a function of how elevated these factors become, which is determined mainly by exercise intensity. A higher intensity of exercise (figure 6.2*b*) causes greater changes in HR, ventilation, body temperature, and hormone levels. It is important to understand that very high intensities also result in a

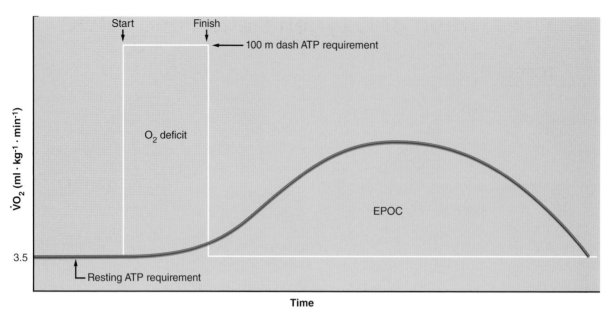

Figure 6.4 Oxygen deficit and EPOC for a 100 m dash.

much greater deficit and EPOC. Indeed, when anaerobic pathways cover the majority of ATP requirement during the exercise bout (maximal efforts of <2 min), the majority of the bout is considered to be oxygen deficit (figure 6.2*b*). An even more extreme example would be a 100 m dash. Figure 6.4 illustrates a prediction of the anaerobic and aerobic contributions during and after an intense anaerobic bout of exercise.

Note that oxygen consumption contributes little to energy production during the 100 m dash; however, $\dot{V}O_2$ continues to rise following the bout of exercise, and it may take a while to return to resting baseline. Again, a large part of the EPOC is due to the replenishment of PCr and oxygen stores. Because this bout of exercise, which lasts about 10 to 13 s, relies mostly on the ATP-PC pathway, the reestablishment

of PCr would contribute to the majority of the EPOC. But remember that almost every exercise situation imaginable involves a mix of the three energy pathways: ATP-PC, fast or anaerobic glycolysis, and oxidative phosphorylation. This holds true for other seemingly anaerobic sports such as weightlifting and for intermittent sports such as football, soccer, and rugby. Also contributing to EPOC is the muscle repair that follows muscle-damaging exercise, which can contribute to elevated oxygen consumption for up to 48 h or more (1).

Go to the web study guide to access electronic versions of the individual data sheet, the question set from the laboratory activity, and case studies for the laboratory activity.

Calculation of Oxygen Deficit and EPOC

Equipment

- Platform scale (or digital or other scale)
- Stadiometer (wall or freestanding) or physician's scale with attached anthropometer
- Metabolic cart, including gas flow meter and O_2 and CO_2 analyzers
- Hans Rudolph valve, mouthpiece, and noseclips
- Treadmill
- Calculator
- Individual data sheet

> This laboratory activity is supported by a virtual lab experience in the web study guide.
>
> www

6.1

One or two students volunteer as subjects; ideally, the subjects have differing fitness levels, but you can also compare intensities in similarly fit subjects. The remaining students split into two groups and assist in the measurement of one of the subjects. Report data for the subject with whom you assisted. Begin by gathering basic data (e.g., age, height, weight) for the individual data sheet as described in laboratory activity 1.1.

Steady-State Oxygen Consumption

Step 1: Have the subject sit quietly for 5 min to attain steady state at rest.

Step 2: Outfit the subject with the breathing apparatus from the calibrated metabolic cart for a 5 min measurement period. Obtain sitting values (i.e., with the subject seated) for the individual data sheet every minute.

Step 3: Have the subject straddle the treadmill. Select a speed and grade that will be submaximal for the subject. (Choose a submaximal intensity so that the subject can reach steady state. If you are comparing two subjects with different fitness levels, they should exercise at the same intensity. If you choose instead to compare intensity, then one subject should go at an easy to moderate intensity, whereas the other subject's pace should be more intense.)

Step 4: When technicians are ready with a stopwatch and metabolic cart, have the subject jump on the treadmill and begin running for 10 min. Begin the stopwatch.

Step 5: Immediately from the start of exercise, record the metabolic data every 15 s for the first minute, every 30 s for the second minute, and then every minute according to the individual data sheet.

Step 6: Have the subject immediately sit for 8 min.

Step 7: Immediately upon the cessation of exercise, record the metabolic data every 15 s for the first minute, every 30 s for the second minute, and then every minute according to the individual data sheet.

Calculating Oxygen Deficit and EPOC

Typically, you would use calculus to calculate the area under a curve (AUC) in order to know the magnitude of deficit and EPOC. However, not all exercise physiology students have this mathematical background, so we will use a simpler method to estimate this calculation.

Step 1: Calculate the sitting mean and steady state exercise mean for the individual data sheet. Determine the average steady-state exercise $\dot{V}O_2$ by taking the mean of the $\dot{V}O_2$ measurements of minutes 5 to 10. In the example used for table 6.1, the average steady-state $\dot{V}O_2$ is 25 ml · kg⁻¹ · min⁻¹.

Step 2: Subtract each of the measurements from minutes 0.25 to 4 during the rest-to-exercise transition from the average steady-state $\dot{V}O_2$.

Step 3: Sum these differences to obtain an estimate of the oxygen deficit before reaching steady state. In table 6.1, this is 54.5 ml · kg⁻¹ · min⁻¹.

Step 4: The EPOC is the reverse of this example—or the sum of the oxygen uptake during recovery (minutes 0.25-4 during postexercise sitting)—minus the average preexercise sitting oxygen uptake (which is approximately 3.5 ml · kg⁻¹ · min⁻¹).

Step 5: Compare the magnitude of the deficit with the EPOC.

Table 6.1 Sample Estimation of Oxygen Deficit

Exercise minute	Measured $\dot{V}O_2$ (ml · kg⁻¹ · min⁻¹)	Average steady-state $\dot{V}O_2$* – measured $\dot{V}O_2$
0.25	6.5	25 – 6.5 = 18.5
0.5	11.7	25 – 11.7 = 13.3
0.75	15.7	25 – 15.7 = 9.3
1	19.5	25 – 19.5 = 5.5
1.5	21.3	25 – 21.3 = 3.7
2	22.8	25 – 22.8 = 2.2
3	23.5	25 – 23.5 = 1.5
4	24.5	25 – 24.5 = 0.5
Estimated total oxygen deficit =		54.5 ml · kg⁻¹ · min⁻¹

*In this example, the average steady-state $\dot{V}O_2$ from minutes 5 to 10 is 25 ml · kg⁻¹· min⁻¹.

Although this is not a true AUC by calculus or trapezoidal method, the timing of the measurements being equal between deficit and EPOC (first 4 min of exercise and recovery) will give an idea of the magnitude of each.

Question Set 6.1

1. What was the steady-state O_2 uptake during exercise for your subject?

2. Calculate the O_2 deficit (energy not accounted for by $\dot{V}O_2$ or the first 4 min of rest-to-exercise transition).

3. Which ATP pathways contribute to the energy needed for the O_2 deficit? What would you measure to determine the source of ATP?

4. Calculate the EPOC (total $\dot{V}O_2$ above rest postexercise).

5. Is the EPOC greater than the deficit? Why might they be equal in some exercise tests, whereas in others the EPOC might be greater than the deficit?

6. EPOC has two components—a fast component (large drop in O_2 uptake values in the first few minutes) and a slow component (gradual return to rest). What explanations are given for the existence of the two components?

7. Using graph paper or a spreadsheet program, graph the values collected in class, identify the areas representing the O_2 deficit and EPOC, and shade in those areas. Your graph should look similar to figure 6.2*a*. Include rest, exercise and recovery intervals, oxygen debt and deficit, and the fast and slow components of oxygen deficit.

8. How does the steady-state exercise oxygen consumption compare with the predicted oxygen consumption for the chosen speed and grade on the treadmill? Use the metabolic prediction equations found in appendix B.

LABORATORY ACTIVITY 6.1

INDIVIDUAL DATA SHEET

Name or ID number: _____ Date: _____

Tester: _____ Time: _____

Sex: M / F (circle one) Age: _____ y Height: _____ in. _____ m

Weight: _____ lb _____ kg Temperature: _____ °F _____ °C

Barometric pressure: _____ mmHg Relative humidity: _____ %

Submax exercise at _____ mph and _____ % grade

	Time (min)	$\dot{V}O_2$ (ml · kg^{-1} · min^{-1})	$\dot{V}CO_2$ (ml · kg^{-1} · min^{-1})	RER	HR (beats · min^{-1})
Sitting	1				
	2				
	3				
	4				
	5				
Sitting mean					
	0.25				
	0.5				
	0.75				
	1				
	1.5				
	2				
	3				
	4				
	5				

	Time (min)	$\dot{V}O_2$ (ml · kg⁻¹ · min⁻¹)	$\dot{V}CO_2$ (ml · kg⁻¹ · min⁻¹)	RER	HR (beats · min⁻¹)
	6				
	7				
	8				
	9				
	10				
Steady state exercise mean					
Postexercise sitting	0.25				
	0.5				
	0.75				
	1				
	1.5				
	2				
	3				
	4				
	5				
	6				
	7				
	8				

Submaximal Exercise Testing

Objectives

- Reinforce the terminology and techniques associated with measuring blood pressure at rest and during exercise.
- Learn the skills needed for administering three types of submaximal exercise tests.
- Reinforce the terminology and techniques associated with measuring heart rate at rest and during exercise.
- Estimate $\dot{V}O_2$max by the measure of HR response to submaximal exercise.

DEFINITIONS

absolute work rate—Intensity of exercise measured on an ergometer by a unit of power (i.e., watts [W] or kg · m · min⁻¹); independent of fitness.

maximum heart rate (HRmax)—Maximum rate at which an individual's heart can beat per minute; often estimated by subtracting age in years from 220.

relative work rate—Intensity of exercise measured relative to an individual's maximal abili-ties, such as percent $\dot{V}O_2$max, percent HRmax, or percent of work rate (power) maximum; dependent on fitness.

submaximal—Exercise intensity below 85% of maximal effort, allowing steady state to be attained.

rating of perceived exertion (RPE)—Scale to assess the difficulty of exercise perceived by the subject.

Knowing a subject's cardiorespiratory capacity, or $\dot{V}O_2$max, can be useful for exercise prescription or for diagnostic purposes in at-risk populations. Tests for estimating $\dot{V}O_2$ max can be maximal (involving incremental workloads to the point of failure; see laboratory 10) or **submaximal** (ending before extreme exertion). Submaximal estimates of cardiorespiratory capacity can be more practical because they do not require expensive automated metabolic laboratory equipment. However, they do require some knowledge and skill to carry them out properly. For this reason, such estimates are often included in the practical portion of certification examinations (e.g., ACSM Certified Exercise Physiologist; for more information about certifications, see appendix F).

Many types of submaximal tests exist, and they differ by target population, number of stages, and modality. Quantifying work performed is necessary in order to determine submaximal effort; thus, common modalities include several approaches that facilitate quantification of work rate, such as bench-stepping, treadmill, and cycle ergometer tests. In this lab, you will perform each of these testing modalities and include a one-stage and a multiple-stage test. Submaximal tests require monitoring of HR (heart rate), BP (blood pressure), patient symptoms, time, ergometer work rates, and RPEs. These lab activities do not involve direct measurement of respiratory gases.

Because of the linear nature of the response of HR to work rate (see figure 7.1), submaximal work rates can be extrapolated to the subject's age-predicted maximum HR. Doing so has become one of the most popular assessment techniques for estimating aerobic capacity, cardiorespiratory fitness, and $\dot{V}O_2$max.

Determining HR response to exercise is one of the most critical skills an exercise physiology student can develop. The usefulness of HR response in predicting cardiorespiratory fitness involves the relationship between HR, age, and fitness level. As people become fitter, their HR decreases at a given absolute submaximal work rate (see figure 7.2).

This response is one of the hallmarks of adaptation to training. Crucially, the reduction in HR at submaximal intensities following training results not from a decrease in $\dot{V}O_2$ or CO (car-

diac output) but from an increase in SV (stroke volume), which allows a lower HR to accomplish the same CO and $\dot{V}O_2$. The predicted oxygen uptake is extrapolated from the HR response to submaximal intensities up to the predicted **maximum heart rate (HRmax)**. Thus, a lower HR response to a given **absolute work rate** extrapolates to a higher intensity at the predicted HRmax and therefore a higher predicted oxygen uptake (see line A in figure 7.2). Another cardiovascular adaptation to training is the recovery of HR following an exercise bout. After exercise, HR returns to normal resting HR more quickly in a trained person than in an untrained one. Thus, accurately timed postexercise HR can be used as a predictor of fitness, as in the step test in laboratory activity 7.1.

It is important to consider the assumptions of submaximal testing in many situations.

1. It is assumed that a subject attains steady-state exercise at each submaximal work rate so that the outcome measures of HR and BP are at steady state.

2. It is assumed that efficiency does not differ significantly between subjects; in other words, it is assumed that all subjects expend

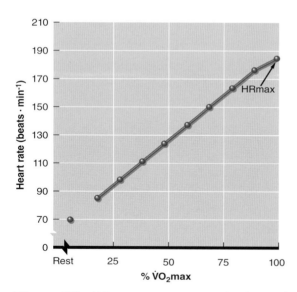

Figure 7.1 HR response to exercise intensity.

Reprinted, by permission, from W.L. Kenney, J.H. Wilmore, and D.L. Costill, 2015, *Physiology of sport and exercise*, 6th ed. (Champaign, IL: Human Kinetics), 197.

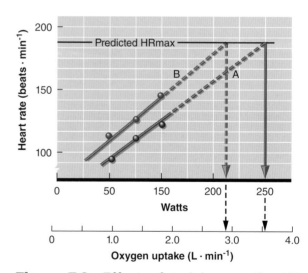

Figure 7.2 Effect of training on the HR response.

Reprinted, by permission, from P.O. Åstrand et al., 2003, *Textbook of work physiology*, 4th ed. (Champaign, IL: Human Kinetics), 285.

the same amount of energy ($\dot{V}O_2$) at a given absolute work rate independent of fitness. However, subjects can be at vastly different percentages of their individual maximal capacity, or **relative work rates**.

3. The most important assumption may be that HRmax is the same across a given age. Error can be significant in the estimation of HRmax according to the formula in which HR = 220 − age; 1 SD is ±12 beats per minute (beats · min⁻¹). A more recent and accurate equation has been developed in which maximal HR = 208 − (0.7 × age) (10).

Although error can arise from these assumptions in prediction of maximum capacity, error can also be attributed to inaccurate pedaling cadence (work rate), inaccurate load on the ergometer, and imprecise measures of HR and BP—all of which are controlled by the testers.

MEASUREMENT OF HEART RATE

HR during exercise can be measured by palpation, auscultation, ECG, or an HR monitor. Measuring HR by palpation, as you will practice in this lab, is an important skill. HR can be palpated at several sites, including the radial, brachial, carotid, temporal, and femoral arteries. This is relatively easy to do when the subject is at rest but can be challenging during exercise due to extraneous noise. With a stopwatch, count the number of beats over 6, 10, or 15 s and multiply the number of beats by 10, 6, or 4, respectively, to determine the number of beats per minute. HR can also be measured by timing the number of seconds for 30 beats and then calculating beats per minute using the following equation:

$$\text{HR (beats} \cdot \text{min}^{-1}) = 1{,}800 \, / \, \text{time (s)}$$
$$\text{for 30 beats}$$

This method is thought to be more accurate because it removes error due to partial beats. With either method, start counting heartbeats with zero when you start timing.

RATING OF PERCEIVED EXERTION

When subjects are performing an exercise test, it is useful to know how they are feeling, and testers often address this need using the **rating of perceived exertion (RPE)**. This type of measure can be used to monitor progress toward maximal exertion. Two subjective scales have been developed (9), and both are appropriate for maximal exercise testing. Both scales were developed by Gunnar Borg; one ranges from 0 to 10 whereas the other ranges from 6 to 20 (8).

The RPE scales include instructions to read to the subject before the start of the test (8):

During the test, we want you to pay close attention to how hard you feel the exercise work rate is. This feeling should reflect your total

Rating of Perceived Exertion Scales

The 15-point Borg scale starts with 6, which is usually associated with sedentary behavior. Descriptors of exertion are assigned to every odd number on the scale, with 7 corresponding to "extremely light," 9 to "very light," 11 to "light," 13 to "somewhat hard," 15 to "hard," 17 to "very hard," and 19 to "extremely hard." Anecdotally, the numbers 15 and 16 are often associated with the anaerobic or ventilatory threshold. The scale continues up to 20, which represents exertion at maximal intensity. The 15-point Borg scale ranges from 6 to 20 in the belief that adding a 0 to the RPE number approximates the HR; however, the previously mentioned differences in maximal and resting HR between individuals call the accuracy of this prediction into question. The Borg category-ratio scale, which can also be used to measure exertion, has values ranging from 0 ("nothing at all") to 10 ("extremely strong") and above, but it has no fixed number indicating the maximum possible exertion (3, 9).

amount of exertion and fatigue, combining all sensations and feelings of physical stress, effort, and fatigue. Don't concern yourself with any one factor, such as leg pain, shortness of breath, or exercise intensity, but try to concentrate on your total, inner feeling of exertion. Try not to underestimate or overestimate your feelings of exertion; be as accurate as you can.

The following laboratory activities involve submaximal testing using a bench step, treadmill, and cycle ergometer. These protocols were selected because of their common use in occupational, clinical, and apparently healthy populations. In addition, this approach exposes you to three measures of work output by humans.

Submaximal testing is generally safe when appropriate procedures are followed, but some circumstances call for terminating a test before the predetermined end point has been reached. Some indications for stopping an exercise test include angina (heart pain), drop in HR or BP with an increase in intensity, or volitional fatigue. For more information on test termination, see *ACSM's Guidelines for Exercise Testing and Prescription* (2).

Go to the web study guide to access electronic versions of individual and group data sheets, the question sets from each laboratory activity, and case studies for each laboratory activity.

Submaximal Bench Step Test

Equipment

- Platform scale (or digital or other scale)
- Stadiometer (wall or freestanding) or physician's scale with attached anthropometer
- Step—41.25 cm (16.25 in.) high for men and women
- Metronome
- HR monitor (optional)
- Stopwatch
- Individual data sheet
- Group data sheet for laboratory 7

Step Box

Stairstepping is an easy and inexpensive alternative to using an exercise ergometer to calculate work and power. You can make a step box from two-by-fours and plywood. A longer bench allows several subjects to be tested at the same time. Some protocols call for different heights for men and women. In this case, the box or bench can be made such that it can rotate from the male height to the female height. You can also use a supplemental step to shorten the step for women. Take care to measure accurately; any error in step height modifies the amount of work accomplished and thus the metabolic response. The metabolic cost of stairstepping may be too high for the unfit or diseased, particularly if the person is overweight (the work of stepping increases with body weight). The estimated $\dot{V}O_2$ of stairstepping can range from 25 to 40 ml $\cdot$ kg^{-1} $\cdot$ min^{-1}. Work and power during step testing can easily be calculated knowing body weight and the step rate and height.

$$\text{Work (kg} \cdot \text{m)} = \text{body weight (kg)} \times \text{distance (m)} \cdot \text{step}^{-1} \times \text{steps} \cdot \text{min}^{-1} \times \text{min}$$

$$\text{Power (kg} \cdot \text{m} \cdot \text{min}^{-1}) = \text{body weight (kg)} \times \text{distance (m)} \cdot \text{step}^{-1} \times \text{steps} \cdot \text{min}^{-1}$$

Queens College Step Test

Step testing is convenient for both indoor and outdoor settings and for use with one person or multiple people. Step tests come in many types, and perhaps one of the most popular is the Queens College Step Test (6, 7). Like most step tests, this test uses the measurement of recovery HR to estimate the subject's level of fitness. (Recall that HR returns to resting values more quickly following submaximal exercise in fitter people.) Many of the available step tests were developed to estimate the fitness necessary for firefighting and other physically demanding occupations. However, they are no longer used for occupational screening because participants sometimes used drugs (e.g., beta-blockers) to lower their HR and thus inflate their apparent fitness. (You would not have been able to get one of these jobs unless your estimated $\dot{V}O_2$max was greater than 45 ml $\cdot$ kg^{-1} $\cdot$ min^{-1} [10].) The test remains useful, however, especially for groups of apparently healthy individuals participating in an exercise program.

Step 1: Because the accuracy of the test relies on the HR response, attempt to eliminate factors that might alter this outcome measure. Ideally, subjects will have avoided exercise for the previous 24 h, fasted for at least 2 h, and avoided the use of foods and drugs that alter HR (e.g., coffee, soda, energy drinks, diet pills, beta-blockers).

Step 2: Pair up with another student and find an appropriate space in which to conduct the test. Either you or your partner will start as the tester, and the other person will serve as the subject. You will then reverse these roles.

Step 3: Gather basic data (e.g., age, height, weight) for the individual data sheet as described in laboratory activity 1.1. The subject then puts on the HR monitor strap, sits on the bench step, and rests for 3 min, after which the tester palpates the radial pulse for 15 s and records the resting HR. Record the simultaneous HR from the HR monitor on the individual data sheet.

Step 4: Set the metronome at 88 beats · min^{-1} to allow the subject to make contact with a foot on each beep in an up-up-down-down manner. This cadence results in the necessary 22 steps · min^{-1} for women. For men, set the metronome at 96 beats · min^{-1} and thus 24 steps · min^{-1}. Allow a few steps for practice.

Step 5: When the subject is ready, begin the 3 min test and start the stopwatch (figure 7.3a).

Step 6: To avoid muscle fatigue, the subject should switch the leading leg at least once during the test.

Step 7: After exactly 3 min of stepping, the subject should stop. The tester should palpate for the radial pulse (figure 7.3b). Begin counting at exactly 3:05 and count for 15 s (i.e., to 3:20). Record the pulse count, recovery HR, and simultaneous HR from the HR monitor on the individual data sheet.

Step 8: Calculate the predicted $\dot{V}O_2$ max by using the recovery HR in the following equations, where HR is beats · min^{-1}.

Men: $\dot{V}O_2$ max (ml · kg^{-1} · min^{-1}) = 111.33 − (0.42 × HR)

Women: $\dot{V}O_2$ max (ml · kg^{-1} · min^{-1}) = 65.81 − (0.1847 × HR)

Step 9: Record your own data on the individual data sheet and on the group data sheet. Include your percentile rank from table 7.1.

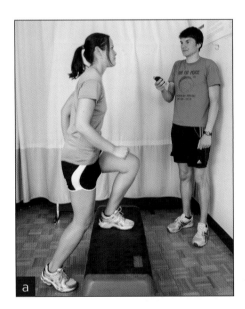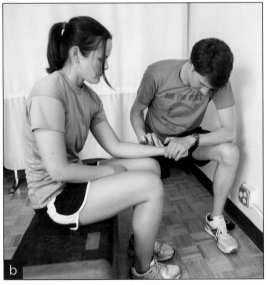

Figure 7.3 Step test: (a) starting position and (b) taking the pulse at the conclusion of the test.

Table 7.1 Reference Standards for Cardiorespiratory Fitness Measured With Treadmill Ergometry

Percentile	Rank	20-29	30-39	40-49	50-59	60-69	70-79
				AGE GROUP (y)			
				MEN			
90	Superior	61.8	56.5	52.1	45.6	40.3	36.6
80	Excellent	57.1	51.6	46.7	41.2	36.1	32.6
70	Good	53.7	48.0	43.9	38.2	32.9	29.5
60		50.2	45.2	40.3	35.1	30.5	26.8
50	Fair	48.0	42.4	37.8	32.6	28.2	24.4
40		44.9	39.6	35.7	30.7	26.6	22.5
30	Poor	41.9	37.4	33.3	28.4	24.6	20.6
20		38.1	34.1	30.5	26.1	22.4	18.9
10	Very poor	32.1	30.2	26.8	22.8	19.8	17.1
				WOMEN			
90	Superior	51.3	41.4	38.4	32.0	27	23.1
80	Excellent	46.5	37.5	34	28.6	24.6	21.9
70	Good	43.2	34.6	31.1	26.8	23.1	20.5
60		40.6	32.2	28.7	25.2	21.2	19.3
50	Fair	37.6	30.2	26.7	23.4	20.0	18.3
40		34.6	28.2	24.9	21.8	18.9	17.0
30	Poor	32.0	26.4	23.3	20.6	17.9	15.9
20		28.6	24.1	21.3	19.1	16.5	14.8
10	Very poor	23.9	20.9	18.8	17.3	14.6	13.6

Based on Kaminsky, Arena, and Myers 2015; Kaminsky et al. 2017.

Question Set 7.1

1. Complete the individual data sheet and the applicable section of the laboratory 7 group data sheet with the data collected in the lab.

2. How did palpated HR results compare with the HR monitor? What explains some of these differences?

3. Using the metabolic equations found in appendix B, what was the oxygen cost for your stepping test?

4. If the step test were being used for occupational screening and required a fitness score of 45 ml · kg^{-1} · min^{-1}, how many of your labmates would pass the test?

5. Do you think using a fitness score as a screening tool for physically taxing occupations is fair to all participants? Why or why not? If yes, which occupations? Why?

6. How do the percentile ranks for $\dot{V}O_2$max differ between tables 7.1 and 7.3? Which do you feel is more appropriate for step testing?

7. In what other settings do you see a step test being useful? Why?

LABORATORY ACTIVITY 7.1

INDIVIDUAL DATA SHEET

Name or ID number: _____ Date: _____

Tester: _____ Time: _____

Sex: M / F (circle one) Age: _____ y Height: _____ in. _____ m

Weight: _____ lb _____ kg Temperature: _____ °F _____ °C

Barometric pressure: _____ mmHg Relative humidity: _____ %

Raw Data

Age-predicted HRmax: _____ beats · min^{-1}

Resting 15 s pulse count: _____   Resting HR: _____ beats · min^{-1}
(HR monitor: _____ beats · min^{-1})

3:05 to 3:20 pulse count: _____   Recovery HR: _____ beats · min^{-1}
(HR monitor: _____ beats · min^{-1})

$\dot{V}O_2$max Determination

Men

$$111.33 - (0.42 \times \underset{\text{recovery HR (beats} \cdot \text{min}^{-1})}{\underline{\hspace{3cm}}}) = \underset{\dot{V}O_2\text{max (ml} \cdot \text{kg}^{-1} \cdot \text{min}^{-1})}{\underline{\hspace{3cm}}}$$

Women

$$65.81 - (0.1847 \times \underset{\text{recovery HR (beats} \cdot \text{min}^{-1})}{\underline{\hspace{3cm}}}) = \underset{\dot{V}O_2\text{max (ml} \cdot \text{kg}^{-1} \cdot \text{min}^{-1})}{\underline{\hspace{3cm}}}$$

Percentile rank based on table 7.1: _____

$\dot{V}O_2$ max classification based on table 7.1: _____

Percentile rank based on table 7.3: _____

$\dot{V}O_2$ max classification based on table 7.3: _____

Submaximal Treadmill Test

Equipment

- Platform scale (or digital or other scale)
- Stadiometer (wall or freestanding) or physician's scale with attached anthropometer
- Treadmill
- Stethoscope and sphygmomanometer
- HR monitor (optional)
- Stopwatch
- Individual data sheet
- Group data sheet for laboratory 7

Ebbeling Submaximal Treadmill Test

The Ebbeling submaximal test is a single-stage treadmill test used to estimate cardiorespiratory fitness in low-risk men and women between 20 and 59 years of age (3). This test has shown high predictability ($R^2 = .86$) through multiple regression for men and women over a wide age range. It is also flexible in its chosen walking speed, thus allowing for a wide variety of fitness levels, and because it does not take much time, multiple tests may be conducted in a short time.

Step 1: Because the accuracy of the test relies on the HR response, try to eliminate factors that might alter this outcome measure. Ideally, subjects will have avoided exercise for the previous 24 h, fasted for at least 2 h, and avoided the use of foods and drugs that alter HR (e.g., coffee, soda, energy drinks, diet pills, beta-blockers).

Step 2: Gather basic data (e.g., age, height, weight) for the individual data sheet as described in laboratory activity 1.1. Have the subject put on the HR monitor strap to compare with palpated HR.

Step 3: Determine the subject's age-predicted HRmax using the following equation: $208 - (0.7 \times age\ in\ y)$. Record the age-predicted HRmax on the individual data sheet.

Step 4: Take resting blood pressure and record this on the individual data sheet.

Step 5: Select a walking speed between 2 and 4.5 mph (3.2-7.2 km · h^{-1}) based on the subject's age, sex, and anticipated fitness level.

Step 6: Have the subject warm up at the selected speed and 0% grade for 4 min. At minutes 2 and 4, take palpated HR, BP, and RPE and record on the individual data sheet. The warm-up should elicit an HR of 50% to 70% of age-predicted maximum. If the HR response is too low or too high, adjust the treadmill speed and repeat the warm-up.

Step 7: The 4 min single-stage test begins by increasing the grade to 5% at the same speed established in the warm-up. At minutes 2 and 4 of this stage, collect palpated HR, BP, and RPE and record on the individual data sheet.

Step 8: Have the subject cool down by lowering the grade to 0% (as in the warm-up). At minutes 2 and 4 of the cool-down, collect HR, BP, and RPE and record on the individual data sheet.

Step 9: Calculate $\dot{V}O_2$ max using the following equation:

$$\dot{V}O_2 \text{ max} = 15.1 + (21.8 \times \text{speed in mph}) - (0.327 \times \text{HR}) - 0.263 \times$$
$$(\text{speed in mph} \times \text{age in y}) + 0.00504 \times (\text{HR} \times \text{age in y}) +$$
$$5.98 \times (\text{sex where } 0 = \text{female and } 1 = \text{male})$$

Question Set 7.2

1. Complete the individual data sheet and the applicable section of the group data sheet for laboratory 7 showing the data collected in the lab.

2. How do your values from the tests compare with the norms? $\dot{V}O_2$ max norms are provided in table 7.1.

3. How do the predictions of $\dot{V}O_2$ max compare between the step test and the treadmill test? What could explain the differences? Which test do you feel is the most accurate? Why?

4. How did palpated HR results compare with the HR monitor? What explains some of these differences?

5. Did your subject reach steady state in the three stages (warm-up, stage 1, cool-down)? Justify your answer.

LABORATORY ACTIVITY 7.2

INDIVIDUAL DATA SHEET

Name or ID number: _____ Date: _____

Tester: _____ Time: _____

Sex: M / F (circle one) Age: _____ y Height: _____ in. _____ m

Weight: _____ lb _____ kg Temperature: _____ °F _____ °C

Barometric pressure: _____ mmHg Relative humidity: _____ %

Raw Data

Age-predicted HRmax: _____ beats · min^{-1}

Resting BP: _____ / _____ mmHg   Resting HR: _____ beats · min^{-1}

50%-70% age-predicted HRmax: _____ – _____ beats · min^{-1}

	Time (min)	Speed (mph)	Grade (%)	HR palpation (beats · min^{-1})	HR monitor (beats · min^{-1})	BP (mmHg)	RPE
Warm-up	2		0				
Warm-up	4		0				
Stage 1	6		5				
Stage 1	8		5				
Cool-down	10		0				
Cool-down	12		0				

$\dot{V}O_2$max Determination

$$15.1 + \left(21.8 \times \underline{\hspace{2cm}}_{\text{speed in mph}}\right) - \left(0.327 \times \underline{\hspace{1.5cm}}_{\text{HR}}\right)$$

$$- 0.263 \times \left(\underline{\hspace{2cm}}_{\text{speed in mph}} \times \underline{\hspace{1.5cm}}_{\text{age (y)}}\right)$$

$$+ 0.0050 \times \left(\underline{\hspace{1.5cm}}_{\text{HR}} \times \underline{\hspace{1.5cm}}_{\text{age (y)}}\right)$$

$$+ \left(5.98 \times \underline{\hspace{3cm}}_{\text{sex where 0 = female and 1 = male}}\right)$$

$$= \underline{\hspace{3cm}}_{\dot{V}O_2 \text{max (ml} \cdot \text{kg}^{-1} \cdot \text{min}^{-1})}$$

Percentile rank based on table 7.1: _____

$\dot{V}O_2$ max classification based on table 7.1: _____

Submaximal Cycle Ergometer Test

Equipment

- Platform scale (or digital or other scale)
- Stadiometer (wall or freestanding) or physician's scale with attached anthropometer
- Monark cycle ergometer
- Stethoscope and sphygmomanometer
- HR monitor (optional)
- Metronome
- Stopwatch
- Individual data sheet
- Group data sheet for laboratory 7

This laboratory activity is supported by a virtual lab experience in the web study guide.

ACSM Bicycle Ergometer Test

The ACSM protocol is a multistage submaximal cycle ergometer test used to estimate cardiorespiratory fitness in men and women (1). This test uses the near-linear relationship between HR and work output to extrapolate an age-predicted HRmax. The protocol involves four 3 min stages of continuous exercise designed to raise the steady-state HR of the subject to between 110 beats · min^{-1} and 85% of the age-predicted HRmax (170 beats · min^{-1} for a 20-year-old) for at least two consecutive stages. Each work rate is performed for 3 min, and HR is recorded during the final 30 s of each minute. A HR that varies by more than 6 beats · min^{-1} within the last minute of a stage indicates non-steady state; in this case, the work rate should be maintained for an additional minute. The test administrator should recognize the error associated with age-predicted maximal HR and monitor the subject throughout the test to ensure the test remains submaximal. The HR measured during the last minute of each steady-state stage is plotted against work rate, which allows you to extrapolate maximum power output at the age-predicted HRmax. This allows you to predict maximum $\dot{V}O_2$.

Test administrators should use the protocol outlined in the following steps. Cadence must remain at 50 rev · min^{-1} for the entire test. Subjects should continue without interruption until the desired HR response is elicited (two HRs from two different stages between 110 and about 170 beats · min^{-1}).

Step 1: One person from each group acts as the subject and performs the submaximal test while the remaining group members administer the test.

Step 2: Gather basic data (e.g., age, height, weight) for the individual data sheet as described in laboratory activity 1.1. Have the subject put on the HR monitor strap to compare with palpated HR.

Step 3: Determine the subject's age-predicted HRmax using the following equation: 208 − (0.7 × age in y). Record the age-predicted HRmax on the individual data sheet. Determine the 85% age-predicted HRmax, and record on the individual data sheet.

Laboratory Activity 7.3 | 171

Step 4: Protocols A, B, and C (figure 7.4) differ by body weight and anticipated fitness. The cutoff to determine whether subjects are "very active" is if they exceed the physical activity recommendations of 150 min of moderate exercise per week or 75 min of vigorous exercise per week. From this information determine the A, B, or C protocol and fill out your individual data sheet accordingly.

Step 5: Set the bicycle seat height so the knee is bent 5° to 15° when the leg is extended (see figure 4.3*a*).

Step 6: Have the subject rest for 3 min. Then, with the subject sitting on the bike, record resting HR and BP before beginning the warm-up (figure 7.5*a*). The warm-up should include pedaling at 50 rev · min^{-1} at a resistance of 0.0 to 0.5 kg.

Step 7: To begin the test, the subject pedals at 50 rev · min^{-1} for stage 1. If the cycle ergometer is not equipped with cadence indicators, use a metronome set at either 50 or 100 beats · min^{-1} for the subject to follow with either one leg or both legs, respectively.

Step 8: Set the appropriate resistance on the ergometer for stage 1. Start the stopwatch to monitor stage time.

Step 9: Follow table 7.2 for the procedures within each stage. Take HR by palpation at each minute of the 3 min stages using a 15 s count. If available, use an HR monitor and compare its results with your palpation HR. Record each in the individual data sheet.

Step 10: Take BP beginning at 2 min into the stage (figure 7.5*b*).

Step 11: Throughout the stage, be aware of the subject's adherence to the 50 rev · min^{-1} cadence.

Step 12: Also, be aware of the subject's subjective symptoms (e.g., pallor, dyspnea).

Step 13: If this is your first attempt at conducting the test, it may be difficult to complete all the necessary tasks within the 3 min stage. Thus, in this early phase of skill development, it is acceptable to extend the stage time in order to complete the necessary measurements. As you become more practiced, strive to complete the data collection within the intended stage time.

Step 14: Following stage 1, adjust the resistance in accordance with the test protocols (A, B, C) provided in figure 7.4.

Step 15: Complete data collection of HR, BP, and RPE for stages 2 through 4 in the same manner used for stage 1.

Step 16: Throughout the stages, be aware of the subject's adherence to the 50 rev · min^{-1} cadence, as well as any subjective symptoms (e.g., pallor, dyspnea). If at any time the subject exceeds 85% of age-predicted HRmax, stop the test.

Step 17: Following stage 4, confirm that two steady-state HRs are obtained between 110 beats · min^{-1} and 85% of age-predicted HRmax (e.g., 170 beats · min^{-1} for a 20-year-old).

Step 18: Have the subject cool down at 0.5 kg resistance at 50 rev · min^{-1}. Record HR, BP, and RPE every 2 min on the individual data sheet. Continue the cool-down for 4 min or until the subject's HR is <120 beats · min^{-1}.

Step 19: Attempt to conduct two or three tests during the class. Rotate testing duties to practice all the necessary skills.

Step 20: Follow table 7.2 and figure 7.4, and record all data on the individual data sheet and where applicable on the group data sheet.

PROTOCOL SELECTION CRITERIA		
Body weight in kg (lb)	Very active: No	Very active: Yes
<73 (160)	A	A
74-90 (161-199)	A	B
>91 (200)	B	C

	TEST PROTOCOL		
Stage	A	B	C
1	150 kg · m · min⁻¹ (0.5 kg; 25 W)	150 kg · m · min⁻¹ (0.5 kg; 25 W)	300 kg · m · min⁻¹ (1.0 kg; 50 W)
2	300 kg · m · min⁻¹ (1.0 kg; 50 W)	300 kg · m · min⁻¹ (1.0 kg; 50 W)	600 kg · m · min⁻¹ (2.0 kg; 100 W)
3	450 kg · m · min⁻¹ (1.5 kg; 75 W)	600 kg · m · min⁻¹ (2.0 kg; 100 W)	900 kg · m · min⁻¹ (3.0 kg; 150 W)
4	600 kg · m · min⁻¹ (2.0 kg; 100 W)	900 kg · m · min⁻¹ (3.0 kg; 150 W)	1200 kg · m · min⁻¹ (4.0 kg; 200 W)

Figure 7.4 Guide to setting workloads on a bicycle ergometer for the ACSM bicycle ergometer test.

Based on Golding, 2000, *YMCA fitness testing and assessment manual*, 4th ed. (Champaign, IL: Human Kinetics).

Figure 7.5 ACSM bicycle ergometer test: *(a)* taking resting HR before the test begins and *(b)* taking RPE after the first stage.

Table 7.2 Suggested Stage Procedures for the Submaximal Cycle Ergometer Test

Stage time (min:s)	Procedure
0:00-0:45	Monitor the subject's cadence, ergometer resistance, and symptoms (e.g., pallor, dyspnea). Idle chatter can relax the subject, but stay on task!
0:45-1:00	Take 15 s HR by palpation.
1:00-1:45	Monitor the subject's cadence, ergometer resistance, and symptoms (e.g., pallor, dyspnea). Idle chatter can relax the subject, but stay on task!
1:45-2:00	Take 15 s HR by palpation.
2:00-2:30	Take the subject's BP for this stage. Be sure the subject relaxes the grip on the handlebars. The subject also needs to continue the 50 rev · min^{-1} cadence.
2:30-2:45	Take the subject's RPE for this stage.
2:45-3:00	Take 15 s HR by palpation.

Based on Golding, 2000, *YMCA fitness testing and assessment manual*, 4th ed. (Champaign, IL: Human Kinetics).

Formulas

Calculating Power on a Monark Cycle Ergometer

$$\text{Power (kg} \cdot \text{m} \cdot \text{min}^{-1}) = \text{rev} \cdot \text{min}^{-1} \times \text{resistance (kg)} \times 6 \text{ m} \cdot \text{rev}^{-1}$$

Predicting Oxygen Consumption on Cycle Ergometer

$$\dot{V}O_2 \text{ (ml} \cdot \text{kg}^{-1} \cdot \text{min}^{-1}) = (1.8 \times \text{work rate [kg} \cdot \text{m} \cdot \text{min}^{-1}] / \text{BW [kg])} + 7 \text{ ml} \cdot \text{kg}^{-1} \cdot \text{min}^{-1}$$

Watts Conversion

$$W = (\text{kg} \cdot \text{m} \cdot \text{min}^{-1}) / 6.12$$

Estimating $\dot{V}O_2$max

Body mass is less of a factor in work on a cycle ergometer. As a result, oxygen consumption is often expressed in L · min^{-1}. However, maximal oxygen consumption is best expressed as ml · kg^{-1} · min^{-1} when comparing data between subjects. We will try two methods for estimating $\dot{V}O_2$max from the data you collected.

Method A

First, using graph paper or a computer spreadsheet, plot the last HR from each of the four stages (dependent variable, or *y*-axis) and work rate (independent variable, or *x*-axis). Draw a straight line of best fit through the HRs in the desired range (110 beats · min^{-1} and 85% HRmax). This may mean eliminating one or more of the HR data points if they fall outside the desired range. Extrapolate this line to the age-predicted HRmax. From this, drop a vertical line to the *x*-axis to establish the estimated maximum work rate. Use the following equation to estimate maximum oxygen uptake ($\dot{V}O_2$max) from the estimated maximum work rate.

$$\dot{V}O_2\text{max (ml} \cdot \text{kg}^{-1} \cdot \text{min}^{-1}) = (1.8 \times \text{max work rate [kg} \cdot \text{m} \cdot \text{min}^{-1}] / \text{BW [kg]})$$
$$+ 7 \text{ ml} \cdot \text{kg}^{-1} \cdot \text{min}^{-1}$$

The result is the estimated $\dot{V}O_2$max.

Method B

This method estimates $\dot{V}O_2$max numerically from the protocol data without the need for graph paper.

First, determine submaximal $\dot{V}O_2$ for two of the steady-state work rates that elicited an HR response in the desired range (110 beats $\cdot$ min^{-1} and 85% HRmax). Use the ACSM metabolic equation for leg cycling (for cycle workloads between 300 and 1,200 kg $\cdot$ m $\cdot$ min^{-1}):

$$\dot{V}O_2 \text{ (ml} \cdot \text{kg}^{-1} \cdot \text{min}^{-1}) = (1.8 \times \text{work rate [kg} \cdot \text{m} \cdot \text{min}^{-1}] / \text{BW [kg]})$$
$$+ 7 \text{ ml} \cdot \text{kg}^{-1} \cdot \text{min}^{-1}$$

Second, determine the slope (m) of the *linear* HR and $\dot{V}O_2$ relationship:

$$\text{Slope (m)} = (\text{SMA} - \text{SMB}) / (\text{HRA} - \text{HRB})$$

where

SMA = submaximal predicted $\dot{V}O_2$ from stage A (ml $\cdot$ kg^{-1} $\cdot$ min^{-1}),

SMB = submaximal predicted $\dot{V}O_2$ from stage B (ml $\cdot$ kg^{-1} $\cdot$ min^{-1}),

HRA = steady-state HR from stage A (beats $\cdot$ min^{-1}), and

HRB = steady-state HR from stage B (beats $\cdot$ min^{-1}).

Note: Stage A and stage B refer not to the first and second stages but to the two stages where target HR (between 110 beats $\cdot$ min^{-1} and 85% of age-predicted HR-max) was attained.

Finally, solve the following equation for $\dot{V}O_2$max (ml $\cdot$ kg^{-1} $\cdot$ min^{-1}):

$$\dot{V}O_2\text{max} = \text{SMB} + \text{m} \times (\text{HRmax} - \text{HRB})$$

The equations that estimate $\dot{V}O_2$ from work rate are valid for estimating oxygen consumption at submaximal steady-state workloads from 300 to 1,200 kg $\cdot$ m $\cdot$ min^{-1} (50-200 W); therefore, caution must be used if extrapolating to workloads outside of this range. Another source of error involved in estimating O_2 from submaximal HR responses occurs because the formula to estimate HRmax (220 − age) can provide only a rough estimate (±12 beats $\cdot$ min^{-1}).

Table 7.3 Reference Standards for Cardiorespiratory Fitness Measured With Cycle Ergometry

Percentile	Rank	AGE GROUP (y)					
		20-29	30-39	40-49	50-59	60-69	70-79
MEN							
90	Superior	55.5	41.7	37.1	34.0	29.9	28.1
80	Excellent	51.4	36.2	34.2	30.7	26.7	24.5
70	Good	47.9	33.9	30.4	28.2	24.5	21.9
60		44.5	31.1	28.6	26.3	23.2	20.4
50	Fair	41.9	30.1	27.1	24.8	22.4	19.5
40		38.3	28.1	25.4	23.6	21.4	18.5
30	Poor	36.2	26.9	24.0	22.6	20.2	17.5
20		33.2	25.4	22.2	21.5	19.0	16.7
10	Very poor	29.5	21.8	20.6	20.4	17.3	19.3
WOMEN							
90	Superior	42.6	30.0	26.2	22.6	20.5	18.0
80	Excellent	38.8	26.0	23.4	20.7	18.8	16.9
70	Good	35.6	24.2	22.0	19.3	17.8	16.1
60		33.6	22.5	20.7	18.2	16.7	15.4
50	Fair	31.0	21.6	19.4	17.3	16.0	14.8
40		28.1	20.1	18.4	16.6	15.4	14.2
30	Poor	25.6	18.8	17.1	15.7	14.7	13.6
20		21.6	17.0	15.8	14.9	14.0	12.8
10	Very poor	19.3	20.9	14.6	13.7	13.0	12.0

Reprinted from *Mayo Clinical Proceeding*, Vol. 92(2), L.A. Kaminsky et al., "Reference standards for cardiorespiratory fitness measured with cardiopulmonary exercise testing using cycle ergometry: data from the fitness registry and the importance of exercise national database (FRIEND) registry," pgs. 228-33, Copyright 2017, with permission from Elsevier.

Question Set 7.3

1. Calculate the estimated $\dot{V}O_2$max (both absolute and relative) for the ACSM cycle test using both methods A and B.

2. Construct a graph illustrating the relationship between workload (kg · m · min^{-1}) and $\dot{V}O_2$ (ml · kg^{-1} · min^{-1}) on the cycle ergometer. Include SMA, SMB, and predicted $\dot{V}O_2$max.

3. Prepare a graph illustrating the relationship between SBP, DBP, and workload during the exercise test.

4. What was your estimated $\dot{V}O_2$ from the ACSM cycle test? What were your percentile ranks and classification using the two methods? Do you think this is an accurate measure of your fitness? Why or why not?

5. Give normative $\dot{V}O_2$max values for both males and females of the following groups: elite marathon runners, sprinters, Olympic weightlifters, college-aged males, and college-aged females. You will have to use a reference source other than the lab manual (e.g., your text or the web).

6. How do the values differ between techniques for calculating $\dot{V}O_2$max (i.e., plotting versus numerical, methods A versus B)? For each method, name at least one source of error.

7. What are some sources of error in estimating maximal aerobic capacity from the submaximal HR response? (Hint: Think about the assumptions inherent in the test.)

LABORATORY ACTIVITY 7.3

INDIVIDUAL DATA SHEET

Name or ID number: _____ Date: _____

Tester: _____ Time: _____

Sex: M / F (circle one) Age: _____ y Height: _____ in. _____ m

Weight: _____ lb _____ kg Temperature: _____ °F _____ °C

Barometric pressure: _____ mmHg Relative humidity: _____ %

Raw Data

Very active? Yes / No

Age-predicted HRmax: _____ beats · min^{-1}

Resting BP: _____ / _____ mmHg

Resting HR: _____ beats · min^{-1}

85% age-predicted HRmax: _____ beats · min^{-1}

	Time (min)	Workload (kg)	HR palpation (beats · min^{-1})	HR monitor (beats · min^{-1})	BP (mmHg)	RPE
1st stage	1				_____	_____
	2					_____
	3				_____	
2nd stage	4				_____	_____
	5					_____
	6				_____	
3rd stage	7				_____	_____
	8					_____
	9				_____	
4th stage (if necessary)	10				_____	_____
	11					_____
	12				_____	

Recovery

Time (min)	Workload (kg)	HR palpation (beats · min⁻¹)	HR monitor (beats · min⁻¹)	BP (mmHg)	RPE
2	0.5				
4	0.5				

Work rate (kg · m · min^{-1}) = workload (kg) × rev · min^{-1} × 6 m · rev^{-1}

Stage 1: _____ kg × 50 rev · min^{-1} × 6m · rev^{-1} = _____ kg · m · min^{-1}

Stage 2: _____ kg × 50 rev · min^{-1} × 6m · rev^{-1} = _____ kg · m · min^{-1}

Stage 3: _____ kg × 50 rev · min^{-1} × 6m · rev^{-1} = _____ kg · m · min^{-1}

Stage 4: _____ kg × 50 rev · min^{-1} × 6m · rev^{-1} = _____ kg · m · min^{-1}

$\dot{V}O_2$max Determination

Method A

$$1.8 \times \underset{\text{estimated max work rate (kg · m · min}^{-1})}{\underline{\hspace{3cm}}} / \underset{\text{body weight (kg)}}{\underline{\hspace{2cm}}} + 7 \text{ ml} \cdot \text{kg}^{-1} \cdot \text{min}^{-1}$$

$$= \underset{\dot{V}O_2\text{max (ml · kg}^{-1} \cdot \text{min}^{-1})}{\underline{\hspace{4cm}}}$$

Percentile rank from table 7.3: _____

$\dot{V}O_2$max classification from table 7.3: _____

Method B

$$1.8 \times \underset{\text{stage A work rate (kg · m · min}^{-1})}{\underline{\hspace{3cm}}} / \underset{\text{body weight (kg)}}{\underline{\hspace{2cm}}} + 7 \text{ ml} \cdot \text{kg}^{-1} \cdot \text{min}^{-1} = \underset{\text{SMA}}{\underline{\hspace{1.5cm}}}$$

$$1.8 \times \underset{\text{stage B work rate (kg · m · min}^{-1})}{\underline{\hspace{3cm}}} / \underset{\text{body weight (kg)}}{\underline{\hspace{2cm}}} + 7 \text{ ml} \cdot \text{kg}^{-1} \cdot \text{min}^{-1} = \underset{\text{SMB}}{\underline{\hspace{1.5cm}}}$$

$$(\underset{\text{SMA}}{\underline{\hspace{1.5cm}}} - \underset{\text{SMB}}{\underline{\hspace{1.5cm}}}) / (\underset{\text{HRA}}{\underline{\hspace{1.5cm}}}) - \underset{\text{HRB}}{\underline{\hspace{1.5cm}}} = \underset{\text{slope (m)}}{\underline{\hspace{1.5cm}}}$$

$$\underset{\text{SMB}}{\underline{\hspace{1.5cm}}} + \underset{\text{slope (m)}}{\underline{\hspace{1.5cm}}} \times (\underset{\text{HRmax}}{\underline{\hspace{1.5cm}}} - \underset{\text{HRB}}{\underline{\hspace{1.5cm}}}) = \underset{\dot{V}O_2\text{max (ml · kg}^{-1} \cdot \text{min}^{-1})}{\underline{\hspace{3cm}}}$$

Percentile rank from table 7.3: _____

$\dot{V}O_2$max classification from table 7.3: _____

Aerobic Power Field Assessments

Objectives

- Explain how to perform several field assessments for aerobic power.
- Conduct the Cooper 1.5-mile run/walk test, the Cooper 12-minute run/walk test, and the Rockport fitness walking test.
- Estimate maximal aerobic power using several types of field tests.
- Evaluate the results of a field test for aerobic power.

DEFINITIONS

aerobic power—Maximal capacity for resynthesizing ATP with aerobic means; best determined with a GXT to exhaustion ($\dot{V}O_2$max).

field test of aerobic power—Field test (not requiring a laboratory) in which the subject either performs a timed completion of a set distance or completes a maximal distance in a set time.

graded exercise test (GXT)—Test in which intensity is increased progressively to test cardiovascular health, wellness, and function (where intensity is increased by increasing speed, resistance, incline, or some combination thereof).

intraclass correlation (ICC)—Indicator of the reliability of a measure.

maximal testing—Test requiring the subject to go to volitional failure in order to evaluate maximal heart rate, maximal oxygen consumption, or aerobic power.

palpation—Examination by touch; typically used to estimate heart rate when a heart rate monitor is not available.

r—Multiple regression correlation coefficient that represents how well a dependent variable can be predicted from a combination of independent variables; ranges between −1 and 1.

regression equation—Statistical method used to relate two or more variables.

standard error of the estimate (SEE)—Measure of the accuracy of a predictive equation; a smaller value indicates a better estimate.

submaximal aerobic power test—Test used to estimate $\dot{V}O_2$max; typically performed in a laboratory setting and involves walking, jogging, running, or cycling.

$\dot{V}O_2$max—Maximal oxygen consumption or aerobic power.

Many consider the assessment of **aerobic power,** or fitness, to be one of the more important methods for determining a person's overall cardiorespiratory fitness. You can choose from multiple methods for determining aerobic power by directly measuring **$\dot{V}O_2$max** (maximal oxygen consumption, see lab 10) or estimating aerobic power from submaximal testing in the laboratory (see lab 7) or field-based setting (4).

The best method for assessing overall cardiorespiratory capacity is generally considered to be direct measurement of oxygen consumption during a **graded exercise test (GXT)** (5). However, though this method is viewed as the gold standard, it requires specialized equipment and trained personnel who can administer the test under controlled conditions (16). In addition, **maximal testing** may be contraindicated for some individuals (2).

An alternative method for determining aerobic power is submaximal testing. These procedures typically require less medical supervision, personnel, and equipment than maximal tests (2). Submaximal tests also tend to require less time, and, because they use fixed workloads, they are ideal for assessing heart rate (HR) and blood pressure (BP) at predetermined time increments. At the same time, a **submaximal aerobic power test** has its own limitations:

- $\dot{V}O_2$ max is predicted rather than directly measured.
- Many of the equations used in conjunction with submaximal tests involve a 10% to 20% prediction error.
- This type of testing offers limited diagnostic capabilities for certain tests.
- Because these tests do not measure a maximal HR, they are of limited usefulness in formulating exercise prescriptions.

The theoretical basis for submaximal testing centers on the relationship between HR, oxygen consumption ($\dot{V}O_2$), and workload (5). Specifically, the linear relationship between HR, $\dot{V}O_2$, and workload (see figure 7.1) makes it is possible to predict $\dot{V}O_2$ from the workload and HRs achieved during a submaximal test (see lab 7). Numerous submaximal tests can be employed based on these relationships, but the need for specialized equipment makes them impractical in some situations.

An alternative method for estimating aerobic power is to use either a maximal or submaximal **field test of aerobic power**. These tests are easy alternatives to laboratory assessments because they require minimal equipment and can be used with large groups of people at one time (1, 2). They fall in two major categories—those that measure the time needed to complete a set distance and those that measure the distance covered in a set amount of time (2). The results of a field test of aerobic power can be used to estimate the $\dot{V}O_2$ max by means

of specialized prediction equations. Major disadvantages of field-based assessment include the following:

- For some people, this type of test could be a maximal test.
- The individual's motivation and pacing strategy during the test can have a profound effect on the final outcome.
- This type of testing does not allow comprehensive monitoring of both HR and BP during the test.

Field tests are generally not recommended for sedentary individuals who have been identified in pretesting screenings to be at moderate or high risk of cardiorespiratory or musculoskeletal complications (6).

The most common field tests of aerobic power are the Cooper 1.5-mile (2.4 km) run/walk test, the Cooper 12-minute run/walk test, and the Rockport fitness walking test. These tests can be used by a variety of individuals with minimal equipment (11). The Cooper 1.5 mi and Cooper 12 min tests are considered to be maximal tests, whereas the Rockport test is a submaximal field test (13). They all allow for reasonable estimates of aerobic power.

COOPER 1.5-MILE RUN/ WALK TEST

This test is one of the more popular field tests for estimating aerobic power (1, 2, 13). Because it is classified as a maximal field test, preexercise screening is necessary to determine whether the subject has known heart disease or risk factors indicative of heart disease. The American College of Sports Medicine (ACSM) suggests that this test is not appropriate for unconditioned individuals or those with heart disease or known heart disease risk factors (2).

The Cooper 1.5 mi test estimates oxygen consumption by measuring the time it takes to run or walk 1.5 mi (2.4 km). Even though walking is allowed during the test, the objective is to cover the distance as fast as possible (13)—the faster the subject covers the distance, the higher the estimated $\dot{V}O_2$max. You can estimate the subject's $\dot{V}O_2$max with a sex-specific **regression equation**:

Men

$\dot{V}O_2$max (ml · kg^{-1} · min^{-1}) = 91.736 − (0.1656 × body mass in kg) − (2.767 × 1.5 mi run time in min)

Women

$\dot{V}O_2$max (ml · kg^{-1} · min^{-1}) = 88.020 − (0.1656 × body mass in kg) − (2.767 × 1.5 mi run time in min)

These equations have been shown to be highly reliable, and the **standard error of the estimate (SEE)** represents about 6% of the $\dot{V}O_2$max (9, 12). Both equations are also valid because of the high correlation ($r = .90$) to the actual maximal testing results and the relatively low standard error of the estimate (13).

Another method for estimating $\dot{V}O_2$max from a 1.5 mi run or walk time uses the following equation:

$\dot{V}O_2$max (ml · kg^{-1} · min^{-1}) = 65.404 + (7.707 × sex) − (0.159 × body mass in kg) − (0.843 × time in min)

where sex = 1 if male, 0 if female.

This equation also produces valid results as indicated by the correlation coefficient ($r = .86$) and SEE (3.37 ml · kg^{-1} · min^{-1}). It is considered reliable because of the high **intraclass correlation (ICC)** (ICC = .93).

COOPER 12-MINUTE RUN/ WALK TEST

The Cooper 12-minute run/walk test was originally used in the development of a regression equation by Cooper (6) to estimate $\dot{V}O_2$max. The test was based on the linear relationship determined between a 12 min best-effort run and running velocity (150-300 m · min^{-1}) and steady-state $\dot{V}O_2$ (3). The original investigation reported that the regression equation

was highly correlated ($r = .897$) to the actual $\dot{V}O_2$max determined with the Balke protocol (4). Though the original work clearly demonstrated that the test was valid, more recent work has reported greater variability in the validity coefficient, ranging from $r = .28$ to $r = .94$ for the measured $\dot{V}O_2$max and the estimated $\dot{V}O_2$max determined with a 12-minute run/walk test (16). This greater variability may in part be explained by the type of criterion $\dot{V}O_2$max test used to determine the actual maximum capacity (15). Despite such discrepancy regarding validity, the Cooper 12-minute run/walk test remains widely accepted as accurately predicting $\dot{V}O_2$max (2, 6, 11, 15, 16).

Because the goal of this test is to cover as much distance as possible in the allotted time, it can be considered a maximal field test (11) and thus, like the Cooper 1.5 mi test, is generally not recommended by the ACSM for unconditioned people or those with heart disease or known heart disease risk factors (2). The total distance traveled during the 12 min time period is used to estimate $\dot{V}O_2$max by means of the following equation:

$\dot{V}O_2$max (ml · kg^{-1} · min^{-1}) = (0.0268 × distance in m) − 11.3

ROCKPORT FITNESS WALKING TEST

The Rockport test was developed by the Rockport Institute to assess cardiorespiratory fitness in men and women from ages 20 to 69. The test was further developed by Kline et al. (14) as a submaximal field test for estimating $\dot{V}O_2$max with the use of a 1 mi (1.6 km) walking protocol. This test was found to exhibit high validity coefficients ($r = .93$) and a low SEE (0.325 L · min^{-1}) (13), which suggests that this test yields a valid estimate of $\dot{V}O_2$max (11). Cross-validation of the original equations by Fenstermaker, Plowman, and Looney (8) revealed that the Rockport test was valid and appropriate for 65- to 79-year-old women.

The following equations can be used for men and women between the ages of 30 and 79:

Men (30-79 Years Old)

$$\dot{V}O_2max \ (ml \cdot kg^{-1} \cdot min^{-1}) = 139.168 - (0.3877 \times age) - (0.1692 \times body\ mass\ in\ lb) - (3.2649 \times walk\ time\ in\ min) - (0.1565 \times HR\ in\ beats \cdot min^{-1})$$

Note: This equation has a validity coefficient of .83 to .88 and an SEE of 4.5 to 5.3 ml · kg⁻¹ · min⁻¹.

Women (30-79 Years Old)

$$\dot{V}O_2max \ (ml \cdot kg^{-1} \cdot min^{-1}) = 132.853 - (0.3877 \times age) - (0.1692 \times body\ mass\ in\ lb) - (3.2649 \times walk\ time\ in\ min) - (0.1565 \times HR\ in\ beats \cdot min^{-1})$$

Note: This equation has a validity coefficient of .59 to .88 and an SEE of 2.7 to 5.3 ml · kg⁻¹ · min⁻¹.

Though the Rockport fitness walking test generally produces valid and reliable data, several studies suggest that the original equations overestimate $\dot{V}O_2max$ in college-aged men and women (7, 9). Based on these findings, Dolgener et al. (5) modified the original equation to better estimate the $\dot{V}O_2max$ for 18- to 29-year-old men and women:

Men (18-29 Years Old)

$$\dot{V}O_2max \ (ml \cdot kg^{-1} \cdot min^{-1}) = 97.660 - (0.0957 \times body\ mass\ in\ lb) - (1.4537 \times walk\ time\ in\ min) - (0.1194 \times HR\ in\ beats \cdot min^{-1})$$

Note: This equation has a validity coefficient of .50 to .85 and an SEE of 3.5 to 5.8 ml · kg⁻¹ · min⁻¹.

Women (18-29 Years Old)

$$\dot{V}O_2max \ (ml \cdot kg^{-1} \cdot min^{-1}) = 88.768 - (0.0957 \times body\ mass\ in\ lb) - (1.4537 \times walk\ time\ in\ min) - (0.1194 \times HR\ in\ beats \cdot min^{-1})$$

Note: This equation has a validity coefficient of .38 to .85 and an SEE of 3.0 to 4.8 ml · kg⁻¹ · min⁻¹.

Procedures for Palpating HR

HR can be checked via **palpation** at several arterial sites (11):

- Brachial artery: About 2 to 3 cm above the antecubital fossa on the anteromedial aspect of the arm below the belly of the biceps brachii
- Carotid artery: Lateral to the larynx in the neck
- Radial artery: In line with the base of the thumb on the anterolateral aspect of the thumb
- Temporal artery: At the temple in line with the hairline

When palpating, use the tips of the middle and index fingers; never use the thumb, which has a pulse of its own and can produce erroneous HR counts. The most common palpation sites during exercise and at rest are at the carotid or radial arteries. Be careful when palpating the carotid artery, because applying too much pressure to this site can result in a baroreceptor response that lowers HR dramatically and can cause the subject to pass out. If you start the stopwatch simultaneously with the initiation of the heartbeat count, use 0 as the first beat counted and then count beats for the preset duration. If, on the other hand, the stopwatch is already running, then initiate the count with 1 (1, 11). When using the Rockport fitness walking test, palpate HR for 15 s immediately after cessation of the test. After counting a total number of beats for this 15 s time period, multiply that number by 4 and use the result in the Rockport test equations.

An additional method for calculating the $\dot{V}O_2$max of adults uses the following equation:

$$\dot{V}O_2\text{max (ml} \cdot \text{kg}^{-1} \cdot \text{min}^{-1}) = 132.853 + (\text{sex} \times 3.315) - (0.3877 \times \text{age}) - (0.1692 \times \text{body mass in kg}) - (3.2649 \times \text{walk time in min}) - (0.1565 \times \text{HR in beats} \cdot \text{min}^{-1})$$

For college-aged subjects, the following equation can be used:

$$\dot{V}O_2\text{max (ml} \cdot \text{kg}^{-1} \cdot \text{min}^{-1}) = 88.768 + (\text{sex} \times 8.892) - (0.2109 \times \text{body mass in kg}) - (1.4537 \times \text{walk time in min}) - (0.1194 \times \text{HR in beats} \cdot \text{min}^{-1})$$

In these equations, sex = 1 for males and 0 for females.

The Rockport fitness walking test is relatively easy to perform, but it does require subjects to palpate their HR for 15 s immediately after completing the test. Therefore, it is necessary for test subjects to able to perform this procedure (see the accompanying highlight box).

Go to the web study guide to access electronic versions of individual and group data sheets, the question sets from each laboratory activity, and case studies for each laboratory activity.

Cooper 1.5-Mile Run/Walk Test and 12-Minute Run/Walk Test

Equipment

- Measured 1.5 mi (2.4 km) distance, ideally on a 0.25 mi (400 m) track
- Physician's scale, stadiometer, or equivalent electronic scale
- Stopwatch
- Individual data sheet
- Group data sheet
- Microsoft Excel or equivalent spreadsheet program

Figure 8.1 presents a traditional 440 yd (0.25 mi) or 400 m track, in which each straightaway and curve is 110 yd (100 m) long and half of each length is 55 yd (50 m). Distance traveled may differ slightly depending on the track; therefore, when performing timed distances, make sure to understand the various lengths on the track. For example, because a 400 m track is 2.5 yd (2.3 m) shorter than a 440 yd track, a subject performing a 1.5 mi run on a 400 m track must go 14 m (15 yd) beyond six laps to complete 1.5 mi. Similarly, when performing a 1 mi (1.6 km) Rockport test, the subject must go 9 m (9 yd) beyond four laps to complete a mile.

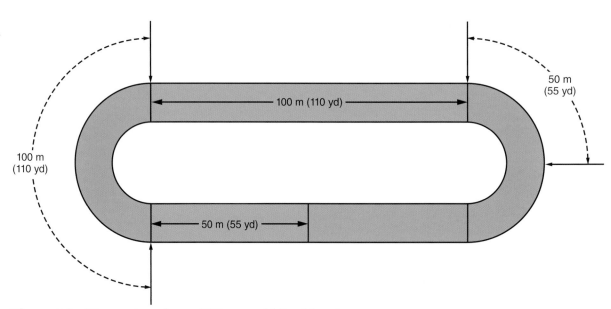

Figure 8.1 Dimensions for a 400 m or 440 yd track.

Warm-Up

As with any performance-based test, the subject should perform a structured warm-up to prepare for the assessment. As a rule, when working with athletes or other fit individuals, devote 5 min to general warm-up activity (e.g., jogging, cycling, jumping rope), and then spend 5 min on dynamic stretching (e.g., high knees, walking lunges, walking knee tucks, butt kicks, inchworms, power skips). With sedentary or untrained individuals, use less rigorous activities (e.g., leg swings, toe touches). After the warm-up, make sure the subject clearly understands that the objective of the test is to complete the distance in as little time as possible.

Cooper 1.5-Mile Run/Walk Test

Step 1: Begin this lab activity by gathering basic data (e.g., age, height, weight) for the individual data sheet as described in laboratory activity 1.1.

Step 2: Have each subject complete a structured warm-up of about 10 min.

Step 3: Prior to starting the test, clearly explain that each individual should walk or run the 1.5 mi (2.4 km) distance as fast as possible.

Step 4: Start a stopwatch at the same time that the walk or run is initiated.

Step 5: When subjects complete the distance, record their time to the nearest second on the individual data sheet.

Step 6: After completing the assessment, each tested individual should perform a cool-down consisting of slow walking followed by stretching.

Step 7: Use the equations presented on the individual data sheet to estimate each individual's $\dot{V}O_2$ max, and then record the results on the group data sheet.

Cooper 12-Minute Run/Walk Test

Step 1: Begin this lab activity by gathering basic data (e.g., age, height, weight) for the individual data sheet as described in laboratory activity 1.1.

Step 2: Have each subject complete a 5 min general warm-up followed by 5 min of dynamic stretching.

Step 3: Clearly explain that the objective of the test is to cover as much distance as possible in the allotted 12 min.

Step 4: Start the stopwatch at the same time that the 12-minute run/walk test is initiated.

Step 5: Provide encouragement to those undertaking the test.

Step 6: Make periodic time checks and provide feedback to those undertaking the test—for example, *"Five minutes to go," "One minute remaining," "Ten seconds . . . Five seconds . . . and stop."*

Step 7: Estimate the distance covered based on the number of laps completed and the place on the track where the subject stopped. Record this distance on individual data sheet.

Step 8: Allow adequate time for a cool-down consisting of slow walking and stretching.

Step 9: Calculate each individual's $\dot{V}O_2$ max using the equations presented on individual data sheet, and record the results on the group data sheet.

Question Set 8.1

1. What is the underlying physiological reason for the relationship between field tests and laboratory measurements of aerobic power?

2. Based on your results, rank your aerobic fitness in relation to the norms and percentile ranks presented in table 8.1 and table 8.2.

3. How did your aerobic fitness results compare with the class averages?

4. Based on the class averages, how would you rate your classmates' overall aerobic fitness?

5. What factors associated with this aerobic power test may result in variations in the values estimated for aerobic power?

Table 8.1 Percentile Ranks for Cooper 1.5-Mile Run/Walk Time (min:s)

AGE (y)	20-29		30-39		40-49		50-59		60-69	
Percentile	Men	Women	Men	Women	Men	Women	Men	Women	Men	Women
90	9:34	10:59	9:52	11:43	10:09	12:25	11:09	13:58	12:10	15:32
80	10:08	11:56	10:38	12:53	11:09	13:38	12:08	15:14	13:25	16:46
70	10:49	12:51	11:09	13:41	11:52	14:33	12:53	16:26	14:33	18:05
60	11:27	13:25	11:49	14:33	12:25	15:17	13:53	17:19	15:20	18:52
50	11:58	14:15	12:25	15:14	13:05	16:13	14:33	18:05	16:19	20:08
40	12:29	15:05	12:53	15:56	13:50	17:11	15:14	19:10	17:19	20:55
30	13:08	15:56	13:48	16:46	14:33	18:26	16:16	20:17	18:39	22:34
20	13:58	17:11	14:33	18:18	15:32	19:43	17:30	21:57	20:13	23:55
10	15:14	18:39	15:56	20:13	17:04	21:52	19:24	23:55	23:27	26:32

Adapted, by permission, from Cooper Institute, *Physical fitness assessments and norms for adults and law enforcement* (Dallas, TX: The Cooper Institute). For more information: http://www.cooperinstitute.org.

Table 8.2 Percentile Ranks for Cooper 12-Minute Run/Walk Distance

AGE (y)	20-29				30-39				40-49				50-59				≥60			
SEX	MEN		WOMEN		MEN		WOMEN		MEN		WOMEN		MEN		WOMEN		MEN		WOMEN	
Percentile	mi	km	mi	km	mi	km	mi	km	mi	km	mi	km	mi	km	mi	km	mi	km	mi	km
90	1.74	2.78	1.54	2.46	1.71	2.74	1.45	2.32	1.65	2.64	1.41	2.26	1.57	2.51	1.29	2.06	1.49	2.38	1.29	2.06
80	1.65	2.64	1.45	2.32	1.61	2.58	1.38	2.21	1.54	2.46	1.32	2.11	1.45	2.32	1.21	1.94	1.37	2.19	1.18	1.89
70	1.61	2.58	1.37	2.19	1.55	2.48	1.33	2.13	1.47	2.35	1.25	2.00	1.38	2.21	1.17	1.87	1.29	2.06	1.13	1.81
60	1.54	2.46	1.33	2.13	1.49	2.38	1.27	2.03	1.42	2.27	1.21	1.94	1.33	2.13	1.13	1.81	1.24	1.98	1.07	1.71
50	1.50	2.40	1.29	2.06	1.45	2.32	1.25	2.00	1.37	2.19	1.117	1.87	1.29	2.06	1.10	1.76	1.19	1.90	1.03	1.65
40	1.45	2.32	1.25	2.00	1.39	2.22	1.21	1.94	1.33	2.13	1.13	1.81	1.25	2.00	1.06	1.70	1.15	1.84	0.99	1.58
30	1.41	2.26	1.21	1.94	1.35	2.16	1.16	1.86	1.29	2.06	1.10	1.76	1.21	1.94	1.02	1.63	1.11	1.78	0.97	1.55
20	1.34	2.14	1.16	1.86	1.29	2.06	1.11	1.78	1.23	1.97	1.05	1.68	1.15	1.84	0.98	1.57	1.05	1.68	0.94	1.50
10	1.27	2.03	1.10	1.76	1.21	1.94	1.05	1.68	1.17	1.87	1.01	1.62	1.09	1.74	0.93	1.49	0.95	1.52	0.89	1.42

Adapted, by permission, from Cooper Institute, *Physical fitness assessments and norms for adults and law enforcement* (Dallas, TX: The Cooper Institute). For more information: www.cooperinstitute.org.

LABORATORY ACTIVITY 8.1

INDIVIDUAL DATA SHEET

Name or ID number: _____ Date: _____

Tester: _____ Time: _____

Sex: M = 1 / F = 0 (circle one)   Age: _____ y

Height: _____ in. _____ m

Weight: _____ lb _____ kg

Temperature: _____°F _____°C

Barometric pressure: _____ mmHg

Relative humidity: _____ %

Location of Testing

❏ Outdoor Field

❏ Indoor Field

❏ Indoor Track

❏ Outdoor Track

❏ Gym

❏ Other _____

Footwear

❏ Jogging shoe

❏ Walking shoe

❏ Tennis shoe

❏ Basketball shoe

❏ Running Shoe

❏ Cross Training Shoe

❏ CrossFit Shoe

❏ Other _____

Cooper 1.5 Mile Run/Walk Test

Men

Equation 1

$$\dot{V}O_2 max\ (ml \cdot kg^{-1} \cdot min^{-1}) = 91.736 - (0.1656 \times \underset{\text{body mass (kg)}}{\underline{\hspace{2cm}}}) - (2.767 \times \underset{\text{time (min)}}{\underline{\hspace{2cm}}})$$

$$\dot{V}O_2 max = \underline{\hspace{4cm}} ml \cdot kg^{-1} \cdot min^{-1}$$

Equation 2

$$\dot{V}O_2 max\ (ml \cdot kg^{-1} \cdot min^{-1}) = 65.404 + (7.707 \times \underset{\text{sex}}{\underline{\hspace{2cm}}}) - 0.159 \times \underset{\text{body mass (kg)}}{\underline{\hspace{2cm}}} - 0.843 \times \underset{\text{time (min)}}{\underline{\hspace{2cm}}}$$

$$\dot{V}O_2 max = \underline{\hspace{4cm}} ml \cdot kg^{-1} \cdot min^{-1}$$ Percentile rank: _____

Women

Equation 1

$$\dot{V}O_2 max\ (ml \cdot kg^{-1} \cdot min^{-1}) = 88.020 - (0.1656 \times \underset{\text{body mass (kg)}}{\underline{\hspace{2cm}}}) - (2.767 \times \underset{\text{time (min)}}{\underline{\hspace{2cm}}})$$

$$\dot{V}O_2 max = \underline{\hspace{4cm}} ml \cdot kg^{-1} \cdot min^{-1}$$

Equation 2

$$\dot{V}O_2 max\ (ml \cdot kg^{-1} \cdot min^{-1}) = 65.404 + (7.707 \times \underset{\text{sex}}{\underline{\hspace{2cm}}}) - 0.159 \times \underset{\text{body mass (kg)}}{\underline{\hspace{2cm}}} - 0.843 \times \underset{\text{time (min)}}{\underline{\hspace{2cm}}}$$

$$\dot{V}O_2 max = \underline{\hspace{4cm}} ml \cdot kg^{-1} \cdot min^{-1}$$ Percentile rank: _____

Cooper 12-Minute Run/Walk Test

Total distance: _____ yd × 0.9144 = _____ m

$$\dot{V}O_2 max\ (ml \cdot kg^{-1} \cdot min^{-1}) = 0.0268 \times \underset{\text{distance (m)}}{\underline{\hspace{2cm}}} - 11.2$$

Percentile rank: _____

Rockport Fitness Walking Test

Equipment

- Measured 1 mi (1.6 km) distance, ideally on a 0.25 mi (400 m) track (see figure 8.1)
- Physician's scale, stadiometer, or equivalent electronic scale
- Heart rate monitor (optional)
- Stopwatch
- Individual data sheet
- Group data sheet
- Microsoft Excel or equivalent spreadsheet program

Warm-Up

Before starting the test, you must teach test subjects how to measure their heart rate (HR) with the palpation method. Make sure they can find their HR and know how to count the heartbeat. It may be prudent to have the subject practice doing so while the tester also palpates to ensure the subject obtains accurate results. Alternatively, you could have the subject wear a HR monitor or have the tester measure the subject's HR. For the sake of this laboratory activity, however, students should palpate their own HR.

Next, measure and record the subject's weight, which will be used in the estimation equations. Then have the subject perform an appropriate warm-up consisting of general and specific activities. When working with athletes or other fit individuals, start with 5 min of general warm-up (e.g., jogging, cycling, jumping rope) before moving on to 5 min of dynamic stretching (e.g., leg swings, walking toe touches, walking lunges, knee tucks, high knees). With sedentary or untrained individuals, use less rigorous activities (e.g., leg swings, toe touches). After the warm-up, make sure the subject clearly understands that the objective is to walk the 1 mi (1.6 km) distance as fast as possible and that jogging or running is not acceptable during this test. Following is a summary of steps to use in administering the test.

One-Mile Rockport Fitness Walking Test

This is an ideal test for assessing cardiorespiratory fitness. It is particularly good for people who are unconditioned, because it uses both walking and HR to estimate aerobic power.

Step 1: Begin this lab activity by gathering basic data (e.g., age, height, weight) for the individual data sheet as described in laboratory activity 1.1.

Step 2: Have each subject complete a 5 min general warm-up followed by 5 min of dynamic stretching.

Step 3:   Clearly explain that the objective of this test is to walk the 1 mi (1.6 km) distance as fast as possible and that jogging or running is not acceptable.

Step 4:   Start the stopwatch at the same time that the walk test is initiated.

Step 5:   Record the time taken to complete the distance on the individual data sheet. Make sure to convert the time to the nearest hundredth of a minute. For example, if the time is 12:35 min:s, convert the value by dividing 35 s by 60 s·min^{-1}, which produces a result of 0.58 min, thus making the time 12.58 min.

Step 6:   Immediately after the subject passes the 1 mi mark, take the HR. Palpate the HR for 15 s and then multiply the result by 4. Alternatively, have the subject wear an HR monitor. Record the HR in the appropriate location on the individual data sheet.

Step 7:   Allow adequate time for a cool-down consisting of slow walking and stretching.

Step 8:   Calculate each individual's $\dot{V}O_2$ max using the equations presented on the individual data sheet, and record the results on the group data sheet. Compare the result with the norms in table 8.3 and table 8.4.

Table 8.3   Norms for 1-Mile Walk Test for Subjects Aged 18 to 29 and 30 to 69 Years (min:s)

AGE (y)	18-29		30-69	
Category	Men	Women	Men	Women
Well above average	>57.8	>45.4	>59.1	>45.0
Above average	51.0-57.8	40.7-45.4	49.2-59.0	37.2-45.0
Average	41.5-50.9	34.1-40.6	35.4-49.2	26.4-37.1
Below average	34.7-41.4	29.4-34.0	25.5-35.3	18.6-26.3
Well below average	<34.7	<29.4	<25.5	<18.6

Data from Bean and Adams 2011; Kline et al 1987; Dolgener et al 1994; George et al. 1993.

Table 8.4   Norms for the 1-Mile Walk Test for Subjects Aged 18 to 30 Years (min:s)

Percentile	Men (*n* = 400)	Women (*n* = 426)
90	11:08	11:45
75	11:42	12:49
50	12:38	13:15
25	13:38	14:12
10	14:37	15:03

Reprinted, by permission, from J. Morrow, A. Jackson, J. Disch, and D. Mood, 2010, *Measurement and evaluation in human performance*, 4th ed. (Champaign, IL: Human Kinetics), 201.

Question Set 8.2

1. Based on the results of your test, rank your aerobic fitness according to the data presented in tables 8.3 and 8.4. How does your ranking compare with your class average?

2. Create a bar graph of your individual results and of the class average. How do your results differ from the class average? Are you above or below average for the class?

3. Do the values determined for the various equations produce different results? If yes, why might this be?

4. What are some limitations associated with the Rockport fitness walking test?

5. Based on the class average, how would rate your classmates' overall fitness?

6. What are some issues related to HR palpation? Are there any procedural concerns? If so, what are they?

LABORATORY ACTIVITY 8.2

INDIVIDUAL DATA SHEET

Name or ID number: _____ Date: _____

Tester: _____ Time: _____

Sex: M = 1 / F = 0 (circle one)   Age: _____ y

Height: _____ in. _____ m

Weight: _____ lb _____ kg

Temperature: _____ °F _____ °C

Barometric pressure: _____ mmHg

Relative humidity: _____ %

Location of Testing
- ❏ Outdoor Field
- ❏ Indoor Field
- ❏ Indoor Track
- ❏ Outdoor Track
- ❏ Gym
- ❏ Other _____

Footwear
- ❏ Jogging shoe
- ❏ Walking shoe
- ❏ Tennis shoe
- ❏ Basketball shoe
- ❏ Running Shoe
- ❏ Cross Training Shoe
- ❏ CrossFit Shoe
- ❏ Other _____

Raw Data

Raw time: _____ min:s   Converted time: _____ min

Body weight: _____ kg × 2.2 = _____ lb

$\dot{V}O_2$max Determination

Men (18-29 Years Old)

$$\dot{V}O_2\text{max (ml} \cdot \text{kg}^{-1} \cdot \text{min}^{-1}) = 97.660 - (0.0957 \times \underset{\text{body mass (lb)}}{\underline{\hspace{1.5cm}}}) - (1.4537 \times \underset{\text{walk time (min)}}{\underline{\hspace{1.5cm}}}) - (0.1194 \times \underset{\text{HR (beats} \cdot \text{min}^{-1})}{\underline{\hspace{1.5cm}}})$$

$\dot{V}O_2$max = _____ ml $\cdot$ kg^{-1} $\cdot$ min^{-1} Ranking: _____   Percentile: _____

Women (18-29 Years Old)

$$\dot{V}O_2\text{max (ml} \cdot \text{kg}^{-1} \cdot \text{min}^{-1}) = 88.768 - (0.0957 \times \underset{\text{body mass (lb)}}{\underline{\hspace{1.5cm}}}) - (1.4537 \times \underset{\text{walk time (min)}}{\underline{\hspace{1.5cm}}}) - (0.1194 \times \underset{\text{HR (beats} \cdot \text{min}^{-1})}{\underline{\hspace{1.5cm}}})$$

$\dot{V}O_2$max = _____ ml $\cdot$ kg^{-1} $\cdot$ min^{-1} Ranking: _____   Percentile: _____

College-Aged Students

$$\dot{V}O_2\text{max (ml} \cdot \text{kg}^{-1} \cdot \text{min}^{-1}) = 88.768 + (\underset{\text{sex}}{\underline{\hspace{1.5cm}}} \times 8.892) - (\underset{\text{body mass (kg)}}{\underline{\hspace{1.5cm}}} \times 0.2109)$$
$$- (\underset{\text{walk time (min)}}{\underline{\hspace{1.5cm}}} \times 1.4537) - (\underset{\text{HR (beats} \cdot \text{min}^{-1})}{\underline{\hspace{1.5cm}}} \times 0.1194)$$

$\dot{V}O_2$max = _____ ml $\cdot$ kg^{-1} $\cdot$ min^{-1} Ranking: _____   Percentile: _____

High-Intensity Fitness Testing

Objectives

- Become familiar with methods for evaluating high-intensity running performance.
- Understand how to conduct the Léger 20 m shuttle test.
- Understand the methods for conducting the Yo-Yo Intermittent Recovery Test (level 1 and level 2).
- Use the 30-15 Intermittent Fitness Test to quantify the V_{IFT}, and apply these results to training.
- Interpret the results of high-intensity intermittent fitness tests.

DEFINITIONS

30-15 Intermittent Fitness Test (30-15$_{IFT-28m}$)—Test performed on a 28 m court that can be used to evaluate high-intensity running performance (22, 23).

30-15 Intermittent Fitness Test (30-15$_{IFT-40m}$)—Test performed on a 40 m field that can be used to evaluate high-intensity running performance (9, 10).

high-intensity interval training (HIIT)—Typically involves repeated short to long bouts of high-intensity exercise interspersed with recovery periods (13).

interval shuttle run test (ISRT)—Submaximal and maximal field test to measure intermittent endurance capacity (32).

Léger 20 m shuttle run test (20mSRT)—Multistage shuttle run test used to estimate $\dot{V}O_2$max and maximal aerobic speed (30, 31).

maximal aerobic speed (MAS)—Lowest speed that elicits $\dot{V}O_2$max (9).

maximal running speed (MRS)—Maximal speed achieved at the end of a high-intensity test (9).

V_{IFT}—Velocity achieved at the end of the 30-15$_{IFT}$ (10).

Yo-Yo Intermittent Recovery Test (Yo-Yo IRT)—Intermittent fitness test that assesses the ability to repeatedly perform high-intensity exercise (3).

Yo-Yo IRT Level 1 (Yo-Yo IRT1)—Evaluates the ability to carry out intermittent exercise, which leads to activation of the aerobic energy system (3).

Yo-Yo IRT Level 2 (Yo-Yo IRT2)—Evaluates the ability to recover from repeated exercise with a high anaerobic demand (3).

Many sports are, for the most part, undertaken in an intermittent fashion (1, 4, 19, 34). It is well documented in the scientific literature that speed, agility, strength, explosive power, and the ability to repeat brief supramaximal exercise is highly related to performance in sports that are dominated by intermittent activities (2, 4). In preparing for intermittent sports, various forms of **high-intensity interval training (HIIT)** are often used to improve cardiorespiratory and metabolic function as well as physical performance (13). Typically employed with running or cycling training, HIIT involves high-intensity exercise performed in short to long bouts interspersed with periods of recovery (5).

Commonly associated with sport performance, HIIT has begun to be seen as an effective training tool for combating cardiometabolic diseases (37, 42), improving vascular function and cardiovascular fitness (35), reducing insulin resistance (25), and improving metabolic health (25) when compared with low- to moderate-intensity continuous endurance training:

Increased

Adiponectin

Availability of nitric oxide

Beta-cell function

Cardiac function

Enjoyment of exercise

High-density lipoproteins

Insulin sensitivity

Maximal rate of Ca^{2+} reuptake

PGC-1α

Quality of life

$\dot{V}O_2$peak

Decreased

BP

Fasting glucose

Fatty acid synthase (FAS)

Fatty acid transport protein (FATP-1)

Inflammation

Oxidative stress

Triglycerides

Adapted from *British Journal of Sports Medicine*, "High-intensity interval training in patients with lifestyle-induced cardiometabolic disease: a systematic review and meta-analysis," K.S. Weston, U. Wisloff, and J.S. Coombes, 48: 1227-1234, ©2014, with permission from BMJ Publishing Group Ltd.

Though HIIT is a useful training tool, it should be used cautiously with clinical populations. The safety of HIIT training is a controversial topic among health professionals who work with clinical patients or at-risk populations (43). In a study by Rognmo et al. (36) examining the cardiovascular risk of HIIT in coronary heart disease patients, both HIIT and moderate-intensity training exhibited a low risk. However, Halle (21) suggests that the data presented by Rognmo et al. (36) should be interpreted with caution because the rates of cardiovascular complications calculated to the number of patient exercise hours were more than five times higher during HIIT. As such, the risks of adverse effects from HIIT may outweigh the benefits for some clinical populations. To avoid contraindications to HIIT, Weston et al. (43) suggest that careful screening must be conducted when working with clinical populations to ensure that HIIT is used in a safe and appropriate fashion. Potential contraindications to HIIT include the following:

- Unstable angina pectoris
- Uncompensated heart failure
- Recent myocardial infarction (<4 wk)
- Recent coronary artery bypass graft or percutaneous coronary intervention (<12 mo)
- Heart disease that limits exercise (valvular, congenital, ischemic, and hypertrophic cardiomyopathy)
- Complex ventricular arrhythmias or heart block
- Severe COPD, cerebrovascular disease, or uncontrolled peripheral vascular disease
- Uncontrolled diabetes mellitus
- Hypertension with BP >180/110 (or uncontrolled)
- Severe neuropathy

Adapted from *British Journal of Sports Medicine*, "High-intensity interval training in patients with lifestyle-induced cardiometabolic disease: a systematic review and meta-analysis," K.S. Weston, U. Wisloff, and J.S. Coombes, 48: 1227-1234, ©2014, with permission from BMJ Publishing Group Ltd.

Regardless of the population using HIIT, it is necessary to use a testing method that allows interval training to be programmed (9, 10). Classically, intense interval training has been programmed based on **maximal aerobic speed (MAS)**, which is the lowest speed that elicits $\dot{V}O_2$max and is typically determined via gas exchange analysis (5, 6, 9). Traditionally composed of shuttle runs, these tests often incorporate accelerations, decelerations, and changes of direction, which reflect the demands associated with many intermittent sports but also can enhance peripheral aspects of cardiorespiratory function (9). To make these tests more practical, numerous field-based tests have been developed to determine the MAS and indirectly reflect $\dot{V}O_2$max (9, 29-31). Typically, these tests are based on continuous linear runs (29) or shuttle tests (30, 31) and are used to determine the **maximal running speed (MRS)**, which is similar to the MAS, at the end of the test (9). However, as noted by Buchheit (9), these tests determine the MRS via efforts that are fundamentally different from intermittent sports, the methods typically used to develop individualized HIIT training programs, and the physiological determinants of performance associated with intermittent or shuttle test efforts.

Careful inspection of the scientific and applied literature suggests that the most effective test protocol for evaluating this type of endurance needs to allow the simultaneous inclusion of intermittent and shuttle runs (9). Two intermittent tests that are often used by practitioners are the **interval shuttle run test (ISRT)** (32) and the Yo-Yo test (26). Although these tests are commonly performed, they only provide an index of intermittent aerobic performance (9, 20, 27) and do not yield an MRS that can be used for developing HIIT programs (9). To address the inability of these tests to yield an MRS that could be used for programming, Buchheit (9, 10) developed the **30-15 Intermittent Fitness Test (30-15$_{IFT-40m}$)**, which is tested on a 40 m field. The test has been modified to be performed on a 28 m court and is called the **30-15 Intermittent Fitness Test (30-15$_{IFT-28m}$)**. The strength of this test is that it incorporates physiological variables similar to those seen in interval training, including explosive expressions of power when changing directions, aerobic qualities, and the ability to recover between efforts. This is accomplished by simultaneously using aspects of intermittent and shuttle tests to establish an MRS.

LÉGER 20 M SHUTTLE RUN TEST

In 1982, the maximal multistage 20 m shuttle run test—the **Léger 20 m shuttle run test (20mSRT)**, also known as the beep test—was first presented by Léger and Lambert (30) as a method for predicting $\dot{V}O_2$max from maximal speed. The original protocol was designed to be a continuous running test performed on a 20 m course, typically in a gymnasium on a nonsliding surface, with 2 min stages (figure 9.1) (30).

The original protocol for the 20mSRT started at 8 km · h⁻¹ and increased by 0.5 km · h⁻¹ every 2 min until the athlete was no longer able to maintain pace with a prerecorded audio signal (24). The maximal speed achieved at the cessation of the test was then placed into a regression equation to predict $\dot{V}O_2$max (ml · kg⁻¹ · min⁻¹):

$$\dot{V}O_2\text{max (ml · kg}^{-1} \cdot \text{min}^{-1}) = 5.857 \text{ (maximal speed in km · h}^{-1}) - 19.458$$

The original test was later modified by Léger et al. (31), reducing the time allotted for each level to 1 min while maintaining the starting speed of 8 km · h⁻¹ and increasing it by 0.5 km · h⁻¹ every minute (level). This continuous running test

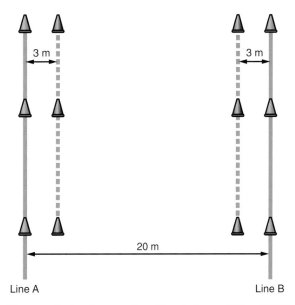

Figure 9.1 Setup for the Léger 20mSRT.

requires the athlete to complete as many levels as possible and is only stopped when the athlete is unable to maintain pace with the audio recording or is unable to reach the 3 m zone in front of each 20 m line for three consecutive times in accordance with the audio signal (9). Upon completion of the shuttle test, the level and number of shuttles is recorded. Normative data have been presented for various populations, including the level and number of shuttles completed (table 9.1 and table 9.2)

Overall this test has been shown to be very reliable for children aged 6 to 19 (31, 33) and adults aged 20 to 45 (31). Buchheit (9) suggests that the 20mSRT yields a MAS score but notes the final velocity could be considered an MRS score (MRS_{20mSRT}).

Regardless of the age group used with this version of the 20mSRT, the subjects' $\dot{V}O_2$max (ml $\cdot$ kg^{-1} $\cdot$ min^{-1}) can be estimated from the MRS_{20mSRT} and the subject's age using the following equation:

$$\dot{V}O_2\text{max (ml} \cdot \text{kg}^{-1} \cdot \text{min}^{-1}) = 31.025 + 3.238$$
$$\text{(maximal speed in km} \cdot \text{h}^{-1}) - 3.248 \text{ (age)} +$$
$$0.1536 \text{ (age} \times \text{maximal speed)}$$

In 2011, Mahar and colleagues (33) reevaluated the 20mSRT in an attempt to improve the fitness classifications and the $\dot{V}O_2$max (ml $\cdot$ kg^{-1} $\cdot$ min^{-1}) predictability for children between the ages of 10 and 16. This study resulted in the following quadratic prediction equation:

$$\dot{V}O_2\text{max (ml} \cdot \text{kg}^{-1} \cdot \text{min}^{-1}) = 41.76799 +$$
$$(0.49261 \times \text{laps}) - (0.0290 \times \text{lap}^2) - (0.61613$$
$$\times \text{BMI}) + (0.34787 \times \text{sex} \times \text{age})$$

where boys = 1 and girls = 0. This new equation has a lower standard error of the estimate (SEE = 6.17 ml $\cdot$ kg^{-1} $\cdot$ min^{-1}) compared with the classic estimation method (33).

Additionally, Mahar et al. (33) created another linear prediction equation:

$$\dot{V}O_2\text{max (ml} \cdot \text{kg}^{-1} \cdot \text{min}^{-1}) = 40.34533$$
$$+ (0.21426 \times \text{laps}) - (0.79462 \times \text{BMI}) +$$
$$(4.27293 \times \text{sex}) + (0.79444 \times \text{age})$$

where boys = 1 and girls = 0. This new equation has a slightly higher standard error of the estimate (SEE = 6.29 ml $\cdot$ kg^{-1} $\cdot$ min^{-1}) compared with quadratic prediction equations.

The 20mSRT has also been modified for use with adults. In 2003, Stickland et al. (39) came up with two equations one for males and one for females aged 18 to 38:

Males

$$\dot{V}O_2\text{max (ml} \cdot \text{kg}^{-1} \cdot \text{min}^{-1}) = 2.75 \times$$
$$\text{(last half} - \text{stage complete)} + 28.8$$

Females

$$\dot{V}O_2\text{max (ml} \cdot \text{kg}^{-1} \cdot \text{min}^{-1}) = 2.85 \times$$
$$\text{(last half} - \text{stage complete)} + 25.1$$

Both equations resulted in accurate predictions of $\dot{V}O_2$ and were more accurate than the ones created by Léger et al. (31).

YO-YO INTERMITTENT RECOVERY TEST

The **Yo-Yo Intermittent Recovery Test (Yo-Yo IRT)** evaluates the ability to repeatedly perform high-intensity exercise (3, 26, 28, 44). The **Yo-Yo IRT Level 1 (Yo-Yo IRT1)** examines the ability to perform intermittent exercise leading to activation of the aerobic system, while the **Yo-Yo IRT Level 2 (Yo-Yo IRT2)** evaluates the ability to recover from exercise with a large contribution of the anaerobic system (3). During each version of the Yo-Yo IRT, HR progressively increases, reflecting increasing oxygen uptake ($\dot{V}O_2$) (26, 28). While both tests display this general trend, the Yo-Yo IRT2 demonstrates a more rapid increase in HR across the test (3). Regardless, both tests can be used to evaluate maximal HR, maximal distance covered, and the physical capacity of people involved in intermittent sports (26, 28).

Both versions of the test require a 25 m area containing a 20 m running zone and 5 m recovery zone (figure 9.2). Two 20 m shuttle runs are performed at increasing speed interspersed with 10 s of active recovery, which consists of 5 m of jogging in the recovery zone (44).

For the Yo-Yo IRT1, the test is initiated at 10 km $\cdot$ h^{-1} and consists of four running bouts between 10 and 13 km $\cdot$ h^{-1} (0-160 m) and another seven bouts between 13.5 and 14 km $\cdot$ h^{-1} (160-440 m) (15, 26). From this point forward, the speed is increased by 0.5 km $\cdot$ h^{-1} after every eight running bouts until exhaustion (table 9.3) (26).

Table 9.1 Normative Data for 20 m Shuttle Test for 12- to 17-Year-Olds

AGE (y)	12		13		14		15		16		17	
Percentile	M	F	M	F	M	F	M	F	M	F	M	F
95	10/5	8/7	11/5	8/8	11/4	8/5	11/9	8/9	12/2	8/9	12/5	9/9
90	9/8	7/9	10/7	8/0	10/8	7/6	11/1	8/0	11/3	8/3	11/7	8/9
80	8/9	6/4	9/8	7/1	10/0	7/1	10/3	7/2	10/5	7/3	10/7	8/1
70	8/1	5/8	9/1	6/4	9/5	6/3	9/9	6/4	9/9	6/6	9/8	7/1
60	7/4	5/3	8/2	5/7	9/0	5/8	9/1	5/4	9/2	5/8	9/2	6/2
50	7/0	5/1	7/8	5/2	8/4	5/3	8/5	4/9	8/7	5/2	8/6	5/7
40	6/2	4/5	7/2	4/8	7/5	5/0	7/6	4/4	8/1	4/9	8/1	5/0
30	5/5	4/0	6/4	4/3	6/8	4/3	6/8	4/1	7/5	4/3	7/5	4/4
20	4/9	3/4	5/8	3/8	6/1	3/8	6/1	3/7	6/7	3/8	6/9	4/1
103	3/8	3/1	4/4	3/2	5/1	3/1	5/2	3/2	5/7	3/4	6/4	3/4

Level/shuttle; for example, 10/5 = 10th level, 5 shuttles. M = male and F = female.

Adapted, by permission, from 20-m Shuttle Run Booklet. ©Australian Sports Commission 2005.

Table 9.2 Normative Data for 20 m Shuttle Test for Various Sports

Sport	Sex	Squad level	Age/position	n	LEVEL/SHUTTLE Mean ± SD	LEVEL/SHUTTLE Range
Australian football	Males	AIS	~18 y	214	13/5 ± 1/0	10/2-15/7
Hockey	Females	National	Striker	5	11/9 ± 0/6	11/1-12/4
		National	Midfielder	8	12/0 ± 0/6	11/1-12/8
		National	Defender	6	12/6 ± 0/11	11/3-13/7
		National	U21	21	11/9 ± 1/1	9/2-13/9
		National	U17	13	10/8 ± 1/2	8/9-13/3
		State	Open	10	10/7 ± 0/9	9/4-12/1
	Males	National	Striker	6	14/9 ± 0/7	14/1-15/8
		National	Midfielder	8	14/12 ± 0/7	13/11-15/11
		National	Defender	7	14/11 ± 1/1	13/2-16/4
		National	U21	18	14/4 ± 0/10	13/5-16/3
		State	Open	11	12/12 ± 2/0	10/5-15/2
Tennis	Females	AIS/National Academy	16+ y	18	11/2 ± 1/2	8/9-12/8
		National Academy	15-16 y	26	10/1 ± 1/5	7/2-12/8
		National Academy	13-14 y	49	10/2 ± 1/4	6/7-13/2
		National Academy	11-12 y	25	10/1 ± 1/4	6/2-12/2
	Males	AIS/National Academy	16+ y	55	13/4 ± 1/0	11/1-15/4
		National Academy	15-16 y	41	13/2 ± 1/1	8/9-15/5
		National Academy	13-14 y	55	11/9 ± 1/1	9/1-14/5
		National Academy	11-12 y	30	10/6 ± 1/8	6/9-13/9

Reprinted, by permission from S. Woolford et al., 2013, Field testing principles and protocols. In *Physiological tests for elite athletes*, edited by R.K. Tanner and C.J. Gore (Champaign, IL: Human Kinetics), 244.

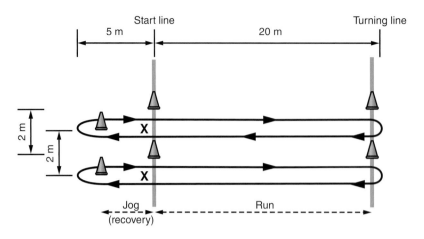

Figure 9.2 Setup for the Yo-Yo IRT.

Reprinted, by permission, from S. Woolford et al., 2013, Field testing principles and protocols. In *Physiological tests for elite athletes*, edited by R.K. Tanner and C.J. Gore (Champaign, IL: Human Kinetics), 245.

Table 9.3 Yo-Yo IRT1 and IRT2 Protocols

		YO-YO IRT1					YO-YO IRT2		
Stage	Speed (km · h⁻¹)	Shuttle bouts (2 × 20 m)	Split distance (m)	Accumulated distance (m)	Stage	Speed (km · h⁻¹)	Shuttle bouts (2 × 20 m)	Split distance (m)	Accumulated distance (m)
1	10	1	40	40	1	11.5	10	200	200
2	12	1	40	80	2	12	11	220	420
3	13	2	80	160	3	12.5	11	220	640
4	13.5	3	120	280	4	13	11	220	860
5	14	4	160	440	5	13.5	12	240	1,100
6	14.5	8	320	760	6	14	12	240	1,340
7	15	8	320	1,080	7	14.5	13	260	1,600
8	15.5	8	320	1,400	8	15	13	260	1,860
9	16	8	320	1,720	9	15.5	13	280	2,140
10	16.5	8	320	2,040	10	16	14	280	2,420
11	17	8	320	2,360	11	16.5	14	280	2,700
12	17.5	8	320	2,680	12	17	15	300	3,000
13	18	8	320	3,000	13	17.5	15	300	3,300
14	18.5	8	320	3,320	14	18	15	300	3,600
15	19	8	320	3,640					

Reprinted, by permission, from C. Castagna et al., 2006, "Aerobic fitness and yo-yo continuous and intermittent tests performances in soccer players: A correlation study," *Journal of Strength Conditioning Research* 20: 320-325.

The main difference between the two levels is that the Yo-Yo IRT2 is initiated at a higher speed (i.e., 11.5 km · h^{-1}) (14). After the first stage, the speed is increased 0.5 km · h^{-1}, resulting in increased distances being covered. In general, the Yo-Yo IRT1 lasts 10 to 20 min for a trained person, but it can also be used with less trained individuals because of the slower starting speeds (3). The Yo-Yo IRT2 lasts 5 to 15 min and is generally used with more advanced athletes (44).

Regardless of which Yo-Yo IRT is performed, the test is terminated if the subject is unable to reach the front line in time for two consecutive trials or is unable to cover another shuttle (14, 15). The total distance covered is the primary performance measure, and the speed attained at the end of the last 2 × 20 m bout is considered the maximal velocity (V_{max}) (15).

The results of either test can also be used to estimate $\dot{V}O_2$max from the distance covered using the following formulas (3):

Yo-Yo IRT1

$$\dot{V}O_2\text{max (ml} \cdot \text{kg}^{-1} \cdot \text{min}^{-1}) = \text{IR1 distance (m)} \times 0.0084 + 36.4$$

Yo-Yo IRT2

$$\dot{V}O_2\text{max (ml} \cdot \text{kg}^{-1} \cdot \text{min}^{-1}) = \text{IR2 distance (m)} \times 0.01364 + 45.3$$

Although $\dot{V}O_2$max can be estimated from the test, the results are generally not very accurate because they are also examining the anaerobic response during exercise and the recovery process (3). However, both Yo-Yo IRT tests are rapid, low-cost ways to estimate $\dot{V}O_2$max with large groups; up to 30 people can be tested in <20 min in a field-based setting (3).

Overall, both the Yo-Yo IRT1 and Yo-Yo IRT2 are highly reliable. For example, the Yo-Yo IRT1 typically displays a high correlation between repeated trials (r = .93-.95) and relatively low coefficients of variation (CV% = 4.9%-8.7%). The Yo-Yo IRT2 typically displays high correlations between repeated trials (r = .97-.99) and relatively low CV% (7.1%-10.4%) (3). Both test levels are highly dependent on effort, so in order to ensure accurate results, subjects must be encouraged to reach the highest possible level before ceasing the test (44).

30-15 INTERMITTENT FITNESS TEST

The 30-15$_{\text{IFT-40m}}$ was first presented by Buchheit (7-9) as an alternative to the Léger-Boucher track test, also known as the Montreal Track Test (29), and the Léger 20mSRT (29). It is highly specific to the training sessions performed by athletes to prepare for intermittent sports and thus can easily be used to determine HIIT plans (10). Recently, this test has become popular for athletes involved in multidirectional intermittent team sports such as soccer (17, 41), rugby union (18), rugby league (38), and team handball (7, 8). The test consists of 30 s shuttle runs interspersed with 15 s of recovery, and it is typically initiated from a velocity of 8 km · h^{-1} and increased by 0.5 km · h^{-1} for each subsequent stage (10, 11). Advanced athletes can start at 10 or even 12 km · h^{-1} and increase velocity in the same 0.5 km · h^{-1} pattern in order to save time (11).

The test is traditionally performed on a field that has three lines on it. Line A and line C are 40 m apart and define the extremities of the testing area. Line B defines the middle of the testing area (i.e., 20 m from lines A and C) (see figure 9.3). A 3 m zone is placed at the extremity of the field (i.e., near lines A and C) and around line B.

A prerecorded beep is used to pace subjects during the test so they can adjust their running speed when they enter one of the 3 m zones at the middle or end of the field (7, 10). During the 15 s recovery period, subjects walk forward to the closest line, either the end or middle of the running area depending on where they completed their last run (10). This line is where they will initiate their next sprint. Because the test is influenced by the ability to change direction, 0.7 s is subtracted from the running period for each change of direction (9, 16).

For example, an athlete running at 8.5 km · h^{-1} would cover a linear distance of 70.8 m in 30 s. If using a 40 m shuttle, the athlete would initiate the test from line A, cross line B to line C, and return (10). This would result in one change of direction (1 × 0.7 s) and a total distance of 69.2 m during the 30 s time period.

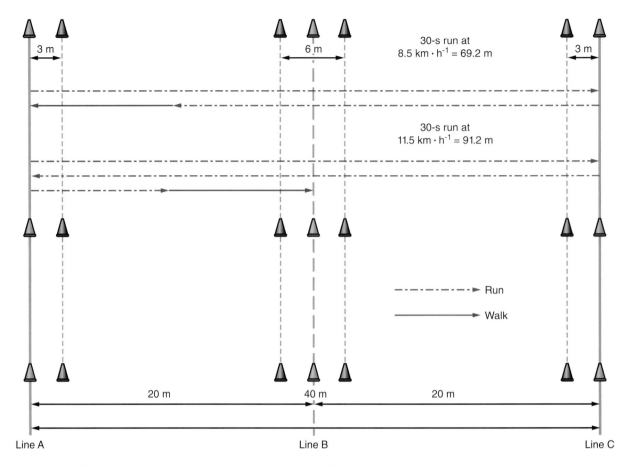

Figure 9.3 Setup for 30-15$_{\text{IFT-40m}}$ on a 40 m field.

Adapted, by permission, from M. Buchheit, 2008, "The 30-15 intermittent fitness test: accuracy for individualizing interval training of young intermittent sport players," *Journal of Strength Conditioning and Research* 22: 365-374.

The athlete would stop this run 8.5 m after line B and then walk to line A during the 15 s recovery period and initiate the next stage of the test from line A. An athlete running at 11.5 km · h^{-1} for a total distance of 91.4 m would start at line A, make one complete round-trip, and stop 9.5 m from line A while moving toward line B. The athlete would then walk to line B during the 15 s recovery period, from which the next stage of the test would be initiated (10). The athlete should complete as many stages of the test as possible, terminating the test when no longer able to maintain pace with the recording or when unable to reach one of the 3 m zones in time with the beep for three consecutive times (10). Upon completion of the test, the velocity achieved during the last stage is recorded as the athlete's **V$_{\text{IFT}}$** (7, 10).

Overall the 30-15$_{\text{IFT}}$ has been determined to be very reliable, with a typical error of measurement of 0.3 km · h^{-1} (95% confidence limit, 0.26-0.48) (9, 10). This suggests that a change of about one stage (i.e., 0.5 km · h^{-1}) would represent a worthwhile change (10). The 30-15$_{\text{IFT}}$ has also been shown to achieve $\dot{V}O_2$max when the V$_{\text{IFT}}$ is achieved (12), which allows the V$_{\text{IFT}}$ to be used to estimate $\dot{V}O_2$max with the following formula (10):

$$\dot{V}O_2\text{max}_{30\text{-}15\text{IFT}} \ (ml \cdot kg^{-1} \cdot min^{-1}) = 28.3 - 2.15(S) - 0.714(A) - 0.0357(W) + 0.0586(A) \times V_{\text{IFT}} + 1.03(V_{\text{IFT}})$$

where S = sex (1 = male; 2 = female), A = age, and W = weight in kg.

Traditionally the 30-15$_{IFT}$ has been completed on a 40 m field; however, it has been modified to be performed in a smaller area such as a basketball court (22, 40). To accomplish this, the shuttle length has been reduced to 28 m in the modified 30-15 Intermittent Fitness Test (30-15$_{IFT-28m}$), which corresponds to the length of a basketball court (figure 9.4) (22, 23).

The main difference between the 30-15$_{IFT-28m}$ and the 30-15$_{IFT-40m}$ is that the 30-15$_{IFT-28m}$ has more changes in direction. Otherwise both tests are conducted with the same work durations (i.e., 30 s), speed increments (i.e., 0.5 km · h^{-1}), and recovery periods (i.e., 15 s) (23).

Once the 30-15 test is complete and the V$_{IFT}$ has been determined, the results can be used to establish individualized training sessions (11). For example, if a 15 s–15 s HIIT running session (run = 15 s; recover = 15 s) is being designed for an intensity of 95% of V$_{IFT}$, the distance would be determined as 75 m. This is accomplished by converting the speed in km · h^{-1} into m · s^{-1} and then multiplying the time by the %V$_{IFT}$ (11). For convenience, dividing the V$_{IFT}$ by 3.6 would be used to convert from km · h^{-1} into m · s^{-1}. Therefore, the 75 m distance would be determined as follows:

$$V_{IFT} \ (m \cdot s^{-1}) = \frac{V_{IFT} \ (km \cdot h^{-1})}{3.6} = \frac{19}{3.6} = 5.28$$

$$\text{Distance (m)} = V_{IFT} \ (m \cdot s^{-1}) \times \%V_{IFT} \times \\ \text{time (s)} = 5.28 \times 0.95 \times 15 = 75$$

This calculation could be completed for a group of subjects, and then individualized training sessions could be constructed. For example, if we had a group that needed to perform a 15 s–15 s running session, the distance covered by each person would be based on individual V$_{IFT}$ (11). Figure 9.5 offers an example of how this could be organized.

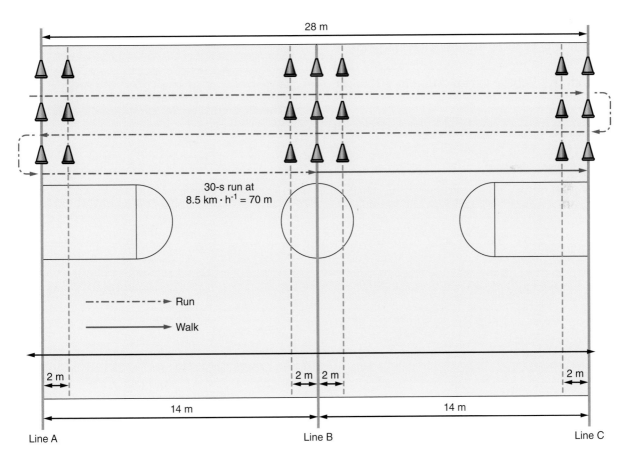

Figure 9.4 Setup for 30-15$_{IFT-28m}$ on a 28 m basketball court.

Based on Haydar and Buchheit 2009.

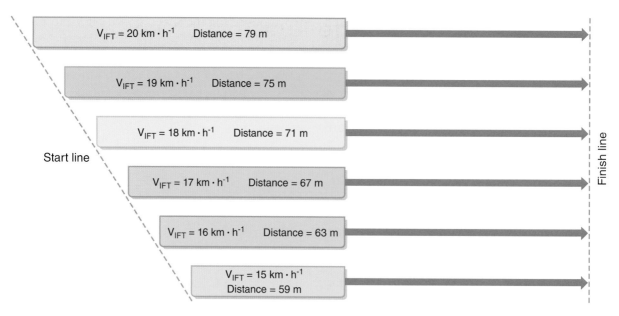

Figure 9.5 Sample running area for straight-line HIIT training (15 s–15 s at 95% V_{IFT}).

Adapted, by permission, from M. Buchheit, 2008, "The 30-15 intermittent fitness test: accuracy for individualizing interval training of young intermittent sport players," *Journal of Strength Conditioning and Research* 22: 365-374.

With this type of organization, individuals would be working at the same percentage of their V_{IFT} but would have to run different distances.

Go to the web study guide to access electronic versions of individual and group data sheets, the question sets from each laboratory activity, and case studies for each laboratory activity.

Léger 20 m Shuttle Run Test (20mSRT)

Equipment

- 23 m court (20 m needed for the test)
- Measuring wheel or 50+ m tape measure
- Cones (12 cones, 6 in one color and 6 in another)
- 20mSRT audio file
- Audio system (including extension cords if not battery powered)
- Individual data sheet
- Group data sheet

> The 20mSRT audio file is available for purchase at http://shop.ausport.gov.au/12-030.

Figure 9.1 depicts the 20 m setup for the Léger 20mSRT used in this laboratory activity. The test requires a space in which a 20 m shuttle run can be performed. Typically, it is set up on a standard gymnasium court such as a basketball or volleyball court. Using a measuring wheel or tape measure, determine the boundaries for the test. Place a same-colored cone at 0 m and 20 m to define the end points of the shuttle test (lines A and B). Then place a different-colored cone 3 m from each of the lines that define the end points of the shuttle run course.

Warm-Up

As with any performance-based test, the subject should perform a structured warm-up to prepare. As a rule, when working with athletes or other fit subjects, devote 5 min to general warm-up activity (e.g., jogging, cycling, jumping rope), and then use 5 min for dynamic activity (e.g., high knees, walking lunges, walking knee tucks, butt kicks, inchworms, power skips). With sedentary or untrained individuals, use less rigorous activities (e.g., leg swings, toe touches).

Léger 20mSRT

Step 1: Assemble the testing area in accordance with figure 9.1. Place cones that are the same color at line A and line B. Then place a different-colored cone 3 m from lines A and B.

Step 2: Place the audio system close to where the test is being conducted. If needed, use extension cords to supply power to the audio system. Make sure to test the volume of the audio system to ensure the subjects can hear it from all parts of the defined testing area.

Step 3: Gather basic data (e.g., age, height, weight) for the individual data sheet as described in laboratory activity 1.1.

Step 4: Have each subject complete a structured warm-up of about 10 min.

Step 5: Clearly explain how the test will be conducted and what the subjects must do. Specify that when they hear the sound that defines the end of the 1 min time period, they must be within the 3 m zone situated ahead of one of the 20 m lines. Make sure they understand that if they are unable to maintain pace or are not within the 3 m zone for three consecutive time shifts, the test is terminated.

Step 6: Remind the subjects of the goals and requirements of the test. At this time, ask them if they are ready to perform the test or if they have any questions. If there are no questions, have the subjects assemble at line A and prepare to begin the test when the signal sounds on the audio system.

Step 7: Once the subjects are ready to initiate the test, start the audio file. Upon the initiation, signal that the test will begin with the first stage of the test, which is performed at 8 km · h^{-1} and increased by 0.5 km · h^{-1} for each subsequent stage.

Step 8: Record the levels and shuttle numbers the subject completes on the individual data sheet. If the subject completes a shuttle, strike out (or cross out) the number; if the subject misses a recovery zone, circle the number to delineate this as a missed recovery zone.

Step 9: If the subject cannot maintain the running pace or misses three recovery zones in a row, terminate the test and record the velocity of the final stage the subject completed as the MRS_{20mSRT} on the individual and group data sheets.

Step 10: Using the equation presented on the individual data sheet, estimate each subject's $\dot{V}O_2$ max, and then record the results on the individual and group data sheets.

Question Set 9.1

1. Based on your estimated $\dot{V}O_2$max rank, compare your aerobic fitness in relation to the norms and percentile ranks presented in table 7.1.

2. Compare your estimated $\dot{V}O_2$max values from the classic and modified formulas. Are there differences in the values estimated?

3. Discuss how the 20mSRT determines MRS.

LABORATORY ACTIVITY 9.1

INDIVIDUAL DATA SHEET

Name or ID number: _____ Date: _____

Tester: _____ Time: _____

Sex: M / F (circle one) Age: _____ y

Location of Testing	Footwear
❏ Outdoor Field	❏ Jogging shoe
❏ Indoor Field	❏ Walking shoe
❏ Indoor Track	❏ Tennis shoe
❏ Outdoor Track	❏ Basketball shoe
❏ Gym	❏ Running Shoe
❏ Other _____	❏ Cross Training Shoe
	❏ CrossFit Shoe
	❏ Other _____

Height: _____ in. _____ m

Weight: _____ lb _____ kg

Temperature: _____ °F _____ °C

Barometric pressure: _____ mmHg

Relative humidity: _____ %

Level	Shuttle #	km · h⁻¹	Shuttle number															
1	7	8	1	2	3	4	5	6	7									
2	8	8.5	1	2	3	4	5	6	7	8								
3	8	9	1	2	3	4	5	6	7	8								
4	9	9.5	1	2	3	4	5	6	7	8	9							
5	9	10	1	2	3	4	5	6	7	8	9							
6	10	10.5	1	2	3	4	5	6	7	8	9	10						
7	10	11	1	2	3	4	5	6	7	8	9	10						
8	11	11.5	1	2	3	4	5	6	7	8	9	10	11					
9	11	12	1	2	3	4	5	6	7	8	9	10	11					
10	11	12.5	1	2	3	4	5	6	7	8	9	10	11					
11	12	13	1	2	3	4	5	6	7	8	9	10	11	12				
12	12	13.5	1	2	3	4	5	6	7	8	9	10	11	12				
13	13	14	1	2	3	4	5	6	7	8	9	10	11	12	13			
14	13	14.5	1	2	3	4	5	6	7	8	9	10	11	12	13			
15	13	15	1	2	3	4	5	6	7	8	9	10	11	12	13			
16	14	15.5	1	2	3	4	5	6	7	8	9	10	11	12	13	14		
17	14	16	1	2	3	4	5	6	7	8	9	10	11	12	13	14		
18	15	16.5	1	2	3	4	5	6	7	8	9	10	11	12	13	14	15	
19	15	17	1	2	3	4	5	6	7	8	9	10	11	12	13	14	15	
20	16	17.5	1	2	3	4	5	6	7	8	9	10	11	12	13	14	15	16
21	16	18	1	2	3	4	5	6	7	8	9	10	11	12	13	14	15	16

Final level: _____ Shuttle #: _____ Speed: _____

Classic:

$$\dot{V}O_2 max_{20mSRT} \ (ml \cdot kg^{-1} \cdot min^{-1}) = \text{_____} = 5.857 \times \underbrace{\text{_____}}_{\text{max speed (km · h}^{-1})} - 19.458$$

Modified: $\dot{V}O_2 max_{20mSRT} \ (ml \cdot kg^{-1} \cdot min^{-1}) = \text{_____}$

$$31.025 + 3.238 \times (\underbrace{\text{_____}}_{\text{max speed (km · h}^{-1})}) - 3.248 \times \underbrace{\text{_____}}_{\text{age (y)}} + 0.1536 \times (\underbrace{\text{_____}}_{\text{age (y)}} \times \underbrace{\text{_____}}_{\text{max speed (km · h}^{-1})})$$

Yo-Yo Intermittent Recovery Test (Yo-Yo IRT)

Equipment

- 25 m field or court (20 m needed for running, 5 m for recovery)
- Measuring wheel or 50+ m tape measure
- Cones (16 cones, all the same color)
- Yo-Yo IRT1 and Yo-Yo IRT2 audio files
- Audio system (including extension cords if not battery powered)
- Individual data sheet
- Group data sheet

> The Yo-Yo IRT1 and Yo-Yo IRT2 audio files are available for purchase at http://bangsbosport.com/shop/the-yo-yo-test-175p.html.

Figure 9.2 depicts the 20 m setup for the Yo-Yo IRT that will be used in this laboratory activity. The test requires space in which a 25 m shuttle run can be performed. Typically, it is set up on a standard gymnasium court such as a basketball or volleyball court. Using the measuring wheel or tape measure, determine the boundaries for the Yo-Yo IRT. Place two cones 2 m apart at 0 m and 20 m to define the end points of the test (starting and turning lines). Then place two cones 2 m from each other 5 m in front of the starting line, straddling the cones placed on the starting line to define the end of the recovery zone.

Warm-Up

As with any performance-based test, the subject should perform a structured warm-up to prepare. As a rule, when working with athletes or other fit subjects, devote 5 min to general warm-up activity (e.g., jogging, cycling, jumping rope), and then use 5 min for dynamic activity (e.g., high knees, walking lunges, walking knee tucks, butt kicks, inchworms, power skips). With sedentary or untrained individuals, use less rigorous activities (e.g., leg swings, toe touches).

Prior to performing this laboratory activity, determine the overall fitness of the subject. If the subject is not well trained, choose level 1 of the test. If the subject is well trained, choose level 2.

Yo-Yo IRT1

Step 1: Set up the testing area in accordance with figure 9.2. Measure out a 25 m course with recovery, starting, and turning lines. The recovery line should be 5 m from the starting line, while the turning line should be 20 m from the starting line. Place cones 2 m apart on the starting and turning lines, and place cones on the recovery line aligned with the middle of the zones established by the cones on the starting line.

Step 2:   Place the audio system close to where the test is being conducted. If needed, use extensions cords to supply power to the audio system. Make sure to test the volume of the audio system in order to ensure the subjects can hear it from all parts of the defined testing area.

Step 3:   Gather basic data (e.g., age, height, weight) for the individual data sheet as described in laboratory activity 1.1.

Step 4:   Have each subject complete a structured warm-up of about 10 min.

Step 5:   Clearly explain to the subjects how the test will be conducted and what they must do. Specify that at the first audio signal, they will run from the starting line toward the turning line. At the second audio signal, they should arrive and turn at the turning line, and they should arrive at the starting line at the next audio signal. As they pass the starting line, they continue to move forward at a reduced pace (i.e., jogging) toward the recovery line, where they turn around and return to the starting line. At this point, they will stop and wait for the next audio signal. Remind the subjects that they need to be stationary on the starting line before the commencement of each sprint.

Step 6:   Remind the subjects that they are to place one foot either on or over the starting or turning line at each audio signal.

Step 7:   Turn on the audio track to initiate the test. The first stage of this test will begin at 10 km · h^{-1} and will increase at a predetermined intensity (table 9.3).

Step 8:   As the subject completes a shuttle, strike out (or cross out) the corresponding number on the individual data sheet.

Step 9:   The first time the start time is not reached, the subject is given a warning; the second time this occurs, the subject is withdrawn from the test. Indicate this on the individual data sheet by circling the corresponding shuttle number. Transfer those data to the group data sheet.

Step 10: When the subject withdraws from the test, record the last stage and number of shuttles performed at this level. Using the information provided in table 9.3 determine the total distance achieved. Record the distance travelled on the individual data sheet. (Note that the last 2 × 20 m shuttle is included even if the subject did not complete it at the requisite pace).

Step 11: The intermittent recovery speed and the interval score obtained are then used to calculate the total distance covered in the test.

Step 12: Calculate the subject's $\dot{V}O_2$ max (ml · kg^{-1} · min^{-1}) using the Yo-Yo IRT1 equation on the individual data sheet.

Yo-Yo IRT2

Step 1:   Set up the testing area in accordance with the setup presented in figure 9.2. Measure out a 25 m course with recovery, starting, and turning lines. The recovery line should be 5 m from the starting line, while the turning line should be 20 m from the starting line. Place cones 2 m apart on the starting and turning lines, and place cones on the recovery line aligned with the middle of the zones established by the cones on the starting line.

Step 2: Place the audio system close to where the test is being conducted. If needed, use extension cords to supply power to the audio system. Make sure to test the volume of the audio system to ensure the subjects can hear it from all parts of the defined testing area.

Step 3: Gather basic data (e.g., age, height, weight) for the individual data sheet as described in laboratory activity 1.1.

Step 4: Have each subject complete a structured warm-up of about 10 min.

Step 5: Clearly explain to the subjects how the test will be conducted and what they must do. Specify that at the first audio signal, they will run from the starting line toward the turning line. At the second audio signal, they should arrive and turn at the turning line, and they should arrive at the starting line at the next audio signal. As they pass the starting line, they continue to move forward at a reduced pace (i.e., jogging) toward the recovery line, where they turn around and return to the starting line. At this point they stop and wait for the next audio signal. Remind the subjects that they need to be stationary on the starting line before the commencement of each sprint.

Step 6: Remind the subject that they are to place one foot either on or over the starting or turning line at each audio signal.

Step 7: Turn on the audio track to initiate the test. The first stage of this test will begin at $11.5 \text{ km} \cdot \text{h}^{-1}$ and will increase at a predetermined intensity (table 9.3).

Step 8: As the subject completes a shuttle, strike out (or cross out) the corresponding number on the individual data sheet.

Step 9: The first time the start time is not reached, the subject is given a warning; the second time this occurs, the subject is withdrawn from the test. Indicate this on the individual data sheet by circling the corresponding shuttle number.

Step 10: When the subject withdraws from the test, record the last stage and number of shuttles performed at this level. Using the information provided in table 9.3 determine the total distance achieved. Record the distance travelled on the individual data sheet. (Note that the last 2×20 m shuttle is included even if the subject did not complete it at the requisite pace).

Step 11: The intermittent recovery speed and interval score are then used to calculate the total distance covered in the test.

Step 12: Calculate the subject's $\dot{V}O_2$max ($\text{ml} \cdot \text{kg}^{-1} \cdot \text{min}^{-1}$) using the Yo-Yo IRT2 equation on the individual data sheet.

Question Set 9.2

1. Based on your estimated $\dot{V}O_2$max rank, compare your aerobic fitness in relation to the norms and percentile ranks presented in table 7.1.

2. Explain the main differences between the Yo-Yo IRT1 and the Yo-Yo IRT2.

3. Discuss the difference between the MAS and MRS.

LABORATORY ACTIVITY 9.2

INDIVIDUAL DATA SHEET

Name or ID number: _____ Date: _____

Tester: _____ Time: _____

Sex: M / F (circle one) Age: _____ y   **Location of Testing**   **Footwear**

Height: _____ in. _____ m   ❏ Outdoor Field   ❏ Jogging shoe

Weight: _____ lb _____ kg   ❏ Indoor Field   ❏ Walking shoe

Temperature: _____°F _____°C   ❏ Indoor Track   ❏ Tennis shoe

Barometric pressure: _____ mmHg   ❏ Outdoor Track   ❏ Basketball shoe

Relative humidity: _____ %   ❏ Gym   ❏ Running Shoe

                                          ❏ Other _____   ❏ Cross Training Shoe

                                                             ❏ CrossFit Shoe

                                                             ❏ Other _____

YO-YO IRT1										
Level	Shuttle #	km · h^{-1}	Number of 2 × 20 shuttles							
1	1	10	1							
2	1	12	1							
3	2	13	1	2						
4	3	13.5	1	2	3					
5	4	14	1	2	3	4				
6	8	14.5	1	2	3	4	5	6	7	8
7	8	15	1	2	3	4	5	6	7	8
8	8	15.5	1	2	3	4	5	6	7	8
9	8	16	1	2	3	4	5	6	7	8
10	8	16.5	1	2	3	4	5	6	7	8
11	8	17	1	2	3	4	5	6	7	8
12	8	17.5	1	2	3	4	5	6	7	8
13	8	18	1	2	3	4	5	6	7	8
14	8	18.5	1	2	3	4	5	6	7	8
15	8	19	1	2	3	4	5	6	7	8

Final-stage speed: _____ Yo-Yo IRT1

Distance traveled: _____ m

Number of intervals completed: _____ Yo-Yo IRT1

$$\dot{V}O_2max_{\text{Yo-Yo IRT1}} (ml \cdot kg^{-1} \cdot min^{-1}) = \underline{\hspace{2cm}} = \underset{\text{Yo-Yo IRT1 distance (m)}}{\underline{\hspace{3cm}}} \times 0.0084 + 36.4$$

| YO-YO IRT2 | | | | | | | | | | | | | | | | | | | |
Level	Shuttle #	km · h⁻¹	Number of 2 × 20 shuttles															
1	10	11.5	1	2	3	4	5	6	7	8	9	10					
2	11	12	1	2	3	4	5	6	7	8	9	10	11				
3	11	12.5	1	2	3	4	5	6	7	8	9	10	11				
4	11	13	1	2	3	4	5	6	7	8	9	10	11				
5	12	13.5	1	2	3	4	5	6	7	8	9	10	11	12			
6	12	14	1	2	3	4	5	6	7	8	9	10	11	12			
7	13	14.5	1	2	3	4	5	6	7	8	9	10	11	12	13		
8	13	15	1	2	3	4	5	6	7	8	9	10	11	12	13		
9	13	15.5	1	2	3	4	5	6	7	8	9	10	11	12	13		
10	14	16	1	2	3	4	5	6	7	8	9	10	11	12	13	14	
11	14	16.5	1	2	3	4	5	6	7	8	9	10	11	12	13	14	
12	15	17	1	2	3	4	5	6	7	8	9	10	11	12	13	14	15
13	15	17.5	1	2	3	4	5	6	7	8	9	10	11	12	13	14	15
14	15	18	1	2	3	4	5	6	7	8	9	10	11	12	13	14	15

Final-stage speed: _____ Yo-Yo IRT2

Distance traveled: _____ m

Number of intervals completed: _____ Yo-Yo IRT2

$$\dot{V}O_2max_{\text{Yo-Yo IRT2}}(ml \cdot kg^{-1} \cdot min^{-1}) = \underline{\hspace{2cm}} = \underset{\text{Yo-Yo IRT2 distance (m)}}{\underline{\hspace{3cm}}} \times 0.01364 + 45.3$$

30-15 Intermittent Fitness Test (30-15$_{IFT-40m}$)

Equipment

- 40 m field
- Measuring wheel or 50+ m tape measure
- Cones (21 cones, 9 in one color and 12 in another)
- 30-15$_{IFT-40m}$ audio file
- Audio system (including extension cords if not battery powered)
- Individual data sheet
- Group data sheet

> E-mail the 30-15 Intermittent Fitness Test creator, Martin Buchheit, at mb@martin-buchheit.net to request a download of the audio file for this test.

Figure 9.3 presents the 40 m setup for the 30-15$_{IFT-40m}$, which will be used in this laboratory activity. The test requires a field on which a 40 m shuttle run can be performed. Using the measuring wheel or tape measure, determine the boundaries for the 40 m shuttle test. Place a same-colored cone at 0 m, 20 m, and 40 m to define lines A, B, and C, respectively. Once these lines are defined, place a different-colored cone 3 m from lines A and C and 3 m from each side of line B to define the recovery zones.

Warm-Up

As with any performance-based test, the subject should perform a structured warm-up to prepare. As a rule, when working with athletes or other fit subjects, devote 5 min to general warm-up activity (e.g., jogging, cycling, jumping rope), and then use 5 min for dynamic activity (e.g., high knees, walking lunges, walking knee tucks, butt kicks, inchworms, power skips). With sedentary or untrained individuals, use less rigorous activities (e.g., leg swings, toe touches). After the warm-up, ensure the subjects understand that the objective of the test is to complete as many 30 s stages as possible. The test ends when they are no longer able to maintain the required running speed or are unable to reach the 3 m recovery zone within the 15 s recovery period established by the audio signal for three consecutive attempts. Make sure they understand where the various recovery zones are and what color of cone indicates these zones.

30-15$_{IFT-40m}$

Step 1: Assemble the testing area in accordance with the setup in figure 9.3. Place cones that are the same color at line A, line B, and line C. Then place a different-colored cone 3 m from lines A and C, while also placing cones 3 m from both sides of line B.

Step 2: Place the audio system close to where the test is being conducted. If needed, use extension cords to supply power to the audio system. Make sure to test the volume of the audio system to ensure the subjects can hear it from all parts of the defined testing area.

Sept 3: Gather basic data (e.g., age, height, weight) for the individual data sheet as described in laboratory activity 1.1.

Step 4: Have each subject complete a structured warm-up of about 10 min.

Step 5: Clearly explain to the subjects how the test will be conducted and what they must do. Specifically explain that when they hear the sound that defines the end of the 30 s period, they have 15 s to walk forward to the nearest line before initiating the next stage of the test. Additionally, make sure they know that they must be at the line or within the recovery zone when the next stage of the test is initiated. Also, explain that the test is complete when they are no longer able to maintain pace or they miss three successive recovery zones.

Step 6: Remind the subjects of the goals and requirements of the test. At this time, ask them if they are ready to perform the test or if they have any questions. If there are no questions, have the subjects assemble at line A and prepare to begin the test when the signal sounds on the audio system.

Step 7: Once the subjects are ready to initiate the test, start the audio file. Upon the initiation signal, the test will begin with the first stage of the test, which is performed at 8 km · h^{-1} and increased by 0.5 km · h^{-1} for each subsequent stage. If the subject is very fit or an intermittent-sport athlete, you may consider starting the test at 10 km · h^{-1} or 12 km · h^{-1}.

Step 6: Record which stages the subjects complete within the designated time periods by filling in the box that corresponds to the stage on the individual data sheet. If the subject misses a recovery zone, record that as a missed recovery zone by filling in the box that corresponds to the stage.

Step 7: If the subject cannot maintain the running pace or misses three recovery zones in a row, terminate the test and record the velocity of the final stage the subject completed as the V_{IFT} on the individual data sheet for this laboratory activity.

Step 8: Using the equation presented on the individual and group data sheets, estimate the subject's $\dot{V}O_2$ max, and then record the results on the group data sheet.

Question Set 9.3

1. In detail, explain what the V_{IFT} represents and how it might be used to determine a training program.

2. Based on your results, design a 15 s–15 s (run–recovery) straight-line running session.

3. Based on your results, design a 30 s–30 s (run–recovery) straight-line running session.

4. Based on your estimated $\dot{V}O_2$ max rank, compare your aerobic fitness in relation to the norms and percentile ranks presented in table 7.1.

INDIVIDUAL DATA SHEET

Name or ID number: _____ Date: _____

Tester: _____ Time: _____

Sex: M / F (circle one) Age: _____ y   **Location of Testing**   **Footwear**

Height: _____ in. _____ m   ❑ Outdoor Field   ❑ Jogging shoe

Weight: _____ lb _____ kg   ❑ Indoor Field   ❑ Walking shoe

Temperature: _____°F _____°C   ❑ Indoor Track   ❑ Tennis shoe

Barometric pressure: _____ mmHg   ❑ Outdoor Track   ❑ Basketball shoe

Relative humidity: _____ %   ❑ Gym   ❑ Running Shoe

❑ Other _____   ❑ Cross Training Shoe

❑ CrossFit Shoe

Stage starting speed: ❑ 8 km · h^{-1} ❑ 10 km · h^{-1} ❑ 12 km · h^{-1} ❑ Other _____

	km · h^{-1}	Completed	Missed		km · h^{-1}	Completed	Missed
Stage 1	8			Stage 21	18		
Stage 2	8.5			Stage 22	18.5		
Stage 3	9			Stage 23	19		
Stage 4	9.5			Stage 24	19.5		
Stage 5	10			Stage 25	20		
Stage 6	10.5			Stage 26	20.5		
Stage 7	11			Stage 27	21		
Stage 8	11.5			Stage 28	21.5		
Stage 9	12			Stage 29	22		
Stage 10	12.5			Stage 30	22.5		
Stage 11	13			Stage 31	23		
Stage 12	13.5			Stage 32	23.5		
Stage 13	14			Stage 33	24		
Stage 14	14.5			Stage 34	24.5		
Stage 15	15			Stage 35	25		
Stage 16	15.5			Stage 36	25.5		
Stage 17	16			Stage 37	26		
Stage 18	16.5			Stage 38	26.5		
Stage 19	17			Stage 39	27		
Stage 20	17.5			Stage 40	27.5		

Final-stage speed: _____ V_{IFT}

$$\dot{V}O_2 max_{30\text{-}15IFT} \ (ml \cdot kg^{-1} \cdot min^{-1}) = \underline{\hspace{2cm}}$$

$$28.3 - (2.15 \times \underset{sex}{\underline{\hspace{1.5cm}}}) - (0.714 \times \underset{age}{\underline{\hspace{1.5cm}}}) - (0.0357 \times \underset{weight}{\underline{\hspace{1.5cm}}}) + (0.0586 \times \underset{age}{\underline{\hspace{1.5cm}}} \times \underset{V_{IFT}}{\underline{\hspace{1.5cm}}}) + (1.03 \times \underset{V_{IFT}}{\underline{\hspace{1.5cm}}})$$

Modified 30-15 Intermittent Fitness Test (30-15$_{IFT-28m}$)

Equipment

- 28 m basketball court
- Cones (21 cones, 9 in one color and 12 in another)
- 30-15$_{IFT-28m}$ audio file
- Audio system (including extension cords if not battery powered)
- Individual data sheet
- Group data sheet

> This laboratory activity is supported by a virtual lab experience in the web study guide.
>
> www

> E-mail the 30-15 Intermittent Fitness Test creator, Martin Buchheit, at mb@martin-buchheit.net to request a download of the audio file for this test.

Figure 9.4 presents the 28 m setup for the 30-15$_{IFT-28m}$, which will be used in this laboratory activity. The test requires a basketball court on which a 28 m shuttle run can be performed. Using the measuring wheel or tape measure, determine the boundaries for the 28 m shuttle test. Place a same-colored cone at 0 m, 14 m, and 28 m to define lines A, B, and C, respectively. Once these lines are defined, place a different-colored cone 2 m from line A and line C, as well as 2 m from each side of line B, to define the recovery zones.

Warm-Up

As with any performance-based test, the subject should perform a structured warm-up to prepare. As a rule, when working with athletes or other fit subjects, devote 5 min to general warm-up activity (e.g., jogging, cycling, jumping rope), and then use 5 min for dynamic activity (e.g., high knees, walking lunges, walking knee tucks, butt kicks, inchworms, power skips). With sedentary or untrained individuals, use less rigorous activities (e.g., leg swings, toe touches). After the warm-up, ensure the subjects clearly understand that the objective of the test is to complete as many 30 s stages of the test as possible. The test ends when they are no longer able to maintain the required running speed or are unable to reach the 2 m recovery zone within the 15 s recovery period established by the audio signal for three consecutive attempts. Make sure they understand where the various recovery zones are and what color of cone indicates these zones.

30-15$_{IFT-28m}$

Step 1: Assemble the testing area in accordance with figure 9.4. Place cones that are the same color at lines A, B, and C. Then place different-colored cones 2 m from lines A and C, as well as 2 m from both sides of line B.

Step 2: Place the audio system close to the field where the test is being conducted. If needed, use extension cords to supply power to the audio system. Make sure to test the volume of the audio system to ensure the subjects can hear it from all parts of the defined testing area.

Step 3: Gather basic data (e.g., age, height, weight) for the individual data sheet as described in laboratory activity 1.1.

Step 4: Have each subject complete a structured warm-up of about 10 min.

Step 5: Clearly explain to the subjects how the test will be conducted and what they must do. Specify that when they hear the sound that defines the end of the 30 s period, they have 15 s to walk forward to the nearest line before initiating the next stage of the test. Additionally, make sure they know that they must be at the line or within the recovery zone when the next stage of the test is initiated. Clearly explain that the test is complete when they are no longer able to maintain pace or they miss three successive recovery zones.

Step 6: Remind the subjects of the goals and requirements of the test. At this time, ask them if they are ready to perform the test or if they have any questions. If there are no questions, have the subjects assemble at line A and prepare to initiate the test when the signal sounds on the audio system.

Step 7: Once the subjects are ready to initiate the test, start the audio file. Upon the initiation signal, the test will begin with the first stage of the test, which is performed at 8 km · h^{-1} and increased by 0.5 km · h^{-1} for each subsequent stage. If the subject is very fit or an intermittent-sport athlete, you may consider starting the test at 10 km · h^{-1} or 12 km · h^{-1}.

Step 8: Record which stages the subject completes within the designated time periods by filling in the box that corresponds to the completed stage on the individual data sheet. If the subject misses a recovery zone, record that as a missed recovery zone in the corresponding box.

Step 9: If the subject cannot maintain the running pace or misses three recovery zones in a row, terminate the test and record the velocity of the final stage the subject completed as the V$_{IFT}$ on the individual data sheet.

Step 10: Using the equation presented on the individual data sheet, estimate the subject's $\dot{V}O_2$ max, and then record the results on the group data sheet.

Question Set 9.4

1. In detail, explain what the V$_{IFT}$ represents and how it might be used to determine a training program.

2. Based on your results, design a 15 s–15 s (run–recovery) straight-line running session.

3. Based on your results, design a 30 s–30 s (run–recovery) straight-line running session

4. Based on your estimated $\dot{V}O_2$max rank, compare your aerobic fitness in relation to the norms and percentile ranks presented in table 7.1.

LABORATORY ACTIVITY 9.4

INDIVIDUAL DATA SHEET

Name or ID number: _____ Date: _____

Tester: _____ Time: _____

Sex: M / F (circle one)   Age: _____ y   **Location of Testing**   **Footwear**

Height: _____ in. _____ m   ❏ Outdoor Field   ❏ Jogging shoe

Weight: _____ lb _____ kg   ❏ Indoor Field   ❏ Walking shoe

Temperature: _____ °F _____ °C   ❏ Indoor Track   ❏ Tennis shoe

Barometric pressure: _____ mmHg   ❏ Outdoor Track   ❏ Basketball shoe

Relative humidity: _____ %   ❏ Gym   ❏ Running Shoe

❏ Other _____   ❏ Cross Training Shoe

❏ CrossFit Shoe

Stage starting speed: ❏ 8 km · h⁻¹   ❏ 10 km · h⁻¹   ❏ 12 km · h⁻¹   ❏ Other _____

	km · h⁻¹	Completed	Missed		km · h⁻¹	Completed	Missed
Stage 1	8			Stage 21	18		
Stage 2	8.5			Stage 22	18.5		
Stage 3	9			Stage 23	19		
Stage 4	9.5			Stage 24	19.5		
Stage 5	10			Stage 25	20		
Stage 6	10.5			Stage 26	20.5		
Stage 7	11			Stage 27	21		
Stage 8	11.5			Stage 28	21.5		
Stage 9	12			Stage 29	22		
Stage 10	12.5			Stage 30	22.5		
Stage 11	13			Stage 31	23		
Stage 12	13.5			Stage 32	23.5		
Stage 13	14			Stage 33	24		
Stage 14	14.5			Stage 34	24.5		
Stage 15	15			Stage 35	25		
Stage 16	15.5			Stage 36	25.5		
Stage 17	16			Stage 37	26		
Stage 18	16.5			Stage 38	26.5		
Stage 19	17			Stage 39	27		
Stage 20	17.5			Stage 40	27.5		

Final-stage speed: _____ V_{IFT}

$$\dot{V}O_2 max_{30\text{-}15IFT\text{-}28m} \ (ml \cdot kg^{-1} \cdot min^{-1}) = \underline{\hspace{3cm}}$$

$$28.3 - (2.15 \times \underset{sex}{\underline{\hspace{1.5cm}}}) - (0.714 \times \underset{age}{\underline{\hspace{1.5cm}}}) - (0.0357 \times \underset{weight}{\underline{\hspace{1.5cm}}}) + (0.0586 \times \underset{age}{\underline{\hspace{1.5cm}}} \times \underset{V_{IFT}}{\underline{\hspace{1.5cm}}}) + (1.03 \times \underset{V_{IFT}}{\underline{\hspace{1.5cm}}})$$

Maximal Oxygen Consumption Measurements

Objectives

- Understand the concept of maximal oxygen consumption as a fitness assessment and describe factors that affect $\dot{V}O_2$max.
- Define the terms *maximal oxygen consumption* and *ventilatory threshold*.
- Review units of measurement for work on a treadmill.
- Understand the criteria for determining whether $\dot{V}O_2$max was attained during a graded exercise test.
- Measure $\dot{V}O_2$max during graded treadmill exercise (Bruce treadmill protocol) using open-circuit calorimetry.
- Measure $\dot{V}O_2$max during graded exercise on a cycle ergometer using open-circuit calorimetry.
- Interpret changes in $\dot{V}O_2$, $\dot{V}CO_2$, $\dot{V}_E$, respiratory exchange ratio, and heart rate as a function of power output.

DEFINITIONS

graded exercise test (GXT)—Ergometer protocol for assessing maximal oxygen consumption; increases intensity every 1 to 3 min stage to exhaustion.

indirect calorimetry—Measurement of energy expenditure from oxygen consumption and carbon dioxide production as registered in expired gases.

maximal aerobic power—Maximal rate at which oxygen can be taken up, transported, and used during physical activity.

maximal oxygen consumption ($\dot{V}O_2$max)— Highest rate of oxygen consumption that a person can attain during exercise; commonly expressed in either the absolute units of $L \cdot min^{-1}$ or the relative units of $ml \cdot kg^{-1} \cdot min^{-1}$.

oxygen consumption ($\dot{V}O_2$)—Amount of oxygen taken up and used by an individual at a given time.

respiratory exchange ratio (RER)—Ratio of carbon dioxide production ($\dot{V}CO_2$) to oxygen consumption ($\dot{V}O_2$).

steady state—Point where aerobic metabolism contributes 100% of the energy requirement at a given absolute intensity.

ventilatory threshold—Point during increasing exercise intensity where ventilation begins to increase disproportionately as the body expires excess CO_2.

Maximal oxygen consumption ($\dot{V}O_2$max) is a measure of the maximum rate of **oxygen consumption ($\dot{V}O_2$)** by an individual using oxidative phosphorylation within the mitochondria. Thus, it is the maximal capability of the oxidative energy system to produce ATP during exercise. Moreover, it is a function of the capac-ity of the heart, lungs, and blood to transport oxygen to the working muscle and of the ability of the muscles to use oxidative phosphorylation to create ATP aerobically (5). The term **maximal aerobic power** can be synonymous with $\dot{V}O_2$max, indicating the maximal rate at which oxygen can be taken up, transported, and used during physical activity. Several other terms are also used for this capacity—among them, *cardiorespiratory endurance, aerobic fitness, cardiorespiratory fitness,* and *cardiovascular endurance.* For the purposes of this lab, we refer to this phenomenon as *maximal oxygen uptake* or $\dot{V}O_2max$.

$\dot{V}O_2max$ is usually measured by means of a **graded exercise test (GXT)** on a treadmill or cycle ergometer using **indirect calorimetry** (see figure 10.1). The test begins at a relatively easy work rate and becomes progressively more demanding due to increasing intensity. Each stage represents a period of time (typically 1-3 min) when workload remains constant and the subject is allowed to reach **steady state**. If an automated system (metabolic cart) is being used, then $\dot{V}O_2$ is measured continuously (breath by breath or from a mixing chamber); it can also be measured manually at the end of each stage (i.e., by collecting bags of expired air to be analyzed by hand; see appendix C). The subject is encouraged to exercise until volitional fatigue. The highest $\dot{V}O_2$ value measured during the test is taken as $\dot{V}O_2max$ if two of the following four criteria have been met (2):

1. $\dot{V}O_2$ plateaus (i.e., increases ≤150 ml · min^{-1}) despite an increase in work rate.
2. Heart rate (HR) during the last stage is no more than 10 beats · min^{-1} below the subject's age-predicted maximal heart rate (HRmax).
3. **Respiratory exchange ratio (RER)** is greater than 1.10 in the final work stage.
4. Rating of perceived exertion (RPE) is greater than 17 using the original Borg scale.

If two of these four criteria are not met, then the subject probably did not attain a true $\dot{V}O_2max$, possibly even with volitional fatigue.

The $\dot{V}O_2max$ test is generally considered the best noninvasive measure of cardiorespiratory fitness. It is highly correlated to maximal CO (cardiac output) and therefore provides an excellent index of the heart's capacity to pump blood. It is not, however, a particularly good predictor of performance in endurance events.

Figure 10.2 demonstrates the $\dot{V}O_2$ response to increases in intensity. Note how the initial increases in intensity are still capable of being

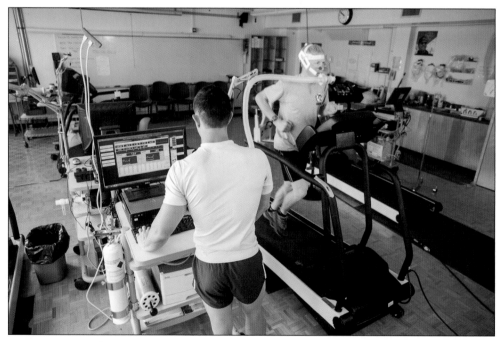

Figure 10.1 $\dot{V}O_2max$ test equipment.

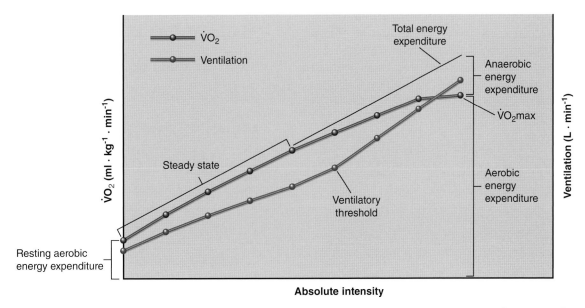

Figure 10.2   Oxygen consumption and ventilation during exercise of increasing intensity.

© Charles Dumke

met aerobically; that is, steady state is attained. As the work rate increases, total energy expenditure remains linear; however, $\dot{V}O_2$ is less able to meet the entire energy requirement. In fact, as a requirement of reaching $\dot{V}O_2$max, the oxygen consumption plateaus with the last increase in work rate. This is only possible by the greater anaerobic energy contribution at these higher intensities. (Ventilatory responses are described later in this laboratory in the Ventilatory Threshold section.)

A subject's $\dot{V}O_2$max can be influenced by several factors:

- Heredity—25% to 50% is considered genetic.
- Sex—Male values are 15% to 20% higher than female values for the same age group.
- Age—Gradual decline of about 8% per decade occurs from 30 years on; this age-related loss can be attenuated with exercise training.
- Training status—A possible 6% to 25% improvement in values may be seen with training (higher values found in aerobically trained individuals).

- Mode of exercise—Outcome is affected by choice of exercise; more muscle recruitment results in greater $\dot{V}O_2$max values.

SELECTING A TEST PROTOCOL

$\dot{V}O_2$max testing can be done on a number of ergometers—for example, on a treadmill, stair-stepper, rowing machine, stationary bicycle, or even a wide treadmill that allows the subject to roller-ski or roller-skate. $\dot{V}O_2$max is generally higher in modalities that recruit greater amounts of muscle. The highest $\dot{V}O_2$max values recorded for a male (94 ml · kg^{-1} · min^{-1}) and for a female (77 ml · kg^{-1} · min^{-1}) were in cross-country skiers (10). Of course, laboratories may be limited by the available ergometers, but when testing athletes you should choose the ergometer that most closely resembles their sport. When this is not possible—or when you are testing an untrained subject—the most popular modes are treadmills and cycle ergometers. See tables 7.1 and 7.3 for percentile norms on a treadmill and cycle ergometer according to sex and age group.

$\dot{V}O_2$max may also depend on the protocol. Many treadmill and cycle ergometer protocols are available, and one area of skill for an exercise physiologist lies in choosing a testing protocol that will give interpretable results. When selecting a protocol, try to stay within the following guidelines:

- The test should be progressive, and each increase in workload should be equal to the previous one. You should not use small increases in the early stages and then switch to larger increases later in the protocol. Equal stage increments can range from 0.5 to 3 METs depending on the subject's anticipated fitness.

- Each stage should be 1 to 3 min long, and the goal is to reach steady state by the end of each stage. This suggests that shorter stages are allowable with less MET increases in intensity from stage to stage.

- Total testing should last from 8 to 15 min. If the duration is too short, it is unlikely that the subject will reach steady state, and the anaerobic component is typically too large. If the duration is too long, other mechanisms of fatigue may cause the subject to quit before reaching $\dot{V}O_2$max.

The maximum measured $\dot{V}O_2$ may depend on the protocol or modality chosen or on the effort put forth by the subject. Because of this, the maximum measured $\dot{V}O_2$ may be either less than or the same as $\dot{V}O_2$max and is often referred to as $\dot{V}O_2peak$. As an example, $\dot{V}O_2$ measured during arm cranking would produce a $\dot{V}O_2peak$ that is less than the whole-body $\dot{V}O_2$max elicited during treadmill testing.

Not everyone should be tested for $\dot{V}O_2$max. Exercising to exhaustion involves inherent risks including light-headedness, pallor, angina (chest pain), nausea, dyspnea (undue shortness of breath), arrhythmia, myocardial infarction, and death. Because of this risk, the ACSM has produced contraindications to maximal exercise testing; for a complete list, see *ACSM's Guidelines for Exercise Testing and Prescription* (1). Participation is voluntary. Subjects should be informed that they can quit at any time before or during the test. For indications for terminating a test once it has started, see *ACSM's Guidelines for Exercise Testing and Prescription* (1).

MONITORING PROGRESS WITH RPE SCALES

In performing a $\dot{V}O_2$max test, a subject is of course unable to talk while wearing a mouthpiece or mask. It is useful to testers, however, to know how the subject is feeling, and thus it is common to use RPE scales. This type of measure can be used to monitor progress toward maximal exertion. Two subjective scales have been developed (6, 9), and both are appropriate for use during maximal exercise testing. The RPE scales include pretest instructions that help clarify their use to the test subject (7, 9). Read these instructions to the subject prior to the start of the test.

During the test, we want you to pay close attention to how hard you feel the exercise work rate is. This feeling should reflect your total amount of exertion and fatigue, combining all sensations and feelings of physical stress, effort, and fatigue. Don't concern yourself with any one factor, such as leg pain, shortness of breath, or exercise intensity, but try to concentrate on your total, inner feeling of exertion. Try not to underestimate or overestimate your feelings of exertion; be as accurate as you can.

Anecdotally, fit subjects are better than untrained subjects at rating their exertion; in fact, you may find that unfit subjects go quickly from moderate RPE to having to quit the test. Even so, the validity of RPE has been tested in both healthy and diseased populations (8). Because RPE can also be affected by external information (e.g., HR, work rate), it is important to keep subjects blind to their own metabolic data; such knowledge could lead them to alter their effort or simply quit.

ESTIMATING FUEL USAGE WITH RER

As discussed in laboratory 5, the ratio of oxygen consumed to carbon dioxide produced is referred to as the *respiratory exchange ratio (RER)*. For the sake of review, and to elaborate, RER (the volume of carbon dioxide expelled divided by the volume of oxygen consumed) can be calculated by measuring $\dot{V}CO_2$ and $\dot{V}O_2$. RER ($\dot{V}CO_2$ / $\dot{V}O_2$) can be used to determine the fuel mixture during exercise, which is typically somewhere between 0.70 (100% fat) and 1.0 (100% carbohydrate), thus indicating that a mixture of these two fuels is contributing to energy production.

To use RER to estimate the amount of fat and carbohydrate being used during exercise, we have to make a few assumptions. First, the breakdown of protein contributes little to the overall energy expenditure during exercise (typically <5%); as a result, the RER that we measure in the laboratory is sometimes referred to as *nonprotein RER*. Also, the subject must be in steady state, meaning that the $\dot{V}O_2$ and RER should not change if there is no change in work rate. RER is affected by the production of CO_2 during exercise. As intensity increases during the $\dot{V}O_2$max test, RER increases due to a shift from greater use of fat to greater reliance on carbohydrate. With this in mind, RER during submaximal steady-state work can be used to determine accurate training zones for clients interested in maximizing absolute fat oxidation. We discussed in laboratory 5 how RER can be converted to kcal · min⁻¹ when $\dot{V}O_2$ in L · min⁻¹ is known. RER is also useful as subjects approach $\dot{V}O_2$max. As we will see shortly, CO_2 production can come from other sources besides macronutrient (fat and carbohydrate) oxidation. This information helps us understand how RER can exceed 1.0; indeed, this is one of the criteria for justifying a valid $\dot{V}O_2$max test (RER > 1.1).

VENTILATORY THRESHOLD

$\dot{V}O_2$ increases as a linear function of power output during a GXT up to $\dot{V}O_2$max. On the other hand, ventilation ($\dot{V}_E$) increases linearly as a function of power output only during exercise requiring up to about 50% to 75% $\dot{V}O_2$max, above which $\dot{V}_E$ increases at a steeper rate (see figure 10.2).

The greater rate of increase in $\dot{V}_E$ above this **ventilatory threshold** coincides with a similar rise in $\dot{V}CO_2$ and a sharp increase in blood lactate as anaerobic energy contribution increases (see figure 10.2). Remember that lactic acid dissociates into a lactate ion and H^+ in the blood, and the H^+ combines with a bicarbonate ion to form carbonic acid, which in turn dissociates into H_2O and CO_2.

$$La^-H^+ + HCO_3^- \leftrightarrow H_2CO_3 \leftrightarrow H_2O + CO_2$$

The accumulation of CO_2 in the blood leads to an increase in CO_2 in the expired air ($\dot{V}CO_2$). The increase in blood CO_2 stimulates an increase in ventilation. Thus, the ventilatory threshold is indicative of an abrupt increase in blood lactate and a subsequent decrease in blood pH. This nonmetabolic or excess CO_2 contributes to the $\dot{V}CO_2$ measured during the test, which is why RER can exceed 1.0 at high intensities. Many people misunderstand the heavy breathing at maximum exercise as an attempt to get more oxygen. In fact, at altitudes of less than 2,000 m (6,562 ft), ventilation is driven more by CO_2 production than by O_2 consumption. Note in figure 10.2 how the increase in ventilation at threshold does not correspond with greater $\dot{V}O_2$. Indeed, it corresponds with $\dot{V}O_2$ no longer meeting the entirety of the energy expenditure (thus *not* at steady state) as anaerobic contribution increases. This concept is discussed further in laboratory 11. Ventilation can be monitored during a $\dot{V}O_2$max test to gauge intensity and anticipate volitional fatigue.

Go to the web study guide to access electronic versions of the individual data sheet, the question sets from each laboratory activity, and case studies for each laboratory activity.

WWW

Graded Treadmill $\dot{V}O_2$max Test

Equipment

- Platform scale (or digital or other scale)
- Stadiometer (wall or freestanding) or physician's scale with attached anthropometer
- Running treadmill
- $\dot{V}O_2$ and $\dot{V}CO_2$ collection system (metabolic cart)
- Breathing valve assembly, mouthpiece (or mask), hoses, noseclip
- Heart rate monitor
- Stopwatch
- Rating of perceived exertion (RPE) scale
- Individual data sheet

> This laboratory activity is supported by a virtual lab experience in the web study guide.
>
> www

Bruce Treadmill Protocol for $\dot{V}O_2$max

At least one subject from each laboratory section will perform the max test on the treadmill. The subject should come prepared to exercise in appropriate shoes and attire, should be well rested, and should have fasted for at least 2 h.

Step 1: Begin this lab activity by gathering basic data (e.g., age, height, weight) for the individual data sheet as described in laboratory activity 1.1. Determine the subject's age-predicted HRmax using the following equation: $208 - (0.7 \times$ age in y). Record the age-predicted HRmax on the individual data sheet.

Step 2: After the lab instructor explains the procedures for using the metabolic cart, proceed to calibrate and operate the cart. Enter the subject's data into the cart.

Step 3: Choose one person to act as a spotter and stand behind the treadmill. (Some subjects get vertigo at higher intensities or catch their foot on the edge.) Others will monitor the subject's HR, RPE, and expired gases, as well as the overall and stage time, and record this information on the data collection sheet.

Step 4: Prepare the subject for testing with the HR monitor and headgear.

Step 5: Before the test begins, read the subject the RPE instructions presented earlier in this laboratory. Allow the subject to ask questions prior to being connected to the metabolic equipment.

Step 6: Connect the subject to the mouthpiece or mask apparatus while the subject stands on the treadmill.

Step 7: Measure the subject's standing HR and metabolic data.

Step 8: Allow a 3 to 5 min warm-up at an intensity no greater than that of the first stage.

Step 9: As soon as the subject begins the first stage, start a stopwatch, because overall time in the Bruce protocol can be used as an accurate estimate of $\dot{V}O_2$max.

Step 10: Follow the Bruce treadmill protocol presented in figure 10.3 to determine the work output at each stage. Collect data during the last 30 s of each stage. Specifically, record HR, $\dot{V}O_2$, $\dot{V}CO_2$, RER, $\dot{V}_E$, and RPE for each stage on the individual data collection sheet.

Step 11: Increase speed and percent grade exactly at the end of every 3 min stage until the subject can go no longer.

Step 12: Observe the exercising individual for signs and symptoms that call for terminating the test. Encouragement is effective in motivating a subject to attain a maximal effort.

Step 13: Because the subject is unable to communicate through the mask or mouthpiece, several common hand signals can be used. For example, a thumbs-up indicates that the subject is doing well and is willing to continue. A raised index finger indicates that the subject will need to stop in about 1 min, allowing the testers to prepare for the end of the test and to collect maximum data for all of the outcome measures.

Step 14: Stop the test when $\dot{V}O_2$max test criteria are reached, when indicators for stopping the test arise, or when the subject stops the test by grabbing the handrail. Slow the treadmill to 3 mph (4.8 km $\cdot$ h^{-1}) and reduce the grade to 0%. Have the subject keep the mouthpiece in for a 5 min cool-down period (record the RER during the cool-down). Immediately record the total time in the protocol, which will be used to predict $\dot{V}O_2$max.

Step 15: The cool-down is important to keep the subject moving to avoid blood pooling in the lower extremities and potential syncope. During the cool-down, which should last at least 5 min, monitor the subject's HR. An appropriate cool-down has the subject attain an HR of ~120 beats $\cdot$ min^{-1} before getting off the ergometer. Continue to monitor the subject's physical appearance and symptoms during the cool-down.

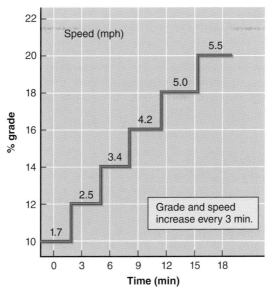

Figure 10.3 Bruce treadmill protocol.

Reprinted, by permission, from J. Hoffman, 2006, *Norms for fitness, performance, and health* (Champaign, IL: Human Kinetics), 24.

Bruce Protocol Formulas

Because the standard Bruce protocol (2) has been used for so long, estimations have been developed from the amount of time spent in the protocol. Thus, if protocol stage times are strictly adhered to, the following equations can be used to estimate $\dot{V}O_2max$ (3):

Men

$$\dot{V}O_2max \ (ml \cdot kg^{-1} \cdot min^{-1}) = 14.76 - (1.379 \times time) + (0.451 \times time^2) - (0.012 \times time^3)$$

Women (6)

$$\dot{V}O_2max \ (ml \cdot kg^{-1} \cdot min^{-1}) = (4.38 \times time) - 3.90$$

These predictions correlate significantly ($r = .98$ and $.91$ for men and women, respectively) with measured $\dot{V}O_2max$ and therefore may be useful if respiratory gas analysis with a metabolic cart is not possible.

Calculate the power output (or work rate) for each work stage on a treadmill using the following equation:

$$\text{Power output} \ (kg \cdot m \cdot min^{-1}) = \text{subject's body weight (kg)} \times \text{treadmill speed} \ (m \cdot min^{-1}) \times \text{fractional grade of the treadmill}$$

where fractional grade = percent grade divided by 100. For example, the fractional grade at a 2% grade = 2% / 100 = 0.02. To convert mph to $m \cdot min^{-1}$, multiply by 26.8.

Question Set 10.1

1. Prepare a data table showing values for HR, $\dot{V}O_2$, $\dot{V}_E$, RPE, and RER for each stage the subject completed.

2. Plot as separate graphs $\dot{V}O_2$, HR, and RER (y-axis) for your subject as a function of the calculated power output ($kg \cdot m \cdot min^{-1}$ or W).

3. Did your subject attain $\dot{V}O_2max$? Justify your answer.

4. Determine relative ($ml \cdot kg^{-1} \cdot min^{-1}$) and absolute ($L \cdot min^{-1}$) $\dot{V}O_2max$ values for your subject. Which is a better index of cardiorespiratory fitness? Why?

5. For the subjects you collected data on, plot $\dot{V}_E$ (y-axis) as a function of HR. Did any of your subjects have a distinct ventilatory threshold? At what absolute HR? At what percentage of HRmax? Why does $\dot{V}_E$ typically increase more sharply at heavy workloads?

6. RER at $\dot{V}O_2max$ is usually greater than 1.0. If an RER of 1.0 is 100% carbohydrate oxidation, how can RER exceed 1.0?

7. Compare the collected data with the normative values in table 7.1. How does your subject rank according to the norms?

8. Using the prediction equations provided for the Bruce protocol, how well does the time in the protocol compare with the measured $\dot{V}O_2max$?

9. How can people increase their $\dot{V}O_2max$? Be specific as possible.

LABORATORY ACTIVITY 10.1

INDIVIDUAL DATA SHEET

Name or ID number: _____ Date: _____

Tester: _____ Time: _____

Sex: M / F (circle one) Age: _____ y Height: _____in. _____m

Weight: _____ lb _____ kg Temperature: _____ °F _____ °C

Barometric pressure: _____ mmHg Relative humidity: _____%

Age-predicted HRmax: _____ beats · min⁻¹

RPE														
RER														
$\dot{V}CO_2$ (L · min⁻¹)														
METs														
$\dot{V}O_2$ (ml · kg⁻¹ · min⁻¹)														
$\dot{V}O_2$ (L · min⁻¹)														
$\dot{V}_E$ (L · min⁻¹)														
HR (beats · min⁻¹)														
Power (kg · m · min⁻¹)	Rest													
Grade (%)	0													
Speed (mph)	0													
Stage time (min)	0													

Cycle Ergometer $\dot{V}O_2$max Test

Equipment

- Platform scale (or digital or other scale)
- Stadiometer (wall or freestanding) or physician's scale with attached anthropometer
- Cycle ergometer
- $\dot{V}O_2$ and $\dot{V}CO_2$ collection system (metabolic cart)
- Breathing valve assembly, mouthpiece (or mask), hoses, noseclip
- Heart rate monitor
- Stopwatch
- Rating of perceived exertion (RPE) scale
- Individual data sheet

Cycle Ergometer Protocols

A graded protocol on a bike is relatively easy. If your lab has an electronically braked bike, then the subject can use a self-selected cadence and the ergometer will adjust resistance accordingly. If you are using a mechanically braked cycle ergometer, such as a Monark, the subject needs to cycle at a fixed cadence because rev · min⁻¹ is a component of power on a bike, as explained later. We recommend using a metronome to keep the subject at the determined cadence (between 60 and 100 rev · min⁻¹). Stage time should be either 2 or 3 min.

The main considerations in determining a protocol are initial work output, time in each stage, and amount of increase between stages. Initial work output is a function of fitness and the subject's familiarity with cycling. Subjects who are less fit should start between 25 and 75 W (about 150 and 450 kg · m · min⁻¹), whereas fitter subjects who are more familiar with cycling can start between 75 and 150 W (about 450 and 920 kg · m · min⁻¹). Similarly, when working with subjects who are less fit, the amount of increase between stages should be smaller (e.g., 10-15 W, or about 60-90 kg · m · min⁻¹); when working with fitter subjects who are experienced in cycling, the increase can be as much as 25 to 50 W (about 150-300 kg · m · min⁻¹).

To serve as an accurate $\dot{V}O_2$max test, the session should last 8 to 15 min. As with the Bruce treadmill test, several continuous cycle ergometer tests have been published, and two of the most popular are the Åstrand and the McArdle protocols (4). However, there is nothing particularly special about these protocols, and in fact their required cadences are lower than those preferred by many subjects (50 and 60 rev · min⁻¹, respectively). The Åstrand test for men starts at 600 kg · m · min⁻¹ (about 100 W) and increases by 300 kg · m · min⁻¹ (about 50 W) every 2 min while pedaling at 50 rev · min⁻¹; for women, it starts at 300 kg · m · min⁻¹ and increases by 150 kg · m · min⁻¹ (about 25 W) every 2 min at 50 rev · min⁻¹. The McArdle protocol, in contrast, is not gender specific. It starts at 900 kg · m · min⁻¹ (about 150 W) and increases by 240 kg · m · min⁻¹ (about 40 W) every 2 min at 60 rev · min⁻¹ (4). Thus, these protocols should be used only for fit individuals who have experience in cycling. They can be used literally or as a gauge for students in designing their own protocol individualized for their subject. Once a decision has been made on the stage time (2 or 3 min) and cadence (60-90 rev · min⁻¹), these should be the same throughout the test.

Cycle Ergometer V̇O₂max Test

At least one subject from each laboratory section will perform the max test on the cycle ergometer. The subject should come prepared to exercise in appropriate shoes and attire, should be well rested, and should have fasted for at least 2 h.

Step 1: Begin this lab activity by gathering basic data (e.g., age, height, weight) for the individual data sheet as described in laboratory activity 1.1. Determine the subject's age-predicted HRmax using the following equation: $208 - (0.7 \times$ age in y$)$. Record the age-predicted HRmax on the individual data sheet.

Step 2: After the lab instructor explains the procedures for using the metabolic cart, proceed to calibrate and operate the cart. Enter the subject's data into the cart.

Step 3: Prepare subject for testing with the HR monitor, chest strap, and head-gear.

Step 4: Before the test begins, read the subject the RPE instructions presented earlier in this laboratory. Allow the subject to ask questions prior to being connected to the metabolic equipment.

Step 5: Connect the subject to the mouthpiece or mask apparatus while the subject sits on the cycle ergometer.

Step 6: Measure the subject's resting HR and metabolic data.

Step 7: Allow a 3 to 5 min warm-up at an intensity no greater than that of the first stage.

Step 8: Follow one of the cycle ergometer protocols (Åstrand, McArdle, or self-created) to determine the work output at each stage.

Step 9: Collect data during the last 30 s of each stage. Record HR, V̇O₂, RER, V̇ₑ, and RPE for each stage.

Step 10: Increase resistance in accordance with the protocol exactly at the end of every 2 or 3 min stage until the subject can go no longer. Throughout the test, monitor the cadence to be sure the subject is meeting the prescribed work rate.

Step 11: Observe the exercising individual for signs and symptoms that call for terminating the test. Encouragement is effective in motivating a subject to attain maximal effort.

Step 12: Because the subject is unable to communicate through the mask or mouthpiece, several common hand signals can be used during max testing. For example, a thumbs-up indicates that the subject is doing well and is willing to continue. A raised index finger indicates that the subject will need to stop in about 1 min, allowing the testers to prepare for the end of the test and to collect maximum data for all of the outcome measures.

Step 13: Stop the test when V̇O₂max test criteria are reached, when indicators for stopping the test arise, or when the subject is no longer able to sustain the prescribed cadence. Reduce the workload immediately (to the kg · m · min⁻¹ of the first stage) so that the subject can begin the cool-down, during which the subject should keep the mouthpiece in (note RER during the cool-down).

Step 14: The cool-down is important to keep the subject moving to avoid blood pooling in the lower extremities and potential syncope. During the cool-down, which should last at least 5 min, monitor the subject's HR. An appropriate cool-down has the subject attain an HR of ~120 beats · min^{-1} before getting off the ergometer. Continue to monitor the subject's physical appearance and symptoms during the cool-down.

Cycle Ergometer Power Output Formulas

Calculate the power output (or work rate) for each work stage on a stationary bicycle using the following equation:

$$\text{Power output } (kg \cdot m \cdot min^{-1}) = \text{pedal cadence } (rev \cdot min^{-1}) \times \text{flywheel distance } (m \cdot rev^{-1}) \times \text{resistance on the flywheel } (kg)$$

Power on a bike is often expressed in watts (W). To convert $kg \cdot m \cdot min^{-1}$ to W, divide by 6.12 (1 W = 6.12 $kg \cdot m \cdot min^{-1}$). Flywheel distance is the distance the bike would travel for one pedal revolution if it could travel. For a multigear bicycle, this would change with every gear; for a stationary bicycle, it is fixed. To measure flywheel distance, first measure the circumference of the flywheel. Then count the number of revolutions of the flywheel for one complete pedal stroke (a piece of tape on the flywheel can be helpful here). Many commonly used laboratory cycle ergometers such as the Monark have a flywheel distance of 6 $m \cdot rev^{-1}$. For more information on calibrating your cycle ergometer, see appendix E.

Question Set 10.2

1. Prepare a data table showing values for HR, $\dot{V}O_2$, $\dot{V}_E$, RPE, and RER for each stage the subject completed.

2. Plot as separate graphs $\dot{V}O_2$, HR, and RER (y-axis) for your subject as a function of the calculated power output ($kg \cdot m \cdot min^{-1}$ or W).

3. Did your subject attain $\dot{V}O_2$max? Justify your answer.

4. Determine relative ($ml \cdot kg^{-1} \cdot min^{-1}$) and absolute ($L \cdot min^{-1}$) $\dot{V}O_2$max values for your subject. Which is a better index of cardiorespiratory fitness? Why?

5. For the subjects you collected data on, plot $\dot{V}_E$ (y-axis) as a function of HR. Did any of your subjects have a distinct ventilatory threshold? At what absolute HR? At what percent of HRmax? Why does $\dot{V}_E$ typically increase more sharply at heavy workloads?

6. RER at $\dot{V}O_2$max is usually greater than 1.0. If an RER of 1.0 is 100% carbohydrate oxidation, how can RER exceed 1.0?

7. Compare the collected data with the normative values table (see table 7.3). How does your subject rank according to the norms?

8. How does the measured $\dot{V}O_2$max on the cycle compare with the predicted $\dot{V}O_2$ from the prediction equations in appendix B? Why might they differ?

9. How can people increase their $\dot{V}O_2$max? Be specific.

10. Why are males expected to have higher $\dot{V}O_2$max values? What does this suggest is important for attaining a high $\dot{V}O_2$max?

LABORATORY ACTIVITY 10.2

INDIVIDUAL DATA SHEET

Name or ID number: _____ Date: _____

Tester: _____ Time: _____

Sex: M / F (circle one) Age: _____ y Height: _____ in. _____ m

Weight: _____ lb _____ kg Temperature: _____ °F _____ °C

Barometric pressure: _____ mmHg Relative humidity: _____ %

Age-predicted HRmax: _____ beats · min^{-1}

Stage time (min)	Speed (mph)	Grade (%)	Power (kg · m · min^{-1})	HR (beats · min^{-1})	$\dot{V}_E$ (L · min^{-1})	$\dot{V}O_2$ (L · min^{-1})	$\dot{V}O_2$ (ml · kg^{-1} · min^{-1})	METs	$\dot{V}CO_2$ (L · min^{-1})	RER	RPE
0	0	0	Rest								

Blood Lactate Threshold Assessment

Objectives

- Understand the concept of lactate threshold (LT) and its importance in predicting endurance performance.
- Learn the process of measuring LT by lactate analysis and noninvasive respiratory gas analysis.
- Compare methods of assessing LT.

DEFINITIONS

fast (or anaerobic) glycolysis—Metabolic pathway that can provide energy quickly; hydrolysis of six-carbon glucose to two three-carbon lactate molecules; does not require oxygen.

lactate—Anion that remains after lactic acid dissociates a proton.

lactate threshold (LT)—Point during exercise of increasing intensity at which the rate of lactate production exceeds the rate of lactate clearance; the point preceding an increase in lactate of >1 mM with increases in intensity.

lactic acid—Three-carbon molecule ($C_3H_6O_3$) formed from pyruvate; as a carboxylic acid,

it dissociates a proton under most conditions (remaining anion is called *lactate*).

metabolic acidosis—State of lowered pH resulting from acids produced in metabolism; thought to be at least in part from lactic acid production from fast glycolysis during high-intensity exercise.

onset of blood lactate accumulation (OBLA)—Point at which blood lactate levels reach 4.0 mM during an exercise protocol of incremental intensity.

Lactic acid is the product of **fast (or anaerobic) glycolysis**. It is formed from pyruvate catalyzed by the enzyme lactate dehydrogenase (LDH) through the following reaction:

$$\text{pyruvate } (C_3H_4O_3) + \text{NADH} + H^+ \leftrightarrow \text{lactic acid } (C_3H_6O_3) + \text{NAD}^+$$

The end point of glycolysis, pyruvate, has two potential fates—lactic acid, as just illustrated, or acetyl-CoA, which is found in the mitochondria and is formed through a reaction catalyzed by the enzyme pyruvate dehydrogenase (PDH).

The fate of pyruvate is important because it involves a commitment to either anaerobic or aerobic metabolism; as a result, exercise physiologists are interested in measuring lactate production.

Lactic acid and *lactate* are often used synonymously—and incorrectly. **Lactate** is the anion of the lactic acid molecule once it has released its proton. Within the muscle-cell cytoplasm and blood, lactic acid typically takes the form of lactate and H^+ (thus we use the term *lactate* in this lab). Because of this loss of a proton, lactic acid has been thought to contribute to changes in pH in muscle and blood during exercise and therefore to **metabolic acidosis**. However, the magnitude of its contribution to metabolic acidosis in muscle is disputed. (For a discussion, see sources 13-16 in this laboratory's reference list.)

Lactate is constantly produced, even in the condition of rest. For many years, it was thought to be a waste product of glycolysis that caused muscle fatigue and soreness. In fact, lactate can be used as a fuel source like many other carbon-containing compounds. However, as exercise intensity increases, blood lactate concentrations increase due to a number of factors, including both increased production and reduced removal. As exercise intensity increases, the demand for ATP production also increases, which can overwhelm the ability of the mitochondria to meet the need aerobically through oxidative phosphorylation of macronutrients. In addition, the recruitment of fast-twitch fibers (with fewer mitochondria) increases, which facilitates greater lactate production. At these higher intensities, less fat and more carbohydrate are being used for fuel, and this increase in glycolytic flux contributes to greater lactate production. Hormones (epinephrine and norepinephrine) increase glycolytic flux at higher intensities by stimulating glycogen breakdown (through the enzyme glycogen phosphorylase). Many exercise physiology texts also report reduced oxygen as a reason for greater lactate production, but this is not the best explanation when at normal altitudes (1, 8). In reality, increasing lactate production can be attributed to the inability of oxidative phosphorylation to meet the energy demand of exercise. As the rate of ATP demand exceeds the capacity of the mitochondria to produce ATP aerobically, greater anaerobic contribution to ATP synthesis is required.

Removal of blood lactate (and its subsequent oxidation) can be performed by almost any metabolically active tissue (e.g., muscle, heart, liver, kidneys). Blood lactate removal is reduced at high exercise intensities due to the reduction in blood flow away from lactate-clearing tissues (e.g., nonworking muscle, liver, kidneys). There-fore, when lactate production exceeds removal, this typically indicates that exercise intensity is above that where the ATP requirement can be met aerobically. This fact makes blood lactate a useful measure in assessing the ability to produce energy aerobically without significant lactate accumulation in the blood, which can be best demonstrated by performing repeated blood draws during an incremental test to volitional fatigue. The point where lactate production exceeds clearance is referred to as the *lactate deflection point*. Blood lactate concentration (measured in mmol · L⁻¹ or mM) at a given absolute intensity can therefore be used as an indicator of an individual's ability to produce energy aerobically, making it an effective measure of fitness for aerobic athletes.

Blood lactate is not only of interest to aerobic athletes. Fast glycolysis provides ATP at a high rate, which makes it a major contributor of energy during anaerobic efforts. Anaerobic systems are thought to contribute the majority of energy for all-out efforts of <3 min (12). The classic example of a fast glycolytic effort is a 400 m (437 yd) dash. Table 11.1 illustrates the effects of a 400 m run on blood lactate levels. Other examples of anaerobic efforts that rely on fast glycolysis are the Wingate test (WAnT) (15) and the Kansas squat test (9). As you will see during lab activity 11.3 as well as in table 11.1, blood lactate following these anaerobic efforts may exceed 15 mmol · L⁻¹. Blood is rarely sampled during anaerobic tests due to the effort involved. In fact, blood lactate is not thought to peak in the blood until 2 to 10 min postexercise. This is the time needed for lactate produced in muscle to spill into the bloodstream. Table 11.1 helps in this understanding by demonstrating elevated muscle lactate compared with blood lactate following a 400 m run.

DETERMINING THE DEFLECTION POINT

Over the years, a number of methods have been developed to define and describe the lactate deflection point from an incremental

Table 11.1 Blood and Muscle pH and Lactate Concentration 5 min After a 400 m Run

		MUSCLE		BLOOD	
Runner	Time (s)	pH	Lactate (mmol · kg⁻¹)	pH	Lactate (mmol · L⁻¹)
1	61.0	6.68	19.7	7.12	12.6
2	57.1	6.59	20.5	7.14	13.4
3	65.0	6.59	20.2	7.02	13.1
4	58.5	6.68	18.2	7.10	10.1
Average	60.4	6.64	19.7	7.10	12.3

Reprinted, by permission, from W.L. Kenney, J.H. Wilmore, and D.L. Costill, 2015, *Physiology of sport and exercise,* 6th ed. (Champaign, IL: Human Kinetics), 217.

test that best corresponds to endurance performance (1, 8, 10). This point can be crucial because it indicates when one switches from mostly aerobic metabolism—and therefore an effort that should be sustainable for some time—to mostly anaerobic metabolism, which may hasten fatigue. At submaximal exercise intensities, blood lactate values are similar to resting values (0.9-2.0 mmol · L⁻¹). As intensity continues to increase, blood lactate concentrations demonstrate an accelerated increase

above resting values (see figure 11.1). The point at which blood lactate concentrations increase nonlinearly is known as the **lactate threshold (LT)**.

Endurance performance depends on the ability to perform for extended periods of time at the highest possible intensity without experiencing the effects of fatigue and lactate accumulation (2, 5, 10, 13). These factors are dependent on the LT because they occur at a certain percentage of $\dot{V}O_2$max and determine the point at which lactate concentrations begin to rise. The LT occurs at different percentages of $\dot{V}O_2$max for different groups of people:

| Untrained, sedentary individuals | ~50%-60% of $\dot{V}O_2$max |
| Well-trained distance runners | >75% of $\dot{V}O_2$max |

Studies show that endurance race performance correlates more closely with LT than with $\dot{V}O_2$max (4, 8, 10, 13, 15), which may explain why athletes with the same $\dot{V}O_2$max often perform differently during endurance events. It also helps explain continued performance improvement without increases in $\dot{V}O_2$max. Determining the LT can aid in the design of training programs to improve race times for endurance athletes. If athletes know the percentage of $\dot{V}O_2$max at which they can train before experiencing lactate accumulation, they can improve the threshold through interval training. Training at or near the LT shifts the curve to the right; in order to do so, of course, it is necessary to know when LT occurs and then

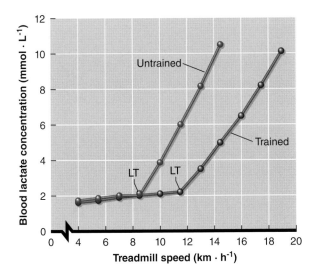

Figure 11.1 Blood lactate concentrations increase with exercise of increasing intensity.

Adapted, by permission, from W.L. Kenney, J.H. Wilmore, and D.L. Costill, 2015, *Physiology of sport and exercise,* 6th ed. (Champaign, IL: Human Kinetics), 277.

to train at that percentage of $\dot{V}O_2$max and HR. More information on training associated with LT can be found in *Lactate Threshold Training* by Peter Janssen (11).

The most clearly defined methods for determining the lactate deflection point are the **onset of blood lactate accumulation (OBLA)** and the anaerobic threshold or LT. OBLA is the workload associated with 4 mmol $\cdot$ L^{-1} during an incremental exercise test (17). The LT is the workload where a nonlinear increase of >1 mmol $\cdot$ L^{-1} is observed in successive workloads (6, 8). It is thought that OBLA typically occurs at a higher intensity and percentage of $\dot{V}O_2$max than LT (see figure 11.1 and figure 11.2). Therefore, LT is considered to better predict performances lasting 60 to 75 min, whereas OBLA better predicts those lasting 20 to 40 min (7).

SELECTING A TEST METHOD

Several methods are available for measuring blood lactate. A small sample of blood needs to be drawn, but this rarely involves more than a drop from a finger stick. Lactate and glucose can be measured in a matter of minutes by automated machines such as the YSI 2300 STAT (Yellow Springs Incorporated, Yellow Springs, OH). Because of the application to the training of endurance athletes, portable analyzers are also commercially available (for examples, see www.lactate.com). These units work like glucose analyzers, where a drop of blood is applied to a strip that engages the machine, which produces results in a matter of seconds. Lactate can also be measured by other automated machines, such as the i-STAT (see www. abbottpointofcare.com/products-services/

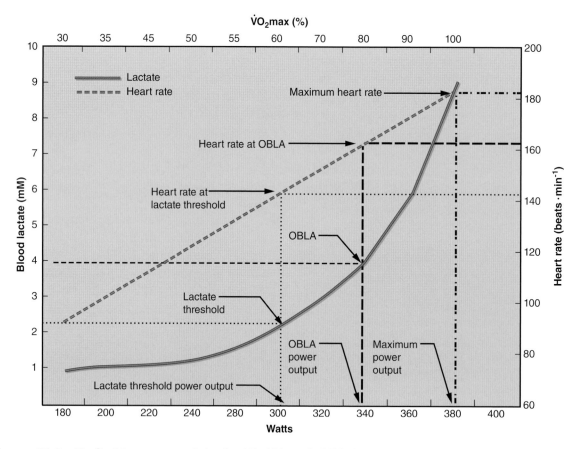

Figure 11.2 Definitions associated with LT and OBLA.

Adapted, by permission, from T. Bompa and G.G. Haff, 2009, *Periodization: Theory and methodology of training*, 5th ed. (Champaign, IL: Human Kinetics), 295.

istat-handheld). Figure 11.3 is an example of a portable lactate analyzer. Finally, lactate can be analyzed using a spectrophotometer, which is considered the most accurate method but requires more blood and takes longer to complete the analysis. More than likely, your lab has a portable analyzer that your lab instructor will show you how to use.

ROLE OF THE VENTILATORY THRESHOLD

As mentioned in laboratory 10, $\dot{V}O_2$ increases as a linear function of power output during a GXT up to $\dot{V}O_2$max. On the other hand, ventilation ($\dot{V}_E$) increases linearly as a function of power output only during exercise requiring up to about 50% to 75% $\dot{V}O_2$max, above which $\dot{V}_E$ increases at a steeper rate (see figure 10.2).

The greater rate of increase in $\dot{V}_E$ above this ventilatory threshold coincides with a similar rise in $\dot{V}CO_2$ and a sharp increase in blood lactate. Remember that lactic acid dissociates into a lactate ion and H^+ in the blood, and the H^+ combines with a bicarbonate ion to form carbonic acid, which in turn dissociates into H_2O and CO_2.

$$La^-H^+ + HCO_3^- \leftrightarrow H_2CO_3 \leftrightarrow H_2O + CO_2$$

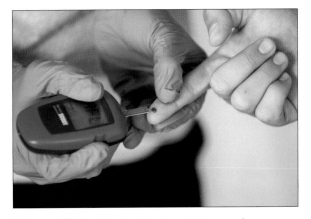

Figure 11.3 Portable lactate analyzer.

The accumulation of CO_2 in the blood leads to an increase in CO_2 in the expired breath ($\dot{V}CO_2$). The increase in blood CO_2 stimulates an increase in ventilation. Thus, the ventilatory threshold indicates an abrupt increase in blood lactate and a subsequent decrease in blood pH. This nonmetabolic or excess CO_2 from the buffering of lactic acid to combat the reduced blood pH contributes to the $\dot{V}CO_2$ from the catabolism of macronutrients. Remembering that RER = $\dot{V}CO_2 / \dot{V}O_2$ explains why RER exceeds 1.0 at high intensities. See figure 11.4 for the relationships between lactate, ventilation, oxygen consumption and the anaerobic and aerobic contributions during exercise of increasing intensity.

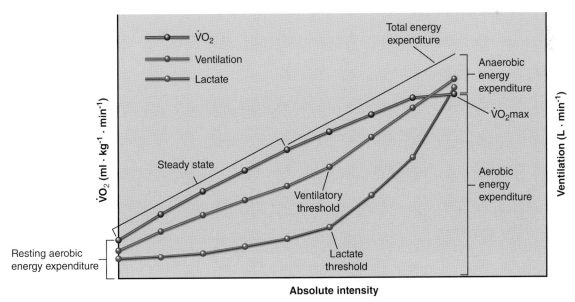

Figure 11.4 Active versus passive recovery on blood lactate clearance postexercise.

© Charles Dumke

It is important to understand that the bicarbonate is not clearing the lactate ion; it is simply buffering the proton. As mentioned, to be cleared, the lactate must be taken up by tissues such as muscle, liver, or kidneys. This can occur either during or following the bout of exercise, depending on the situation.

As you will learn in this lab, measurement of blood lactate complicates an incremental test of increasing intensity to exhaustion. The necessity of taking even a small drop of blood during every stage increases the number of people needed to carry out the test. More supplies are also needed (e.g., lactate analyzer, gloves, gauze, alcohol swabs, bandages). For this reason, the ventilatory threshold is often used to approximate the LT. In this lab, you will also experience how the ventilatory threshold relates to the LT.

Go to the web study guide to access electronic versions of individual and group data sheets, the question sets from each laboratory activity, and case studies for each laboratory activity.

Blood Lactate Measurement at Rest

Equipment

- Lactate analyzer with strips
- Lancets
- Exam gloves
- Alcohol swabs
- Sterile gauze, bandages
- Group data sheet

Laboratory Orientation

Collecting a drop of blood is a simple, relatively noninvasive measure. However, because blood is a bodily fluid, you must take care in handling it. You should wear gloves when handling blood and be careful to dispose of bloodstained materials into bins designated for biological waste. It is a common mistake for students to spread blood contamination onto other equipment or elsewhere in lab because of residual blood on their gloves. Be conscious of potential contamination. Some institutions may require you to pass a bloodborne pathogens safety quiz prior to participation in this lab. Your laboratory instructor will know more about this safety requirement at your institution.

As mentioned at the beginning of this lab, lactate can be measured in blood samples via several methods. Because of the availability and affordability of portable lactate analyzers, the following steps assume that you will be using one. Note that calibration of portable analyzers may require that a calibration strip be used for each batch of lactate test strips. Follow the manufacturer's guidelines for calibration of your lactate monitor. It is possible to collect blood from earlobes or fingertips. However, fingertips are both more common and easier for students, so we will assume you are using fingertip blood draws in this lab. Your laboratory instructor will clarify the method to use for your particular lactate analyzer.

Lactate Analyzer Test

Step 1:   Your laboratory instructor will demonstrate the procedure for taking blood samples for the lactate analyzer in your lab. Assuming that your lab uses a portable analyzer, calibration is relatively easy. Each batch of lactate strips comes with a code that needs to be confirmed in the analyzer, either by inserting a code strip or by manual entry. The machine should then recognize a lactate strip once it is inserted and be ready for the application of a drop of blood.

Step 2:   Have the subject sit for about 3 min before taking a resting blood sample.

Step 3:   While wearing exam gloves, clean the subject's fingertip with an alcohol swab. Wait for the alcohol to evaporate so that the blood is not mixed with the alcohol during sampling.

Step 4:   With a lancet, prick the swabbed finger and if necessary pulse-squeeze (don't just squeeze) it to obtain a drop of blood.

Step 5: Place the lactate analyzer strip in contact with the drop of blood to begin measurement. Capillary action should draw blood into a reservoir in the strip. Be careful not to introduce bubbles of air into the strip reservoir.

Step 6: Wipe any remaining blood off the finger with sterile gauze, and apply a bandage if necessary.

Step 7: Record the data on the group data sheet.

Question Set 11.1

1. Prepare a data table showing resting lactate values for your class.
2. Did resting lactate differ by sex? Why might this be the case?
3. What do these values say about lactate production at rest?
4. What could explain a resting lactate value of >2.0 mM?

LT During an Incremental Cycle Test

Equipment

- Platform scale (or digital or other scale)
- Stadiometer (wall or freestanding) or physician's scale with attached anthropometer
- Cycle ergometer
- Metabolic cart, including gas flow meter and O_2 and CO_2 analyzers
- HR monitor
- Lactate analyzer with strips
- Exam gloves
- Lancets
- Alcohol swabs
- Sterile gauze and bandages
- Individual data sheet

This laboratory activity is supported by a virtual lab experience in the web study guide.

www

Laboratory Orientation

Collecting a drop of blood is a simple, relatively noninvasive measure. However, because blood is a bodily fluid, you must take care in handling it. You should wear gloves when handling blood and be careful to dispose of bloodstained materials into bins designated for biological waste. It is a common mistake for students to spread blood contamination onto other equipment or elsewhere in lab because of residual blood on their gloves. Be conscious of potential contamination. Some institutions may require you to pass a bloodborne pathogens safety quiz prior to participation in this lab. Your laboratory instructor will know more about this safety requirement at your institution.

As mentioned at the beginning of this lab, lactate can be measured in blood samples via several methods. Because of the availability and affordability of portable lactate analyzers, the following steps assume that you will be using one. Note that calibration of portable analyzers may require that a calibration strip be used for each new batch of lactate test strips. Follow the manufacturer's guidelines for calibration of your lactate monitor. It is possible to collect blood from earlobes or fingertips. However, fingertips are both more common and easier for students, so we will assume you are using fingertip blood draws in this lab. Your laboratory instructor will clarify the method to use for your particular lactate analyzer.

Incremental Cycle Test for LT

One student will serve as the subject to determine LT during an incremental test. Other students will facilitate the testing by performing such tasks as blood sampling, controlling workload on the cycle ergometer, taking RPE and HR measurements, and controlling the metabolic cart.

Step 1: Begin this lab activity by gathering basic data (e.g., age, height, weight) for the individual data sheet as described in laboratory activity 1.1.

Step 2: Fit the subject on the cycle ergometer comfortably with a slight knee angle (5°-15°) when the leg is in the downward position.

Step 3: The lab instructor will explain procedures for using the metabolic cart, but students will operate the cart and conduct the test, including calibration and data recording. Enter the subject's data into the metabolic cart.

Step 4: Prepare subject for testing by helping with the HR monitor chest strap and headgear.

Step 5: Prior to the test, read the subject the RPE instructions (see lab 10). Allow the subject to ask questions before being connected to the metabolic equipment.

Step 6: Connect the subject to the mouthpiece or mask apparatus while the subject sits on the bike.

Step 7: Allow the subject to sit for about 3 min to attain resting steady state.

Step 8: Your laboratory instructor will demonstrate the procedure for taking blood samples for the lactate analyzer. Briefly, while wearing exam gloves, clean the subject's fingertip with an alcohol swab. Wait for the alcohol to evaporate so that the blood is not mixed with the alcohol during sampling. With a lancet, prick the swabbed finger and if necessary pulse-squeeze (don't just squeeze) it to obtain a drop of blood. Place the lactate analyzer strip in contact with the drop of blood to begin measurement. Wipe any remaining blood off the finger with sterile gauze, and apply a bandage if necessary.

Step 9: Measure and record the subject's resting HR, RPE, and data from the metabolic cart.

Step 10: Allow a 3 to 5 min warm-up at an intensity no greater than that of the first stage.

Step 11: Select a cycle protocol appropriate for your subject (see laboratory 10), such as the Åstrand or McArdle. However, these protocols have 2 min stages, which can be challenging for blood draws. Thus, we suggest that noncyclists pedal at a cadence of 75 rev · min⁻¹ at an initial resistance of 0.5 kg (1.1 lb) and then increase it by 0.5 kg every 3 min. If a fit cyclist is the subject, start at 1 or 1.5 kg (2.2 or 3.3 lb).

Step 12: At each stage, collect the following data during the last 30 s of the stage: power, HR, $\dot{V}O_2$, $\dot{V}CO_2$, RER, $\dot{V}_E$, and RPE. Record the results on the individual data sheet.

Step 13: During the last minute of each stage, collect a lactate sample. Be prepared with the alcohol, gauze, strips, and analyzer to obtain the blood sample as quickly as possible. Doing so may take several students. It is ideal not to interrupt the stage timing, but if necessary for the purposes of this lab you may prolong the stage until a good blood sample is obtained for lactate analysis. Do not attempt to get a drop of blood from a previous lancet site—it can delay the blood draw and possibly result in coagulated blood being placed on the strip. Use a fresh lancet and new blood-draw site for each stage.

Step 14: Observe the exercising individual for signs and symptoms indicating that the test should be terminated. Encouragement is effective in motivating a subject to attain maximal effort.

Step 15: Because the subject is unable to communicate through the mask or mouthpiece, several common hand signals can be used during maximal testing. For example, a thumbs-up indicates that the subject is doing well and is willing to continue. A raised index finger indicates that the subject will need to stop in about 1 min, allowing the testers to prepare for the end of the test and to collect maximum data for all of the outcome measures.

Step 16: Stop the test when $\dot{V}O_2$max test criteria are reached, or when indicators for stopping the test arise (see laboratory 7 for criteria for terminating a test), or when the subject stops the test by volitional fatigue. At this point, reduce the workload on the bike to that of the initial stage of the protocol. Have the subject keep the mouthpiece in for a 5 min cool-down (record RER during the cool-down).

Step 17: The cool-down is important to keep the subject moving to avoid blood pooling in the lower extremities and potential syncope. During the cool-down, which should last at least 5 min, monitor the subject's HR. An appropriate cool-down would have the subject attain an HR of 120 beats $\cdot$ min^{-1} before getting off the ergometer.

Step 18: Continue to monitor the subject's physical appearance and symptoms during the cool-down.

Step 19: Record the data from each stage, and complete the individual data sheet for the subject.

Cycle Ergometer Work and Power Formulas

The Monark cycle ergometer allows you to accurately set the work rate by adjusting the tension on a belt around the flywheel while having the subject pedal at a constant number of revolutions per minute (rev $\cdot$ min^{-1}). The circumference of the flywheel is such that for each pedal revolution, a given spot on the flywheel would travel 6 m (20 ft). The flywheel velocity is, therefore, equal to the pedal rev $\cdot$ min^{-1} × 6 m $\cdot$ rev^{-1}. As with the treadmill, no calculated work is done on a cycle ergometer when pedaling at no resistance. Work (force × distance) and power (work / time or force × velocity) can be calculated on a Monark cycle ergometer using the following formulas:

$$\text{Work} = \text{force} \times \text{distance}$$

$$\text{Work on bike} = \text{resistance (kg)} \times (\text{pedal rev} \cdot \text{min}^{-1} \times 6 \text{ m} \cdot \text{rev}^{-1} \times \text{total min of exercise})$$

$$\text{Power (work rate)} = \text{force} \times \text{velocity}$$

$$\text{Power on bike} = \text{resistance (kg)} \times (\text{pedal rev} \cdot \text{min}^{-1} \times 6 \text{ m} \cdot \text{rev}^{-1})$$

Note that the result of the power equation will be expressed in kg $\cdot$ m $\cdot$ min^{-1}, which means you can divide by 6.12 to express power in watts (W).

Question Set 11.2

1. Prepare a graph illustrating the relationship between blood lactate concentration (mM) and work rate (kg · m · min^{-1} or W).

2. Prepare a graph illustrating the relationship between HR (beats · min^{-1}) and work rate (kg · m · min^{-1} or W).

3. Prepare a graph illustrating the relationship between $\dot{V}_E$ (L · min^{-1}) and work rate (kg · m · min^{-1} or W).

4. Recall the laboratory 10 discussion. Did your subject meet the criteria for achieving a true $\dot{V}O_2$max?

5. At what workload, HR, and %HRmax does the subject reach LT? OBLA? Ventilatory threshold? (Hint: A table would work best.)

6. Does the LT occur at the same workload as the ventilatory threshold? Why or why not?

7. What might explain a subject achieving different $\dot{V}O_2$max values on the cycle ergometer and the treadmill?

8. What causes the abrupt increase in blood lactate accumulation, indicated by LT, that occurs during dynamic exercise in an incremental test?

9. Explain the importance of LT for performance prediction.

LABORATORY ACTIVITY 11.2

INDIVIDUAL DATA SHEET

Name or ID number: _____ Date: _____

Tester: _____ Time: _____

Sex: M / F (circle one) Age: _____ y Height: _____ in. _____ m

Weight: _____ lb _____ kg Temperature: _____ °F _____ °C

Barometric pressure: _____ mmHg Relative humidity: _____ %

Mode and protocol: _____ Metabolic cart: _____

	Lactate											
RPE												
RER												
$\dot{V}CO_2$ (ml · kg^{-1} · min^{-1})												
METs												
$\dot{V}O_2$ (ml · kg^{-1} · min^{-1})												
$\dot{V}O_2$ (ml · min^{-1})												
$\dot{V}_E$ (L · min^{-1})												
HR (beats · min^{-1})												
Power (kg · m · min^{-1})	Rest											
Resist-ance (kg)	Rest											
Stage time (min)	0											

Blood Lactate After Anaerobic Exercise

Equipment

- Platform scale (or digital or other scale)
- Stadiometer (wall or freestanding) or physician's scale with attached anthropometer
- Monark cycle or equivalent ergometer equipped to perform the WAnT (e.g., Monark 814E, 824E, or 834E, all of which contain a peg or basket upon which the external load can be applied, as shown in figure 13.9).
- Revolutions counter that counts the number of pedal revolutions performed during the test
- Timing apparatus (e.g., countdown set-timer, stopwatch, timer attached to the cycle ergometer)
- HR monitor
- Lactate analyzer with strips
- Exam gloves
- Lancets
- Alcohol swabs
- Sterile gauze and bandages
- Individual data sheet

Laboratory Orientation

Lactate is typically cleared within minutes following exercise. The removal of lactate from the blood following exercise depends on blood flow to tissues such as muscle, liver, heart, and kidneys, which is typically facilitated by an exercise cooldown (figure 11.5). This process is particularly important following a bout of exercise that is more reliant on fast glycolysis—the more anaerobic the exercise, the greater the lactate production, and thus the more time needed to clear lactate from the blood. In this lab activity, you will use the Wingate test to demonstrate the contribution to energy from fast glycolysis and the ability to clear lactate postexercise.

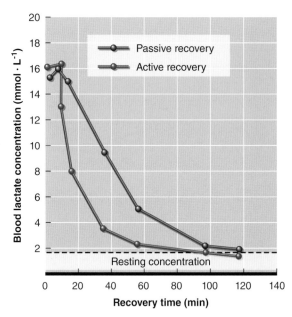

Figure 11.5 Active versus passive recovery on blood lactate clearance postexercise.

Reprinted, by permission, from W.L. Kenney, J.H. Wilmore, and D.L. Costill, 2015, *Physiology sport and exercise*, 6th ed. (Champaign, IL: Human Kinetics), 218.

Blood Lactate Levels Following a Wingate Test

Two similarly sized students will perform the WAnT. Following the test, one subject (referred to as *SS* for "sitting subject") will sit quietly without a cool-down, while the other student (referred to as *CDS* for "cool-down subject") will continue pedaling at a low workload.

Step 1: Set up the cycle ergometer, and check to see if it is in working order.

Step 2: Gather basic data (e.g., age, height, weight) for the individual data sheet as described in laboratory activity 1.1.

Step 3: Calculate the prescribed load for each subject based on the subject's training status (see table 13.3). Record the load on the individual data sheet.

Step 4: Fit the cycle ergometer to the subjects so that while seated on the bike the downward leg has a slight bend of 5° to 15° (see figure 4.3*a*).

Step 5: Completely explain the test protocol to the subjects, and emphasize that this is an all-out effort that lasts 30 s.

Step 6: Have the subjects perform the standardized warm-up and recovery periods outlined in table 13.4.

Step 7: Instruct the subjects to increase their pedaling rate to near maximum. Instruct the force setter to apply the load, tell the timer to start the timer, and yell *"Go"* when the subjects reach a maximal pedaling rate.

Step 8: The counter should begin counting pedal revolutions upon hearing the "Go" command. (If you are using a computerized system, this step need not be performed.)

Step 9: The counter should tell the recorder the number of revolutions completed at the end of each 5 s time interval during the test (i.e., 5, 10, 15, 20, 25, and stop).

Step 10: At the 30 s mark, the timer yells *"Stop"* and the force setter reduces the load to 0. At this point, the SS subject should sit in a chair next to the cycle ergometer. The CDS subject should pedal at a light load for 30 min while lactate is being collected.

Step 11: Following the test, collect blood samples from both the CDS and SS (see details in this laboratory's preceding activities for procedures) every 2 min for the first 10 min, then once every 5 min for the next 30 min. Record blood lactate results in the data table for both subjects.

Step 12: To convert the prescribed load to newtons, multiply the load by 9.80665. Calculate the power output from the total number of revolutions, the distance the flywheel travels per rotation (6 m), and the duration of the interval. Calculate the power output for each time interval, and record it in the appropriate location on the individual data sheet for both subjects.

Step 13: Divide the power output calculated for each time interval by body weight to determine a relative power output ($W \cdot kg^{-1}$). Record these values on the individual data sheet for both subjects.

Step 14: Calculate the absolute and relative work accomplished during the 30 s time interval, and record these values on the individual data sheet for both subjects.

Step 15: Calculate the absolute and relative mean power outputs accomplished during the 30 s test, and record these data on the individual data sheet for both subjects.

Step 16: Complete calculations for absolute power, relative power, work, mean power, and fatigue index for the SS and CDS subjects.

Question Set 11.3

1. What are the physiological reasons for the large amount of lactate produced after the WAnT?

2. Prepare a graph illustrating the relationship between blood lactate concentration and time postexercise for both subjects.

3. At what time point did peak lactate occur following the WAnT? What is the reason for this delayed response?

4. What do your data indicate about the importance of an active cool-down? For which sports is this particularly important?

LABORATORY 11.3

INDIVIDUAL DATA SHEET

Date: _____ Time: _____

Tester: _____

Temperature: _____ °F _____ °C Barometric pressure: _____ mmHg

Relative humidity: _____ %

Wingate Test Subject Data: Cool-Down Subject (CDS)

Name or ID number: _____

Sex: M / F (circle one)   Age:_____ y

Height: _____ in. _____ m Weight: _____ lb _____ kg

Prescribed load = _____ × _____ = _____ kp
 body mass (kg) kp per kg body mass*

Note: As a default, use 0.075 kp · kg^{-1} body mass or see table 13.3.

Force (N) = _____ × 9.80665 = _____ N
 force setting in kp

Wingate Test Subject Data: Sitting Subject (SS)

Name or ID number: _____

Sex: M / F (circle one)   Age:_____ y

Height: _____ in. _____ m Weight: _____ lb _____ kg

Prescribed load = _____ × _____ = _____ kp
 body mass (kg) kp per kg body mass*

Note: As a default, use 0.075 kp · kg^{-1} body mass or see table 13.3.

Force (N) = _____ × 9.80665 = _____ N
 force setting in kp

Pedal Revolutions

Time interval (s)	# of revolutions (CDS)	# of revolutions (SS)
0-5		
5-10		
10-15		
15-20		
20-25		
25-30		

Absolute Power Data for CDS

Time interval (s)	Power (W)
0-5	Power (W) = (_____ kp × _____ # revolutions / 5 s × 12 × 6 m) / 6.12 = _____ W
5-10	Power (W) = (_____ kp × _____ # revolutions / 5 s × 12 × 6 m) / 6.12 = _____ W
10-15	Power (W) = (_____ kp × _____ # revolutions / 5 s × 12 × 6 m) / 6.12 = _____ W
15-20	Power (W) = (_____ kp × _____ # revolutions / 5 s × 12 × 6 m) / 6.12 = _____ W
20-25	Power (W) = (_____ kp × _____ # revolutions / 5 s × 12 × 6 m) / 6.12 = _____ W
25-30	Power (W) = (_____ kp × _____ # revolutions / 5 s × 12 × 6 m) / 6.12 = _____ W

Absolute Power Data for SS

Time interval (s)	Power (W)
0-5	Power (W) = (_____ kp × _____ # revolutions / 5 s × 12 × 6 m) / 6.12 = _____ W
5-10	Power (W) = (_____ kp × _____ # revolutions / 5 s × 12 × 6 m) / 6.12 = _____ W
10-15	Power (W) = (_____ kp × _____ # revolutions / 5 s × 12 × 6 m) / 6.12 = _____ W
15-20	Power (W) = (_____ kp × _____ # revolutions / 5 s × 12 × 6 m) / 6.12 = _____ W
20-25	Power (W) = (_____ kp × _____ # revolutions / 5 s × 12 × 6 m) / 6.12 = _____ W
25-30	Power (W) = (_____ kp × _____ # revolutions / 5 s × 12 × 6 m) / 6.12 = _____ W

Relative Power for CDS

Time interval (s)	Relative power (W · kg^{-1})
0-5	Relative power (W · kg^{-1}) = _____ power (W) / _____ body mass (kg) = _____ W · kg^{-1}
5-10	Relative power (W · kg^{-1}) = _____ power (W) / _____ body mass (kg) = _____ W · kg^{-1}
10-15	Relative power (W · kg^{-1}) = _____ power (W) / _____ body mass (kg) = _____ W · kg^{-1}
15-20	Relative power (W · kg^{-1}) = _____ power (W) / _____ body mass (kg) = _____ W · kg^{-1}
20-25	Relative power (W · kg^{-1}) = _____ power (W) / _____ body mass (kg) = _____ W · kg^{-1}
25-30	Relative power (W · kg^{-1}) = _____ power (W) / _____ body mass (kg) = _____ W · kg^{-1}

Relative Power for SS

Time interval (s)	Relative power (W · kg⁻¹)
0-5	Relative power (W · kg⁻¹) = _____ power (W) / _____ body mass (kg) = _____ W · kg⁻¹
5-10	Relative power (W · kg⁻¹) = _____ power (W) / _____ body mass (kg) = _____ W · kg⁻¹
10-15	Relative power (W · kg⁻¹) = _____ power (W) / _____ body mass (kg) = _____ W · kg⁻¹
15-20	Relative power (W · kg⁻¹) = _____ power (W) / _____ body mass (kg) = _____ W · kg⁻¹
20-25	Relative power (W · kg⁻¹) = _____ power (W) / _____ body mass (kg) = _____ W · kg⁻¹
25-30	Relative power (W · kg⁻¹) = _____ power (W) / _____ body mass (kg) = _____ W · kg⁻¹

Work Data for CDS

Time interval (s)	Work
0-30	Total work = _____ force (N) × _____ total revolutions × 6 m = _____ J
0-30	Relative work = _____ total work (J) / _____ body mass (kg) = _____ J · kg⁻¹

Work Data for SS

Time interval (s)	Work
0-30	Total work = _____ force (N) × _____ total revolutions × 6 m = _____ J
0-30	Relative work = _____ total work (J) / _____ body mass (kg) = _____ J · kg⁻¹

Mean Power Output for CDS

Time interval (s)	Mean power output
0-30	Mean power output = _____ total work (J) / 30 s = _____ J · sec⁻¹
0-30	Relative mean power output = _____ mean power (W) / _____ body mass (kg) = _____ W · kg⁻¹

Mean Power Output for SS

Time interval (s)	Mean power output
0-30	Mean power output = _____ total work (J) / 30 s = _____ J · sec⁻¹
0-30	Relative mean power output = _____ mean power (W) / _____ body mass (kg) = _____ W · kg⁻¹

Calculating the Fatigue Index

Calculate the fatigue index for both the CDS and SS subject.

$$\text{Fatigue index (CDS)} = [(\underline{\hspace{2cm}}_{\text{highest power (W)}} - \underline{\hspace{2cm}}_{\text{lowest power (W)}}) / \underline{\hspace{2cm}}_{\text{highest power (W)}}] \times 100 = \underline{\hspace{2cm}} \%$$

$$\text{Fatigue index (SS)} = [(\underline{\hspace{2cm}}_{\text{highest power (W)}} - \underline{\hspace{2cm}}_{\text{lowest power (W)}}) / \underline{\hspace{2cm}}_{\text{highest power (W)}}] \times 100 = \underline{\hspace{2cm}} \%$$

Lactate Results Following Wingate Test

	CDS	SS
Peak power (W)		
Mean power (W)		
Lactate 2 min post (mM)		
Lactate 4 min post (mM)		
Lactate 6 min post (mM)		
Lactate 8 min post (mM)		
Lactate 10 min post (mM)		
Lactate 15 min post (mM)		
Lactate 20 min post (mM)		
Lactate 25 min post (mM)		
Lactate 30 min post (mM)		

Musculoskeletal Fitness Measurements

Objectives

- Become familiar with various methods of evaluating muscular strength and endurance.
- Differentiate between direct and indirect methods of determining muscular strength.
- Understand the proper methods for performing a 1RM bench press and leg press test.
- Introduce various prediction equations that are useful when attempting to estimate muscular strength.

DEFINITIONS

1-repetition maximum (1RM)—Maximum amount of weight or load that can be lifted one time.

alternated grip—Grip in which the dominant hand is supinated and the other hand is pronated.

muscular endurance—Ability to exert submaximal forces repetitively (2, 4).

muscular fitness—Combination of muscular endurance and maximal strength.

muscular strength—Highest amount of force that can be generated by a muscle or group of muscles during a single contraction (12, 23).

prediction equation—Equation established to predict one variable from one or a series of other variables. (In this laboratory, equations are used that predict 1RM strength from the load lifted and the number of repetitions performed.)

pronated grip—Overhand grip.

repetition maximum (RM)—Maximum amount of weight that can be lifted for a prescribed repetition range (e.g., 1RM is the heaviest weight that can be lifted one time, whereas 10RM is the heaviest weight that can be lifted 10 times).

supinated grip—Underhand grip.

Musculoskeletal or **muscular fitness** is associated with numerous health benefits, including a reduced risk of heart disease, osteoporosis, glucose intolerance, and musculoskeletal injuries (2, 25). A healthy musculoskeletal system is also associated with an improved ability to complete activities of daily living and is directly related to quality of life (25, 50). Ultimately, high levels of musculoskeletal fitness are associated with positive health status, whereas lower levels of musculoskeletal fitness are associated with lower health status (50). Research strongly suggests that muscular fitness is related to a reduced all-cause mortality risk (8). Specifically, it appears that reductions in muscular fitness as indicated by reductions in muscle mass and strength are important risk factors as one ages (20).

Because musculoskeletal fitness is such an important contributor to health and wellness, it should be evaluated as part of a comprehensive health and wellness screening process (23, 44). The ACSM defines *muscular* or *musculoskeletal fitness* as the combination of muscular strength and muscular endurance (2). **Muscular strength** refers to the largest amount of force that a muscle or group of muscles can generate during a single contraction (12, 23), whereas **muscular endurance** refers to the ability to exert submaximal forces repetitively (2, 4). Muscular strength and endurance can be evaluated with several methods, and individual tests are generally selected based on the muscle group being tested, the equipment available, and the subject's capabilities.

ASSESSMENTS OF MUSCULAR STRENGTH

No single assessment evaluates total-body muscular strength or endurance; rather, musculoskeletal fitness testing is specific to the muscle groups tested, the velocity of movement employed, the type of contraction and ROM used, and the type of equipment used to perform the assessment (2). These issues sometimes make it difficult to compare individual results with those reported in the scientific literature.

When conducting assessments of musculoskeletal or muscular fitness, you should famil-

iarize clients or subjects with the equipment and protocol to help ensure that the testing process is as reliable as possible (2). You should also perform all protocols under standardized conditions. The ACSM recommends six steps—including using proper lifting technique and familiarizing individuals with the equipment—to increase the accuracy and reliability of any musculoskeletal fitness assessment (table 12.1).

Muscular strength can be assessed with either dynamic or static methods. Dynamic assessments involve moving an external load or body part, whereas static methods exhibit no overt muscular or limb movement (2). Muscular strength assessment can also be divided into two categories: (1) the **1-repetition maximum (1RM)** test and (2) static or isometric tests.

1RM Testing

Traditionally, the 1RM test has been considered the gold standard when evaluating dynamic muscular strength. It requires the client to exert maximal force dynamically through a ROM in a controlled manner while maintaining proper technique (2, 5). When performing this assessment with a client for the first time, it is often difficult to obtain an accurate and reliable 1RM because of the number of attempts that may be

Table 12.1 ACSM Recommendations for Standardized Conditions for Muscular Strength and Endurance Tests

Guideline	Rationale
Maintenance of proper technique	Maintaining proper technique ensures the exercise is performed correctly and safely.
Consistent repetition duration or movement speed	Maintaining consistent repetition speed allows for easier interpretation of results.
Full range of motion	Using a full range of motion allows for easier interpretation of results and maximizes safety.
Use of spotters	In exercises such as the bench press or back squat, spotters must be used in order to maximize safety.
Familiarization with equipment	All subjects should be familiar with the equipment and protocols in order to maximize the reliability of the assessment.
Proper warm-up	As with all testing, employ a proper warm-up in order to maximize performance capacity while minimizing potential risks that may be inherent in the testing process.

Adapted from ACSM 2010.

used or because the client may have unstable technique (5, 31). However, if the client is familiarized with the exercise and protocol used in the assessment, the 1RM test can be highly reliable. Overall, 1RM tests exhibit high test–retest reliabilities as indicated by ICCs ranging between .79 and .99 (28).

The 1RM procedure can be applied to both free-weight and machine-based exercises through the following methods (2, 28, 45):

1. Have the subject estimate his or her 1RM. This is easier to do with trained individuals because you can form the estimation using the loads they train with. With untrained or novice subjects, it is more difficult to determine the perceived maximum, and the process is rather exploratory. In most cases, body weight is used as the perceived 1RM to obtain these subjects' testing loads.

2. The subject warms up with 5 to 10 repetitions at 40% to 60% of perceived maximum.

3. After 1 min of rest, the subject performs 3 to 5 repetitions with a resistance between 60% and 80% of perceived maximum.

4. After 3 min rest, the subject performs 1 repetition equal to ~90% of perceived maximum.

5. If done correctly, step 4 should take the subject close to the 1RM. At this point, make conservative increases in resistance and have the subject perform 1 repetition. If the lift is completed, the subject should rest for 3 min and then make another attempt with an increased load. Continue this process (resting 3 min, then increasing the load) until the subject is unable to complete the repetition with good technique. To maximize the reliability of the process, the 1RM should be achieved within 3 to 5 sets.

6. The 1RM value is reported as the heaviest weight that the client successfully completed.

Generally, 1RM testing is considered safe for most populations (28); investigations looking at 1RM testing with children, older adults, and athletes have found few or no injuries when testing is conducted according to standard procedures

and is properly supervised (9, 12, 19, 22, 28, 42, 43). Collectively, the literature highlights the fact that 1RM testing is safe for clinical clients and athletes, but it also stresses the importance of familiarizing clients with the exercise being tested and of ensuring proper supervision by qualified testers. The test supervisor must assess the client's technique and determine whether or not the client should continue in the testing process.

Once the 1RM is established, you can represent the results in several ways. The first method is to simply report the heaviest load (in kg) that the person lifted. This method is acceptable when attempting to track a client's musculoskeletal fitness over time, but it is not as useful when making comparisons between individuals (6). The most common method for evaluating or comparing individuals is to divide the 1RM by body mass and compare the results with data from norm tables (e.g., table 12.2 and table 12.3) (23).

The ACSM recommends using the bench press and leg press for assessing upper- and lower-body strength when conducting a health and wellness screening (2). However, you can also assess overall strength fitness by examining the 1RM values for six specific exercises: the bench press, arm curl, latissimus pull-down, leg press, leg extension, and leg curl. Each 1RM is divided by body mass to determine relative strength levels that are converted to points and summed to determine a total strength fitness score, which is then classified (see table 12.4).

Predicting 1RM

In some scenarios, a direct 1RM test may not be feasible or safe. In these cases, a strength endurance test may be recommended as an indirect method of determining muscular strength (9). Prediction or indirect determination of 1RM is based on a negative linear relationship between the percentage of 1RM and the number of repetitions that can be performed (5). This linear relationship suggests that the percentage of 1RM load decreases by about 2% to 2.5% per maximal repetition performed. Thus, the 1RM load is the maximal resistance that can be lifted one

Table 12.2 Age- and Sex-Based Norms for the 1RM Bench Press

AGE (y)		20-29		30-39		40-49		50-59		≥60	
Descriptors	% rank	M	F	M	F	M	F	M	F	M	F
Well above average	90	1.48	0.54	1.24	0.49	1.10	0.46	0.97	0.40	0.89	0.41
	80	1.32	0.49	1.12	0.45	1.00	0.40	0.90	0.37	0.82	0.38
Above average	70	1.22	0.42	1.04	0.42	0.93	0.38	0.84	0.35	0.77	0.36
	60	1.14	0.41	0.98	0.41	0.88	0.37	0.79	0.33	0.72	0.32
Average	50	1.06	0.40	0.93	0.38	0.84	0.34	0.75	0.31	0.68	0.30
	40	0.99	0.37	0.88	0.37	0.80	0.32	0.71	0.28	0.66	0.29
Below average	30	0.93	0.35	0.83	0.34	0.76	0.30	0.68	0.26	0.63	0.28
	20	0.88	0.33	0.78	0.32	0.72	0.27	0.63	0.23	0.57	0.26
Well below average	10	0.80	0.30	0.71	0.27	0.65	0.23	0.57	0.19	0.53	0.25

Values represented are 1RM (kg) / body mass (kg). M = male and F = female.

Adapted, by permission, from V. Heyward and A.L. Gibson, 2014, *Advanced fitness assessment and exercise prescription*, 7th ed. (Champaign, IL; Human Kinetics), 162. Data for women provided by the Women's Exercise Research Center, The George Washington University Medical Center, Washington, D.C., 1998. Data for men provided by The Cooper Institute for Aerobics Research, 2005, *The Physical Fitness Specialist Manual* (The Cooper Institute, Dallas, TX).

Table 12.3 Age- and Sex-Based Norms for the 1RM Leg Press

AGE (y)		20-29		30-39		40-49		50-59		≥60	
Descriptors	% rank	M	F	M	F	M	F	M	F	M	F
Well above average	90	2.27	2.05	2.07	1.73	1.92	1.63	1.80	1.51	1.73	1.40
	80	2.13	1.66	1.93	1.50	1.82	1.46	1.71	1.30	1.62	1.25
Above average	70	2.05	1.42	1.85	1.47	1.74	1.35	1.64	1.24	1.56	1.18
	60	1.97	1.36	1.77	1.32	1.68	1.26	1.58	1.18	1.49	1.15
Average	50	1.91	1.32	1.71	1.26	1.62	1.19	1.52	1.09	1.43	1.08
	40	1.83	1.25	1.65	1.21	1.57	1.12	1.46	1.03	1.38	1.04
Below average	30	1.74	1.23	1.59	1.16	1.51	1.03	1.39	0.95	1.30	0.98
	20	1.63	1.13	1.52	1.09	1.44	0.94	1.32	0.86	1.25	0.94
Well below average	10	1.51	1.02	1.43	0.94	1.35	0.76	1.22	0.75	1.16	0.84

Values represented are 1RM (kg) / body mass (kg). M = male and F = female.

Adapted, by permission, from V. Heyward and A.L. Gibson, 2014, *Advanced fitness assessment and exercise prescription*, 7th ed. (Champaign, IL; Human Kinetics), 162. Data for women provided by the Women's Exercise Research Center, The George Washington University Medical Center, Washington, D.C., 1998. Data for men provided by The Cooper Institute for Aerobics Research, 2005, *The Physical Fitness Specialist Manual* (The Cooper Institute, Dallas, TX).

time and represents a 100% 1RM load, while lifting a load 5 times would be associated with approximately 90% of 1RM load (5).

There are two indirect methods for determining 1RM—the repetition maximum test (27) and the use of **prediction equations** (1, 14, 15, 18, 30, 32, 35, 36, 39, 40, 49).

Estimating 1RM

One method for estimating 1RM is to use a **repetition maximum (RM)** test or strength endurance test. When this test is used in this capacity, it should be performed with a load that ranges between 5 and 10 repetitions (i.e., the

Table 12.4 Relative Strength Ratios* for Selected 1RM Tests in College-Aged Men and Women

Points	BENCH PRESS M	F	LAT PULL-DOWN M	F	LEG PRESS M	F	LEG EXTENSION M	F	LEG CURL M	F	ARM CURL M	F
10	1.50	0.90	1.20	0.85	3.00	2.70	0.80	0.70	0.70	0.60	0.70	0.50
9	1.40	0.85	1.15	0.80	2.80	2.50	0.75	0.65	0.65	0.55	0.65	0.45
8	1.30	0.80	1.10	0.75	2.60	2.30	0.70	0.60	0.60	0.52	0.60	0.42
7	1.20	0.70	1.05	0.73	2.40	2.10	0.65	0.55	0.55	0.50	0.55	0.38
6	1.10	0.65	1.00	0.70	2.20	2.00	0.60	0.52	0.50	0.45	0.50	0.35
5	1.00	0.60	0.95	0.65	2.00	1.80	0.55	0.50	0.45	0.40	0.45	0.32
4	0.90	0.55	0.90	0.63	1.80	1.60	0.50	0.45	0.40	0.35	0.40	0.28
3	0.80	0.50	0.85	0.60	1.60	1.40	0.45	0.40	0.35	0.30	0.35	0.25
2	0.70	0.45	0.80	0.55	1.40	1.20	0.40	0.35	0.30	0.25	0.30	0.21
1	0.60	0.35	0.75	0.50	1.20	1.00	0.35	0.30	0.25	0.20	0.25	0.18

INTERPRETATION

To determine strength fitness category, sum the points for the 6 exercises tested:

Total points	Strength fitness classification
48-60	Excellent
37-47	Good
25-36	Average
13-24	Fair
0-12	Poor

* Relative values are determined by dividing the 1RM in kg by the subject's body mass in kg. M = male and F = female.

Adapted, by permission, from V. Heyward and A.L. Gibson, 2014, *Advanced fitness assessment and exercise prescription*, 7th ed. Champaign, IL: Human Kinetics, 164.

heaviest load that allows the client to complete between 5 and 10 repetitions with proper technique). Once the maximum number of repetitions that can be performed with a specific load is established, it can then be used in conjunction with a 1RM estimation table (e.g., table 12.5) to predict the client's maximal muscular strength with a specific exercise (9).

Using 1RM Prediction Equations

The second indirect method is to conduct a strength endurance test that allows the client to perform fewer than 10 repetitions (1, 14, 16, 37, 51). In this approach, you can choose from numerous prediction equations (see table 12.6) (1, 14, 15, 17, 18, 30, 32, 35, 37, 39, 40, 49). The

predictions are generally more accurate with heavier loads and fewer repetitions (1).

Isometric Strength Tests

You can also assess strength with static or isometric tests, in which there is no visible movement and the length of the active muscle is constant. Isometric tests are usually specific to the muscle group and joint angle being assessed and thus are limited in their ability to describe a person's overall strength (2, 41). The results of any isometric test are affected by factors such as joint angle, testing protocol, feedback, and individual motivation (22, 28). Because force varies across the ROM of any joint, it is

Table 12.5 1RM Estimation Table

Max reps (RM)	1	2	3	4	5	6	7	8	9	10	12	15
%1RM	100	95	93	90	87	85	83	80	77	75	67	65
Load (lb or kg)	10	10	9	9	9	9	8	8	8	8	7	7
	20	19	19	18	17	17	17	16	15	15	13	13
	30	29	28	27	26	26	25	24	23	23	20	20
	40	38	37	36	35	34	33	32	31	30	27	26
	50	48	47	45	44	43	42	40	39	38	34	33
	60	57	56	54	52	51	50	48	46	45	40	39
	70	67	65	63	61	60	58	56	54	53	47	46
	80	76	74	72	70	68	66	64	62	60	54	52
	90	86	84	81	78	77	75	72	69	68	60	59
	100	95	93	90	87	85	83	80	77	75	67	65
	110	105	102	99	96	94	91	88	85	83	74	72
	120	114	112	108	104	102	100	96	92	90	80	78
	130	124	121	117	113	111	108	104	100	98	87	85
	140	133	130	126	122	119	116	112	108	105	94	91
	150	143	140	135	131	128	125	120	116	113	101	98
	160	152	149	144	139	136	133	128	123	120	107	104
	170	162	158	153	148	145	141	136	131	128	114	111
	180	171	167	162	157	153	149	144	139	135	121	117
	190	181	177	171	165	162	158	152	146	143	127	124
	200	190	186	180	174	170	166	160	154	150	134	130
	210	200	195	189	183	179	174	168	162	158	141	137
	220	209	205	198	191	187	183	176	169	165	147	143
	230	219	214	207	200	196	191	184	177	173	154	150
	240	228	223	216	209	204	199	192	185	180	161	156
	250	238	233	225	218	213	208	200	193	188	168	163
	260	247	242	234	226	221	206	208	200	195	174	169
	270	257	251	243	235	230	224	216	208	203	181	176
	280	266	260	252	244	238	232	224	216	210	188	182
	290	276	270	261	252	247	241	232	223	218	194	189
	300	285	279	270	261	255	249	240	231	225	201	195
	310	295	288	279	270	264	257	248	239	233	208	202
	320	304	298	288	278	272	266	256	246	240	214	208
	330	314	307	297	287	281	274	264	254	248	221	215
	340	323	316	306	296	289	282	272	262	255	228	221
	350	333	326	315	305	298	291	280	270	263	235	228
	360	342	335	324	313	306	299	288	277	270	241	234
	370	352	344	333	322	315	307	296	285	278	248	241

(continued)

Max reps (RM)	1	2	3	4	5	6	7	8	9	10	12	15
	380	361	353	342	331	323	315	304	293	285	255	247
	390	371	363	351	339	332	324	312	300	293	261	254
	400	380	372	360	348	340	332	320	308	300	268	260
	410	390	381	369	357	349	340	328	316	308	274	267
	420	399	391	378	365	357	349	336	323	315	281	273
	430	409	400	387	374	366	357	344	331	323	288	280
	440	418	409	396	383	374	365	352	339	330	295	286
	450	428	419	405	392	383	374	360	347	338	302	293
	460	437	428	414	400	391	382	368	354	345	308	299
	470	447	437	423	409	400	390	376	362	353	315	306
	480	456	446	432	418	408	398	384	370	360	322	312
	490	466	456	441	426	417	407	392	377	368	328	319
	500	475	465	450	435	425	415	400	385	375	335	325
	510	485	474	459	444	434	423	408	393	383	342	332
	520	494	484	468	452	442	432	416	400	390	348	338
	530	504	493	477	461	451	440	424	408	398	355	345
	540	513	502	486	470	459	448	432	416	405	362	351
	550	523	512	495	479	468	457	440	424	413	369	358
	560	532	521	504	487	476	465	448	431	420	375	364
	570	542	530	513	496	485	473	456	439	428	382	371
	580	551	539	522	505	493	481	464	447	435	389	377
	590	561	549	531	513	502	490	472	454	443	395	384
	600	570	558	540	522	510	498	480	462	450	402	390

Reprinted, by permission, from National Strength and Conditioning Association, 2016, Resistance training, by J.M. Sheppard and N.T. Triplett. In *Essentials of strength training and conditioning*, 4th ed., edited by G.G. Haff and N.T Triplett (Champaign, IL: Human Kinetics), 455, 456.

Table 12.6 1RM Prediction Equations

Reference	Equation	Constant error*	ICC
Adams (5)	1RM = RepWt / (1 − 0.02 × RTF)	0.7	.90
Brown (13)	1RM = (RTF × 0.0338 + 0.9849) × RepWt	0.9	.95
Mayhew et al. (34)	1RM = RepWt / (0.522 + 0.419 $e^{-0.055 \times RTF}$)	0.2	.96
O'Conner et al. (39)	1RM = 0.025 (RepWt × RTF) + RepWt	−0.8	.96
Reynolds et al. (40)	1RM = RepWt / (0.5551 $e^{-0.0723 \times RTF + 0.4847}$)	0.8	.96
Tucker et al. (49)	1RM = 1.139 × RepWt + 0.352 × RTF + 0.243	0.4	.93

1RM = 1-repetition maximum; RepWt = repetition weight, load < 1RM to perform repetitions; RTF = repetitions to failure.
*Predicted 1RM − actual 1RM.

Adapted from Mayhew et al. 2008.

important that the joint angle assessed with this type of testing is standardized (figure 12.1).

Isometric strength and endurance can be assessed using an isometric dynamometer, which can assess strength in the grip (figure 12.2*a*), the back and legs (figure 12.2*b*), and the upper body. The simplest isometric assessment is done with a handgrip dynamometer, and results obtained this way have been related to muscle mass (24) and to the function of the forearm muscles. As a whole, handgrip tests are very reliable (ICC > .90) (33) and constitute valid methods for assessing strength (5). The strength levels determined for each hand are then summed to create a composite score that represents a level of strength (see table 12.7).

More complex multijoint isometric tests (e.g., midthigh pull, back squat, bench press) can be performed with the combination of a force plate and a custom isometric rack (28, 48) (see figure 12.2*c*). The reliability and validity of this approach are generally related to the standardization of both position and protocol (47). When standardized positions are used, these testing methods have been shown to be very reliable (21, 46, 48).

The most basic measurement of isometric strength reveals the peak force that can be generated and the time to achieve peak force (5). More complex isometric testing procedures yield a detailed force-time curve that can be analyzed (see figure 12.3) (22) in terms of three considerations: overall peak force, sometimes termed *maximal voluntary contraction (MVC)*; rate of force development, sometimes referred to as *explosive strength*; and starting strength (47).

ASSESSMENTS OF MUSCULAR ENDURANCE

Muscular endurance is the ability of the neuromuscular system to perform one of two types of action: (1) repeated contractions over a period of time until fatigue occurs with dynamic exercises and (2) maintenance of a specific percentage of an MVC for a period of time with static exercises (2). Generally, muscular endurance is represented by the total number of repetitions a person can perform with a percentage of 1RM. Methods for evaluating muscular endurance include the YMCA bench press test and the push-up test (2, 4, 23).

YMCA Bench Press Protocol

The YMCA bench press test is a classic protocol for muscular endurance. It assesses the muscular endurance of the upper body, specifically the pectoralis major, anterior deltoid,

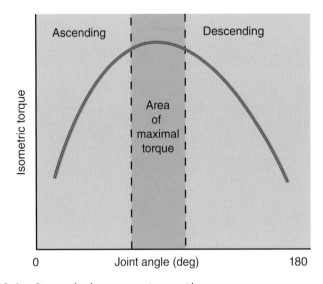

Elbow flexion = 70-120°
Elbow extension = 90-120°
Hip extension = 40-50°
Hip flexion = 145-150°
Knee extension = 80-130°
Knee flexion = 130-170°

Figure 12.1 Sample human strength curve.

Reprinted, by permission, from P.J. Maud and C. Foster, 2005, *Physiological assessment of human fitness*, 2nd ed. (Champaign, IL: Human Kinetics), 131.

Figure 12.2 Isometric dynamometers can be used to assess *(a)* grip strength and *(b)* back and leg strength. *(c)* Force-power racks can be used to test a variety of isometric movements.

Photo *(c)* courtesy of Greg Haff

Table 12.7 Age- and Sex-Based Norms for Combined Handgrip Test

AGE (y)	15-19				20-29				30-39			
SEX	**M**		**F**		**M**		**F**		**M**		**F**	
Classification	kg	kN	kg	kN	kg	kN	kg	kN	kg	kN	kg	kN
Excellent	≥108	≥1.06	≥68	≥0.67	≥115	≥1.13	≥70	≥0.69	≥115	≥1.13	≥71	≥0.70
Very good	98-107	0.96-1.05	60-67	0.59-0.66	104-114	1.02-1.12	63-69	0.62-0.68	104-114	1.02-1.12	63-70	0.62-0.69
Good	90-97	0.88-0.95	53-59	0.52-0.58	95-103	0.93-1.01	58-62	0.57-0.61	95-103	0.93-1.01	58-62	0.57-0.61
Fair	79-89	0.77-0.87	48-52	0.47-0.51	84-94	0.82-0.92	52-59	0.51-0.58	84-94	0.82-0.92	51-57	0.50-0.56
Needs improvement	≤78	≤0.76	≤47	≤0.46	≤83	≤0.81	≤51	≤0.50	≤83	≤0.81	≤50	≤0.49
AGE (y)	**40-49**				**50-59**				**60-69**			
SEX	**M**		**F**		**M**		**F**		**M**		**F**	
Classification	kg	kN	kg	kN	kg	kN	kg	kN	kg	kN	kg	kN
Excellent	≥108	≥1.06	≥69	≥0.68	≥101	≥0.99	≥61	≥0.60	≥100	≥0.98	≥54	≥0.53
Very good	97-107	0.95-1.05	61-68	0.60-0.67	92-100	0.90-0.98	54-60	0.53-0.59	91-99	0.89-0.97	48-53	0.47-0.52
Good	88-96	0.86-0.94	54-60	0.53-0.59	84-91	0.82-0.89	49-53	0.48-0.52	84-90	0.82-0.88	45-47	0.44-0.46
Fair	80-87	0.78-0.85	49-53	0.48-0.52	76-83	0.75-0.81	45-48	0.44-0.47	73-83	0.72-0.81	41-44	0.40-0.43
Needs improvement	≤79	≤0.77	≤48	≤0.47	≤75	≤0.74	≤44	≤0.43	≤72	≤0.71	≤40	≤0.39

kN = kilonewton.

Sources: adapted from Canadian Physical Guidelines 2011, 2012. Used with permission from the Canadian Society for Exercise Physiology and from V. Heyward and A.L. Gibson, 2014, *Advanced fitness assessment and exercise prescription,* 7th ed. (Champaign, IL; Human Kinetics), 157.

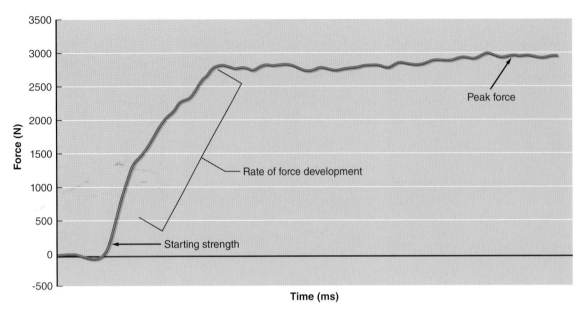

Figure 12.3   Isometric force-time curve.

and triceps. This widely used assessment of muscular endurance requires subjects to lie on their back on a bench and press a constant load at a cadence of 30 reps · min⁻¹ until they experience volitional fatigue or an inability to maintain the prescribed cadence (26). Kim et al. (26) have reported that the YMCA bench press test can also be used to estimate maximal bench press strength with the following prediction equations:

Men 1RM (kg) = 1.55 × (repetitions performed at 30 reps · min⁻¹) + 37.9

Women 1RM (kg) = 0.31 × (repetitions performed at 30 reps · min⁻¹) + 19.2

Once a maximum number of repetitions has been established, you can compare it with data from a norm table (e.g., table 12.8) to classify the client's upper-body muscular endurance.

Push-Up Test

This test of upper-body (pectoralis major, anterior deltoid, and triceps) muscular endurance has no time limit and requires only minimal equipment (23). It is performed in a continuous fashion and requires the subject to maintain a straight back at all times. The test is stopped when proper technique cannot be maintained for two consecutive repetitions. The total number of repetitions is then used to tabulate the score, which is compared with age- and sex-based norms (e.g., table 12.9).

Go to the web study guide to access electronic versions of individual and group data sheets, the question sets from each laboratory activity, and case studies for each laboratory activity.

Table 12.8   Muscular Endurance Norms for the Bench Press

AGE (y)	18-25		26-35		36-45		46-55		56-65		>65	
% rank	M	F	M	F	M	F	M	F	M	F	M	F
90	49	49	48	46	41	41	33	33	28	29	22	22
80	34	30	30	29	26	26	21	20	17	17	12	12
70	26	21	22	21	20	17	13	12	10	9	8	6
60	17	13	16	16	12	10	8	6	4	4	3	2
50	5	2	4	2	2	2	1	0	0	0	0	0

Score is based on the number of repetitions completed in 1 min with an 80 lb (36.3 kg) bar for men and a 35 lb (15.9 kg) bar for women. M = male and F = female.

Adapted, by permission, from V. Heyward and A.L. Gibson, 2014, *Advanced fitness assessment and exercise prescription*, 7th ed. (Champaign, IL: Human Kinetics), 165.

Table 12.9   Age- and Sex-Based Norms for the Push-Up Test

AGE (y)	15-19		20-29		30-39		40-49		50-59		60-69	
Classification	M	F	M	F	M	F	M	F	M	F	M	F
Excellent	≥39	≥33	≥36	≥30	≥30	≥27	≥25	≥24	≥21	≥21	≥18	≥17
Very good	29-38	25-32	29-35	21-29	22-29	20-26	17-24	15-23	13-20	11-20	11-17	12-16
Good	23-28	18-24	22-28	15-20	17-21	13-19	13-16	11-14	10-12	7-10	8-10	5-11
Fair	18-22	12-17	17-21	10-14	12-16	8-12	10-12	5-10	7-9	2-6	5-7	2-4
Needs improvement	≤17	≤11	≤16	≤9	≤11	≤7	≤9	≤4	≤6	≤1	≤4	≤1

M = male and F = female.

Source: *The Canadian Physical Activity, Fitness & Lifestyle Approach: CSEP-Health & Fitness Program's Health-Related Appraisal and Counselling Strategy,* 3rd Edition © 2003. Adapted with permission of the Canadian Society for Exercise Physiology.

Maximal Upper-Body Strength

Equipment

- Bench press and appropriate free weights
- Physician's scale or equivalent electronic scale
- Stadiometer
- Calculator
- Stopwatch
- Individual data sheet
- Group data sheet
- Microsoft Excel or equivalent spreadsheet program

This laboratory activity is supported by a virtual lab experience in the web study guide.

www

Warm-Up

Regardless of which method is used to assess muscular strength, each subject should perform a structured warm-up to prepare for the tests that will be performed. The warm-up should contain a 5 min general warm-up and a 5 min dynamic stretching warm-up (8). Sample general warm-up exercises include light jogging, cycling, and jumping rope. Dynamic stretching activities might include push-ups, arm circles and swings, and other activities targeting the upper body. After completing the warm-up protocol, the subject can begin the testing process.

1RM Bench Press Test

The bench press test requires a flat bench and free weights that provide for a range of possible loads. For example, various weight bars may be needed (e.g., 5 kg, 15 kg, 20 kg) depending on the population being tested. Ideally, you will have a barbell set with loads ranging from 0.5 kg to 25 kg to allow for a wide variety of loads. When performing the bench press test, it is essential to use appropriate spotting techniques (see figure 12.4). Briefly, the spotter should stand behind the bench press and use an **alternated grip** (i.e., one hand using a **supinated grip** and the other hand using a **pronated grip**) when lifting the bar out of the racked position on the bench and when helping rerack the barbell. It is also important for the subject to maintain five points of contact with the floor and bench—head, shoulders and upper back, right foot, left foot, and buttocks. Before starting the 1RM test, consult with the subject to estimate the subject's perceived maximum. This process can be challenging with untrained individuals because it is likely to be speculative at best. One possible method for estimating the 1RM is to estimate it off of the subject's body weight. This is typically done by multiplying the subject's body weight by a correction factor. For example, if a female subject weighs 60 kg and wants to estimate her 1RM for the free weight bench press, we would multiply this value by a correction factor of 0.35 to yield an estimated 1RM of 21 kg. With male subjects, the correction factor utilized for the free weight bench press is 0.60. Based upon this correction factor, the estimated 1RM for the free weight bench press for a 70 kg male would be 42 kg. When using this method, it is generally recommended that the maximum weight used is 79 kg (175 lb) for men and 64 kg (140 lb) for women. With athletes, in contrast, it is relatively easy to determine the perceived maximum because you can likely use

Figure 12.4 Bench press series: *(a)* liftoff, *(b)* starting position, *(c)* downward position, *(d)* upward movement, and *(e)* racking the bar.

the actual 1RM or estimates based on the individual's training weights. This value should be noted on the individual data sheet for laboratory activity 12.1 and used to calculate the 40% to 60% load, 60% to 80% load, and 90% load needed to conduct the test. Once the loads have been determined, use the following steps to conduct the assessment.

Step 1: Set up the bench press area with all the equipment needed to conduct the test.

Step 2: Gather basic data (e.g., age, height, weight) for the individual data sheet as described in laboratory activity 1.1. Have the subject put on his or her shoes.

Step 3: Lead the subject through a 10 min warm-up that includes dynamic warm-up activities such as arms swings to prepare the upper body for the bench press test.

Step 4: Load the bar with the weight corresponding to the subject's warm-up weight (40%-60% of perceived maximum). Move into an appropriate spotting position behind the bench press (figure 12.4*a*).

Step 5: Instruct the subject to lie on his or her back on the bench in a five-point-contact position (figure 12.4*a*). The subject's body should be positioned on the bench such that the racked bar is directly above the eyes. All repetitions should begin in this position.

Step 6: Upon the subject's command, lift the bar off the rack using an alternated grip. Guide the bar to a position over the subject's chest, and then release the bar to the subject (figure 12.4*b*).

Step 7: Have the subject lower the bar in a controlled manner to touch the chest at approximately nipple height while maintaining the five points of contact and inhaling (figure 12.4*c*). During this movement, keep your hands in the alternated position near the bar but not touching the bar.

Step 8: The subject should push the bar upward until the elbows are fully extended while maintaining the five points of contact and exhaling (figure 12.4*d*). At the end of the prescribed number of repetitions (5-10)—or if maximum is achieved—grasp the bar with an alternated grip and place it back on the rack (figure 12.4*e*).

Step 9: After the subject completes 5 to 10 repetitions with the 40% to 60% load, quickly change the load to the predetermined second-set load. The subject should have only 1 min to rest during this change.

Step 10: After the 1 min rest interval, conduct steps 5 through 8 and have the subject perform 3 to 5 repetitions with a load that is 60% to 80% of perceived maximum.

Step 11: After the completion of set 2, the subject rests for 3 min while you change the resistance to a load that is 90% of perceived maximum.

Step 12: After the 3 min rest, conduct steps 5 through 8 and have the subject perform 1 repetition with the 90% load.

Step 13: While the subject rests for 3 min, increase the load. If the attempt appeared relatively easy, increase the load by 5 to 10 kg. If, however, the attempt was more difficult, increase the load by 1 to 5 kg, depending on the weights available.

Step 14: Repeat steps 12 and 13, continuing to have the subject perform only 1 repetition until 1RM is achieved.

Step 15: If an attempt is unsuccessful, reduce the load by 1-10 kg, but keep it above the last successful attempt and have the subject rest for 3 min.

Step 16: After the 3 min rest, have the subject perform another attempt with the procedures outlined in steps 5 to 8.

Step 17: If the attempt is successful, increase the load by a small amount (1-5 kg) and repeat step 16. If the attempt is unsuccessful, the 1RM load is determined as the heaviest load successfully completed. Record the maximum weight lifted in the appropriate location on the individual and group data sheets.

Step 18: Divide the maximal weight lifted by the subject's body mass and record the result in the appropriate location on the individual and group data sheets. This value should then be compared with the data in the norm table for the bench press (see table 12.2). Based on the results, note the subject's strength classification and percent ranking on the individual data sheet.

Bench Press 1RM Prediction Test

When performing an RM test as a tool for predicting the 1RM, use a load that allows fewer than 10 repetitions to be performed (29). This repetition range is selected because it has been established that using heavier loads in the prediction process results in greater accuracy in predicting the 1RM (1). In this lab, you will use the 1RM to calculate a load that is 87% of 1RM, which is chosen as the target intensity because it corresponds to previously established 5RM loads for most people (9, 52). Once the target load has been established, initiate the testing process with the following steps.

Step 1:   Set up the bench press area with the appropriate weights.

Step 2:   Gather basic data (e.g., age, height, weight) for the individual data sheet as described in laboratory activity 1.1. Instruct the subject to put on his or her shoes and move to the test area.

Step 3:   Explain the test to the subject. Then begin a 10 min warm-up that contains a 5 min general warm-up (e.g., cycling, jogging, rope jumping) followed by 5 min of dynamic stretching activities targeting the upper body.

Step 4:   Use the subject's 1RM to determine the test set load. This will be considered the repetition weight (kg) used for the test (RepWt).

Step 5:   Load the barbell with 50% of the calculated test set load.

Step 6:   Instruct the subject to lie on the bench on his or her back with five points of contact (head, shoulders and upper back, buttocks, left foot, and right foot).

Step 7:   Assume the spotting position depicted in figure 12.4*a*.

Step 8:   Instruct the subject to grasp the bar using a pronated grip with hands approximately shoulder-width apart.

Step 9:   Lift the barbell off the rack with an alternated grip while the subject is grasping the bar (see figure 12.4*b*). The subject should hold the barbell at arm's length.

Step 10:   Have the subject lower the barbell in a controlled fashion while inhaling until it touches the chest by moving the elbows down past the torso and slightly away from the body (see figure 12.4*c*). Once the barbell touches the chest, instruct the subject to press the bar upward while exhaling until the arms are fully extended (see figure 12.4*d*). This pattern should be completed six times. While the subject is performing each repetition, follow the bar with an alternated grip in order to assist the subject if needed.

Step 11:   Once the last repetition has been completed for the set, help the subject place the bar onto the rack (see figure 12.4*e*).

Step 12:   Adjust the weight to a load corresponding to 70% of the test weight previously calculated. During this time, the subject rests for 3 min.

Step 13:   Repeat steps 6 through 11.

Step 14:   While the subject rests for 3 min, change the weight to a load corresponding to 90% of the predetermined test weight.

Step 15: Repeat steps 6 through 11.

Step 16: Adjust the weight on the bar to the test set weight (87% of estimated 1RM) while the subject rests for 3 min. After the 3 min, complete steps 6 through 11, but have the subject perform as many repetitions as possible.

Step 17: If the subject completes fewer than 10 repetitions with the predetermined load, the test is complete. Record the total number of repetitions completed (RTF), as well as the load used (RepWt), in the appropriate location on the individual data sheet. Skip to step 21.

Step 18: If the subject performs more than 10 repetitions, increase the load by 2.5% to 5.0% while the subject rests for 3 min.

Step 19: Repeat steps 6 through 11, but have the subject perform as many repetitions as possible.

Step 20: After the subject completes step 19, the tester can then complete step 17. Continue this process until reaching a weight where the subject can lift fewer than 10 times.

Step 21: Using the Adams (5) equation, calculate the subject's predicted 1RM bench press.

$$\text{Predicted 1RM (kg)} = \text{RepWt (kg)} / (1 - 0.02 \times \text{RTF})$$

Place the weight used for the RM test in the RepWt (kg) location and place the number of repetitions (RTF) completed in the appropriate location. Using a calculator, perform the calculation and record the result on the individual and group data sheets.

Step 22: Using the Brown (13) equation, calculate the subject's predicted 1RM bench press.

$$\text{Predicted 1RM (kg)} = (\text{RTF} \times 0.0338 + 0.9849) \times \text{RepWt (kg)}$$

Place the weight used for the RM test in the RepWt (kg) location and place the number of repetitions (RTF) completed in the appropriate location. Using a calculator, perform the calculation and record the predicted 1RM in the appropriate location on the individual and group data sheets.

Step 23: Using the Mayhew et al. (34) equation, calculate the subject's predicted 1RM bench press.

$$\text{Predicted 1RM (kg)} = \text{RepWt (kg)} / (0.522 + 0.419\ e^{-0.055 \times \text{RTF}})$$

Place the weight used for the RM test in the RepWt (kg) location and place the number of repetitions (RTF) completed in the appropriate location. Using a calculator, perform the calculation and record the predicted 1RM in the appropriate location on the individual and group data sheets.

Step 24: Using the O'Conner et al. (39) equation, calculate the subject's predicted 1RM bench press.

$$\text{Predicted 1RM (kg)} = 0.025 \times [\text{RepWt (kg)} \times \text{RTF}] + \text{RepWt (kg)}$$

Place the weight used for the RM test in the RepWt (kg) location and place the number of repetitions (RTF) completed in the appropriate location. Using a calculator, perform the calculation and record the predicted 1RM in the appropriate location on the individual and group data sheets.

Step 25: Using the Reynolds (40) equation, calculate the subject's predicted 1RM bench press.

$$\text{Predicted 1RM (kg)} = \text{RepWt (kg)} / (0.5551\ e^{-0.0723 \times \text{RTF} + 0.4847})$$

Place the weight used for the RM test in the RepWt (kg) location and place the number of repetitions (RTF) completed in the appropriate location. Using a calculator, perform the calculation and record the predicted 1RM in the appropriate location on the individual and group data sheets.

Step 26: Using the Tucker et al. (49) equation, calculate the subject's predicted 1RM bench press.

$$\text{Predicted 1RM (kg)} = 1.139 \times \text{RepWt (kg)} + 0.352 \times \text{RTF} + 0.243$$

Place the weight used for the RM test in the RepWt (kg) location and place the number of repetitions (RTF) completed in the appropriate location. Using a calculator, perform the calculation and record the predicted 1RM in the appropriate location on the individual and group data sheets.

Question Set 12.1

1. Why is it important to test muscular strength?

2. When performing the 1RM bench press tests, what are some important factors to remember as a tester? As a subject?

3. What was your predicted 1RM for the bench press? Did any of the formulas differ from your previously established 1RM? Was this pattern similar for the class as a whole?

4. Why might a 1RM prediction differ from an actual 1RM? (Hint: Is there a physiological reason?)

5. Using Excel or an equivalent spreadsheet program, create a graph of each of the estimated maximum results and the actual 1RM. Which formula best estimates the 1RM? Which formula is the worst? What might explain the differences between the regression equations used to predict 1RM?

6. Explain the pros and cons of 1RM and RM testing. Which test is safer and why?

7. Why is important to use fewer than 10 repetitions when predicting 1RM values? What is the physiological reason for performing fewer than 10 repetitions?

LABORATORY ACTIVITY 12.1

INDIVIDUAL DATA SHEET

Name or ID number: _____ Date: _____

Tester: _____ Time: _____

Sex: M / F (circle one) Age: _____ y Height: _____ in. _____ m

Weight: _____ lb _____ kg Temperature: _____ °F _____ °C

Barometric pressure: _____ mmHg Relative humidity: _____ %

1RM Bench Press Test

Perceived maximum		Load	kg	Completed
Set 1	5-10 repetitions at 40%-60% of perceived maximum			❏
Set 2	3-5 repetitions at 60%-80% of perceived maximum			❏
Set 3	1 repetition at 90% of perceived maximum			❏
Set 4	1RM attempt			❏
Set 5	1RM attempt			❏

Note: 1 min rest between sets 1 and 2; 3 min rest preceding sets 3, 4, and 5.

Absolute 1RM = _____ kg

Relative 1RM = load (kg) / body weight (kg) = _____

Strength classification = % rank = _____

Notes:

Bench Press Predicted Maximum Test

Set	Load	Repetitions	Repetitions Completed	Completed
Warm-up set 1		6		❏
3 min rest				
Warm-up set 2		6		❏
3 min rest				
Warm-up set 3		6		❏
3 min rest				
Test set 1		To failure		❏
3 min rest				
Test set 2*		To failure		❏

*If test set 1 results in more than 10 repetitions, repeat the test after adding weight.

Bench Press 1RM Prediction Equations

Resistance (RepWt)		kg	Note: This resistance is 87% of 1RM.
Number of repetitions (RTF)			
	Adams (5)		
	Brown (11)		
	Mayhew et al. (34)		
1RM prediction	O'Conner et al. (39)		
	Reynolds et al. (40)		
	Tucker et al. (49)		

Maximal Lower-Body Strength

Equipment

- Leg press machine and weights
- Physician's scale or equivalent electronic scale
- Stadiometer
- Calculator
- Stopwatch
- Individual data sheet
- Group data sheet
- Microsoft Excel or equivalent spreadsheet program

Warm-Up

Regardless of which method is used to assess muscular strength, each subject should perform a structured warm-up to prepare for the tests that will be performed. The warm-up should contain a 5 min general warm-up and a 5 min dynamic stretching warm-up (8). Sample general warm-up exercises include light jogging, cycling, and jumping rope. For this test, dynamic stretching activities might include body weight squats, leg swings, walking lunges, high knees, and trunk circles. After completing the warm-up protocol, the subject can begin the testing process.

1RM Leg Press Test

Performance of the 1RM leg press test is similar to that of the bench press test. The subject sits in a leg press machine with his or her feet hip-width apart on the foot platform (figure 12.5*a*). Make sure the subject is properly fitted to the machine. The foot and chair placement should allow the subject to fully extend his or her legs (figure 12.5*b*) and to achieve a bottom position where the knee is bent to about 90° (figure 12.5*c*). Monitor the test to ensure the subject's hips and back remain in contact with the back pad throughout the lift. Consult with the subject to determine a perceived maximum; this load should be recorded in the appropriate location on the individual data sheet and then used to calculate the loads needed for the various sets contained in the 1RM testing protocol.

Step 1: Prepare the leg press machine to perform the test.

Step 2: Gather basic data (e.g., age, height, weight) for the individual data sheet as described in laboratory activity 1.1. Have the subject put on his or her shoes.

Step 3: Direct the subject through a 10 min warm-up that includes dynamic exercises targeting the lower body (e.g., body weight squats, walking knee hugs, toy soldiers, walking lunges).

Step 4: Following the warm-up, have the subject get into the leg press machine. Adjust the machine as needed to ensure the subject is properly seated and the feet are in the appropriate position (figure 12.5*a*). Load the leg press with the warm-up load (40%-60% of perceived maximum).

Figure 12.5 Leg press series: *(a)* foot position, *(b)* starting position, and *(c)* downward and upward movements.

Step 5: Have the subject initiate the leg press (figure 12.5*b*) by removing the support mechanism from the foot platform and then grasping the handles or seat.

Step 6: Instruct the subject to allow the hips and knees to slowly flex, lowering the foot platform. The knees should flex until the tops of the thighs are parallel to the foot platform (figure 12.5*c*). Remind the subject to inhale during this portion of the lift.

Step 7: Having achieved the bottom position, the subject should extend the legs, which will move the foot platform upward (figure 12.5*c*). Encourage the subject to exhale and fully extend the legs during this portion of the lift (figure 12.5*b*).

Step 8: Following the warm-up, increase the load to 60% to 80% of the subject's perceived maximum. After 1 min of rest, have the subject repeat steps 5 through 7.

Step 9: After completion of the second set, instruct the subject to rest for 3 min while you increase the load to correspond to about 90% of the subject's perceived maximum. After the recovery period, have the subject repeat steps 5 through 7, but perform only 1 repetition with an increased load.

Step 10: Based on the how the subject's form looked, increase the load for the next set. For example, if the subject appeared to have difficulty completing the prescribed repetitions, increase the resistance by a smaller amount. If, on the other hand, it seemed relatively easy, increase the weight by more. Be careful to neither increase the load too rapidly nor be overly conservative. After a 3 min rest, the subject repeats steps 5 through 7 while performing only 1 repetition with an increased load.

Step 11: Repeat step 10 until the subject reaches the heaviest weight that can be lifted with appropriate technique.

Step 12: Record the heaviest load properly lifted in the appropriate location on the individual and group data sheets as the subject's 1RM for the leg press.

Step 13: Divide the maximal weight lifted by the subject's body mass and record the result in the appropriate location on the individual and group data sheets. Compare this value with the data in the norm table for the leg press (table 12.3). Based on the results, note the subject's strength classification and percent rank on the individual data sheet.

Leg Press 1RM Prediction Test

The leg press RM test should be performed with an estimated load that allows fewer than 10 repetitions to be performed. In this lab, you will use the 1RM to calculate a load that is 87% of 1RM, which is chosen as the target intensity because it corresponds to previously established 5RM loads for most people (8, 48). Once the target load has been established, initiate the testing process with the following steps.

Step 1: Set up the leg press area with the appropriate weights.

Step 2: Gather basic data (e.g., age, height, weight) for the individual data sheet as described in laboratory activity 1.1. Instruct the subject to put his or her shoes back on and move to the test area.

Step 3: Explain the test to the subject. Then begin a 10 min warm-up that contains a 5 min general warm-up (e.g., cycling, jogging, or rope jumping) followed by 5 min of dynamic stretching activities targeting the lower body.

Step 4: Load the leg press machine with 50% of the calculated test set load. The test set load can be calculated using an actual or predicted 1RM. The weight determined for the test is referred to as the RepWt in all prediction equations.

Step 5: Instruct the subject to get into the leg press machine with the feet on the footpad (figure 12.5a). Adjust the leg press machine to ensure the subject is properly seated in it.

Step 6: To initiate the lift, have the subject engage the leg press machine (figure 12.5b) by removing the support mechanism from the foot platform and then grasping the handles or seat.

Step 7: Have the subject slowly flex the hips and knees to lower the foot platform. The knees should flex until the tops of the thighs are parallel to the foot platform (figure 12.5c). During this portion of the lift, ensure that the subject inhales. After achieving the bottom position, the subject should exhale while extending the legs to move the foot platform upward (figure 12.5c). Once the knees are fully extended, the subject should repeat this step for a total of 6 repetitions.

Step 8: Adjust the leg press weight to a load corresponding to 70% of the test weight previously calculated. During this time, the subject rests for 3 min.

Step 9: Have the subject repeat steps 6 and 7.

Step 10: While the subject rests for 3 min, change the weight to 80% of the pre-determined test weight.

Step 11: Have the subject repeat steps 6 and 7.

Step 12: Adjust the weight on the leg press load to the test set weight (87% of 1RM) while the subject rests for 3 min. After the 3 min, have the subject repeat steps 6 and 7 but complete as many repetitions as possible.

Step 13: If the subject completes fewer than 10 repetitions with the predetermined load, the test is complete. Record the total number of repetitions completed, as well as the load used, in the appropriate location on the individual data sheet. Skip to step 17.

Step 14: If the subject performs more than 10 repetitions, increase the load by 2.5% to 5.0% while the subject rests for 3 min.

Step 15: Have the subject repeat steps 6 and 7 but complete as many repetitions as possible.

Step 16: After the subject completes step 15, the tester can then complete step 13. Continue this process until the subject reaches a weight that can be lifted fewer than 10 times. The total number of repetitions completed is considered the repetitions completed to failure (RTF).

Step 17: Using the Adams (5) equation, calculate the subject's predicted 1RM leg press.

$$\text{Predicted 1RM (kg)} = \text{RepWt (kg)} / (1 - 0.02 \times \text{RTF})$$

Place the weight used for the RM test in the RepWt (kg) location and place the number of repetitions (RTF) completed in the appropriate location. Using a calculator, perform the calculation and record the result in the appropriate location on the individual and group data sheets.

Step 18: Using the Brown (13) equation, calculate the subject's predicted 1RM leg press.

$$\text{Predicted 1RM (kg)} = (\text{RTF} \times 0.0338 + 0.9849) \times \text{RepWt (kg)}$$

Place the weight used for the RM test in the RepWt (kg) location and place the number of repetitions (RTF) completed in the appropriate location. Using a calculator, perform the calculation and record the predicted 1RM in the appropriate location on the individual and group data sheets.

Step 19: Using the Mayhew et al. (34) equation, calculate the subject's predicted 1RM leg press.

$$\text{Predicted 1RM (kg)} = \text{RepWt (kg)} / (0.522 + 0.419 \, e^{-0.055 \times \text{RTF}})$$

Place the weight used for the RM test in the RepWt (kg) location and place the number of repetitions (RTF) completed in the appropriate location. Using a calculator, perform the calculation and record the predicted 1RM in the appropriate location on the individual and group data sheets.

Step 20: Using the O'Conner et al. (39) equation, calculate the subject's predicted 1RM leg press.

$$\text{Predicted 1RM (kg)} = 0.025 \times [\text{RepWt (kg)} \times \text{RTF}] + \text{RepWt (kg)}$$

Place the weight used for the RM test in the RepWt (kg) locations and place the number of repetitions (RTF) completed in the appropriate location. Using a calculator, perform the calculation and record the predicted 1RM in the appropriate location on the individual and group data sheets.

Step 21: Using the Reynolds (40) equation, calculate the subject's predicted 1RM leg press.

$$\text{Predicted 1RM (kg)} = \text{RepWt (kg)} / (0.5551 \ e^{-0.0729 \times \text{RTF} + 0.4847})$$

Place the weight used for the RM test in the RepWt (kg) location and place the number of repetitions (RTF) completed in the appropriate location. Using a calculator, perform the calculation and record the predicted 1RM in the appropriate location on the individual and group data sheets.

Step 22: Using the Tucker et al. (49) equation, calculate the subject's predicted 1RM leg press.

$$\text{Predicted 1RM (kg)} = 1.139 \times \text{RepWt (kg)} + 0.352 \times \text{RTF} + 0.243$$

Place the weight used for the RM test in the RepWt (kg) location and place the number of repetitions (RTF) completed in the appropriate location. Using a calculator, perform the calculation and record the predicted 1RM in the appropriate location on the individual and group data sheets.

Question Set 12.2

1. What is the basic procedure for conducting a 1RM test?
2. Based on your individual results, how do you compare with the class average and with the norms presented in this laboratory activity?
3. Graph the class, male, and female results using Microsoft Excel or an equivalent spreadsheet program. Based on this graph, what can you say about these data?
4. Using the class results, perform a *t*-test comparing the absolute and relative male data with the absolute and relative female data.
5. Using Excel or an equivalent spreadsheet program, create a graph of each of the estimated maximum results and the actual 1RM. Which formula best estimates the 1RM? Which formula is the worst? Is this consistent between males and females?

LABORATORY ACTIVITY 12.2

INDIVIDUAL DATA SHEET

Name or ID number: _____ Date: _____

Tester: _____ Time: _____

Sex: M / F (circle one) Age: _____ y Height: _____ in. _____ m

Weight: _____ lb _____ kg Temperature: _____ °F _____ °C

Barometric pressure: _____ mmHg Relative humidity: _____ %

1RM Leg Press Maximum Test

Perceived maximum		Load	kg	Completed
Set 1	5-10 repetitions at 40%-60% of perceived maximum			❏
Set 2	3-5 repetitions at 60%-80% of perceived maximum			❏
Set 3	1 repetition at 90% of perceived maximum			❏
Set 4	1RM attempt			❏
Set 5	1RM attempt			❏

Note: 1 min rest between sets 1 and 2; 3 min rest preceding sets 3, 4, and 5.

Absolute 1RM = _____ kg

Relative 1RM = load (kg) / body weight (kg) = _____

Strength classification = % rank = _____

Notes:

Leg Press Predicted Maximum Test

Set	Load	Repetitions	Repetitions Completed	Completed
Warm-up set 1		6		❏
3 min rest				
Warm-up set 2		6		❏
3 min rest				
Warm-up set 3		6		❏
3 min rest				
Test set 1		To failure		❏
3 min rest				
Test set 2*		To failure		❏

*If test set 1 results in more than 10 repetitions, repeat the test after adding weight.

Leg Press 1RM Prediction Equations

Resistance (RepWt)		kg	*Note:* This resistance is 87% of 1RM.
Number of repetitions (RTF)			

1RM prediction		
Adams (5)		
Brown (11)		
Mayhew et al. (34)		
O'Conner et al. (39)		
Reynolds et al. (40)		
Tucker et al. (49)		

Maximal Handgrip Strength

Equipment

- Handgrip dynamometer (hydraulic, spring, other)
- Physician's scale or equivalent electronic scale
- Stadiometer
- Stopwatch
- Individual data sheet
- Group data sheet
- Microsoft Excel or equivalent spreadsheet program

Warm-Up

Regardless of which method is used to assess muscular strength, each subject should perform a structured warm-up to prepare for the tests that will be performed. The warm-up should contain a 5 min general portion and a 5 min dynamic stretching warm-up (8). Sample general warm-up exercises include light jogging, cycling, and jumping rope. Dynamic stretching activities might include wrist, finger, and elbow rotations. After completing the warm-up, the subject can begin the testing process.

Isometric Handgrip Strength Test

The assessment of handgrip strength is generally completed with a handgrip dynamometer. Of the many types available, the hydraulic (e.g., Jamar) and spring (e.g., Stoelting, Smedley, Lafayette; figure 12.2a) are the most prevalent. Regardless of type, the basic testing procedures are the same.

Step 1:   Gather all the equipment needed to perform the test and set up a testing area. Ensure that the pointer of the dynamometer is set at zero.

Step 2:   Gather basic data (e.g., age, height, weight) for the individual data sheet as described in laboratory activity 1.1. Instruct the subject to put on his or her shoes.

Step 3:   Lead the subject through a 10 min warm-up.

Step 4:   Direct the subject stand and face straight ahead.

Step 5:   Adjust the dynamometer grip so that the middle portion of the middle finger is at a right angle. The method for adjusting the grip depends on the instrument; refer to the equipment manufacturer's instructions to understand how to adjust the dynamometer.

Step 6:   Record the grip setting in the appropriate box on the individual data sheet. Use this setting for all subsequent tests.

Step 7:   Have the subject place the forearm at an angle of 90° to 180° to the upper arm, which is kept in a vertical position. This puts the arm in a position that is either bent at a right angle or straight.

Step 8:   Ensure the subject keeps the wrist and forearm in a midprone position.

Step 9: Give the subject the following commands (7, 31):

- *"Are you ready?"* Ask just before having the subject initiate the test; when the subject answers *yes*, move to the next command.
- *"Squeeze as hard as possible."* Say this as the subject initiates the test.
- *"Harder! . . . Harder! . . . Relax."* Say this as the subject undertakes and then finishes the test.

Step 10: Have the subject undertake two or three (36, 38) trials for each hand (and alternate between hands). Each test should be separated by 1 min (21).

Step 11: After each test, record the maximum force generated in kilograms in the appropriate location on the individual data sheet.

Step 12: After each test, reset the pointer to zero.

Step 13: Take the best trial for each hand and place them in the appropriate location on the individual and group data sheets.

Step 14: Sum the best right and left hand trials to create a sum score. Divide this sum score by body mass in kg and place in the appropriate location on the individual and group data sheets.

Question Set 12.3

1. What are some important considerations when performing a handgrip dynamometer test?
2. Why is it important to reset the pointer after each test? If it were not reset, how would the test be affected?
3. How do your results compare with those of the class?
4. Using Excel or an equivalent spreadsheet program, graph the class, male, and female results. Make sure to place standard deviations on your graph. Based on the graph, what can you say about these data?
5. Graph the relative and absolute strength of the five strongest men and the five strongest women. Based on this graph, what can you say about the class?

LABORATORY ACTIVITY 12.3

INDIVIDUAL DATA SHEET

Name or ID number: _____ Date: _____

Tester: _____ Time: _____

Sex: M / F (circle one) Age: _____ y Height: _____ in. _____ m

Weight: _____ lb _____ kg Temperature: _____ °F _____ °C

Barometric pressure: _____ mmHg Relative humidity: _____ %

Handgrip Strength

Circle the best of the first two trials; underline the best of the three trials.

Instrument	Grip setting position		TRIAL 1		TRIAL 2		TRIAL 3	
			R	L	R	L	R	L
Hydraulic	1 2 3 4 5 (circle appropriate number)							
Spring		mm	Setting on handle					
Other		Setting						

R = right and L = left.

Best of two trials	Right hand	+	Left hand	=	Sum				Rank
Hydraulic		+		=		kg		kN	
Spring		+		=		kg		kN	
Other		+		=		kg		kN	

Ratio of grip strength to body mass	Grip sum = _____ kN	÷	Body mass = _____ kg	=		Ratio
Ratio of dominant to nondominant hand	Dominant = _____ kN	÷	Nondominant = _____ kg	=		Ratio

Notes:

Upper-Body Muscular Endurance

Equipment

- Bench press with appropriate free weights
- Floor mats (for push-up stations)
- Metronome
- Individual data sheet
- Group data sheet
- Microsoft Excel or equivalent spreadsheet program

Warm-Up

Regardless of which method is used to assess muscular endurance, each subject should perform a structured warm-up to prepare for the tests that will be performed. The warm-up should contain a 5 min general warm-up and a 5 min dynamic stretching warm-up (8). Sample general warm-up exercises include light jogging, cycling, and jumping rope. Dynamic stretching activities might include arm circles and swings, as well as other activities targeting the upper body. After completing the warm-up protocol, the subject can begin the testing process.

YMCA Bench Press Test

This test requires a station with a bench press, barbell (with appropriate weights), and metronome. Men use an 80 lb (36.3 kg) barbell, and women use a 35 lb (15.9 kg) barbell. During the test, a metronome counts 60 beats · min^{-1} to establish a rate of 30 reps · min^{-1}. If the subject cannot maintain this rate, the test is ended.

Step 1: Set up the testing area with the appropriate weight, as well as a bench and a metronome.

Step 2: Set the metronome to 60 beats · min^{-1} to establish the necessary cadence (30 reps · min^{-1}).

Step 3: Gather basic data (e.g., age, height, weight) for the individual data sheet as described in laboratory activity 1.1.

Step 4: Assume a spotting position while directing the subject to lie on his or her back on the bench. Explain that the subject must maintain all five points of contact (e.g., head, shoulders, buttocks, right and left feet) during the test (figure 12.6a).

Step 5: Once the subject is positioned on the bench, lift the bar off the rack using an alternated grip (figure 12.6b) while the subject holds the bar using a pronated grip with the hands slightly farther than shoulder-width apart. Throughout the test, spot the subject with an alternated grip in order to assist when muscular failure is reached.

Step 6: Have the subject lower the bar to the starting position with the bar resting on the chest and the elbows in a flexed position (figure 12.6c).

Step 7: Instruct the subject to press the bar up to a position in which the arms are fully extended (figure 12.6d) and then return bar to the chest at a cadence of 30 reps · min⁻¹. The up-and-down movement will continue as long as the subject can maintain the prescribed cadence. During each repetition, ensure that a full range of motion (ROM) is performed. Count each successful repetition.

Step 8: When the subject can no longer maintain the 30 reps · min⁻¹ cadence, help return the barbell to the rack (figure 12.6*e*).

Step 9: Record the total number of repetitions completed and determine the subject's percent ranking by comparing the results with the data presented in table 12.8. Record the results in the appropriate locations on the individual data sheet.

Step 10: Using the equations presented on page 264, estimate the subject's 1RM bench press. Transfer this value to the individual and group data sheets.

Figure 12.6 Bench press series: *(a)* liftoff, *(b)* starting position, *(c)* downward position, *(d)* upward movement, and *(e)* racking the bar.

Push-Up Test

The push-up test is recommended by the ACSM for evaluating upper-body muscular endurance (2, 3). It is an ideal field test because it requires no special equipment. When conducting the push-up test, make sure the subject maintains a flat back at all times. For men, the subject takes the standard push-up position in which the hands are about shoulder-width apart, the back is straight, the head is up, and the toes are used as a pivot point (figure 12.7a). Women use a modified knee push-up position in which the legs are held together, the lower leg is in contact with a mat or the floor with the ankles plantar-flexed, the back is straight, the head is up, and the hands are approximately shoulder-width apart (figure 12.8a).

Figure 12.7 Push-up test: (a) standard start position and (b) standard lowered position.

Figure 12.8 Push-up test: (a) modified start position and (b) modified lowered position.

Step 1: Set up the testing area and explain the basics of the test to the subject.

Step 2: Gather basic data (e.g., age, height, weight) for the individual data sheet as described in laboratory activity 1.1. Have the subject put his or her shoes back on and move to the test area.

Step 3: Lead the subject through a 10 min warm-up that includes both a general warm-up and a dynamic stretching routine targeting the musculature of the upper body.

Step 4: Instruct the subject to assume the proper starting position (for men, use the standard position in figure 12.7a; for women, use the modified position in figure 12.8a).

Step 5: Have the subject lower the body while maintaining a straight back until the chin touches the mat (figures 12.7b and 12.8b). The stomach should not touch the mat at any time during the assessment.

Step 6: Once the chin touches the mat, the subject should push up by extending the arms until achieving a straight-arm position. Record each successful push-up as it is performed.

Step 7: Without resting, the subject should repeat steps 5 and 6 until no longer able to maintain a straight-back position or complete any more push-ups.

Step 8: Record the total number of push-ups completed in the appropriate location on the individual and group data sheets. Compare the subject's results with the data presented in table 12.9 and record the appropriate classification.

Question Set 12.4

1. What were your individual upper-body muscular endurance ratings for the YMCA bench press test and the push-up test? How did these compare with the results for the class and with the values presented in the norm tables?

2. Did the men and women in the class differ in upper-body muscular endurance? Did you expect this result? Why or why not?

3. How did your push-up test results compare with the normative values? Was the muscular endurance rating similar to the YMCA bench press test results?

4. Using Excel or an equivalent spreadsheet program, perform a correlation analysis between the number of push-ups performed and the number of repetitions performed in the YMCA bench press test. Are these related? If so, how might they be related?

5. Using Excel or an equivalent spreadsheet program, perform a correlation analysis between the number of push-ups performed and the estimated maximal weight determined. Are these related? If so, how might they be related?

LABORATORY ACTIVITY 12.4

INDIVIDUAL DATA SHEET

Name or ID number: _____ Date: _____

Tester: _____ Time: _____

Sex: M / F (circle one) Age: _____ y Height: _____ in. _____ m

Weight: _____ lb _____ kg Temperature: _____ °F _____ °C

Barometric pressure: _____ mmHg Relative humidity: _____ %

YMCA bench press test		Comments
Number of repetitions		
Percent ranking		
1RM prediction	Men	
	Women	

Push-up test		Comments
Number of repetitions		
Classification		

Anaerobic Fitness Measurements

Objectives

- Explain the various anaerobic power and capacity tests.
- Gain exposure to basic procedures used to evaluate anaerobic power and capacity.
- Describe a method for evaluating the stretch–shortening cycle with the use of jumping tests.
- Examine the Wingate anaerobic test (WAnT).
- Examine published norms for various anaerobic power tests.

DEFINITIONS

anaerobic capacity—Mean power output achieved during exercise bouts that last between 10 and 120 s. Tests that assess anaerobic capacity typically stress the ATP-PC and glycolytic energy systems. The term *mean power* is often used synonymously with *anaerobic capacity*.

anaerobic power—Mean or peak power output in exercise lasting 10 s or less. Anaerobic power is typically evaluated with tests that stress the phosphagen or ATP-PC systems and require very high intensity for a short duration.

eccentric utilization ratio (EUR)—Representation of the ability to use the stretch–shortening cycle; estimated by dividing countermovement vertical jump results by static vertical jump results (44) and typically calculated with peak power output and vertical displacement.

fatigue rate—Rate of decline in performance; may be considered as the degree of power drop-off from the highest power output to the value at the end of the test (29), thus often calculated as percent change between these two values.

force plate—Device that measures ground reaction forces; often considered the gold standard when evaluating vertical jump performance.

horizontal power—Power typically estimated by sprinting tasks such as a 40 m sprint.

Margaria-Kalamen test—Stair test used to evaluate anaerobic peak power (40).

mean anaerobic power—Average power output achieved during a performance test that stresses the anaerobic energy supply mechanisms.

mean power—Average power output achieved during a specified time interval or during a test.

peak anaerobic power—Highest power output achieved during a test in which the anaerobic system is the primary supplier of energy (these tests typically last less than 10 s).

peak power—Highest power output achieved during a test (25).

power—Rate of doing work (36); calculated in terms of either work divided by time (power = work / time) or force multiplied by velocity (power = force × velocity).

power endurance—Ability to repetitively achieve high power outputs or to maintain a level of power output.

rate of fatigue—Rate that is represented as the degree of power drop-off during the test; also known as the *fatigue index*.

reactive strength index (RSI)—Typically assessed during a drop jump; quantified as the ratio of flight time to ground-contact time.

stretch–shortening cycle (SSC)—Active stretch of a muscle (i.e., eccentric muscle action) followed by immediate shortening of the same muscle (i.e., concentric muscle action).

switch mat—Device used to measure flight time, which can then be used to determine jump height and power output.

total work—Product of mean power and time.

Wingate anaerobic test (WAnT)—Test of anaerobic power and capacity performed on a cycle ergometer that typically lasts 30 s and uses a resistance equal to $0.075 \text{ kp} \cdot \text{kg}^{-1}$ body mass.

The assessment of anaerobic fitness is an important part of the overall assessment of physical capacity for both athletes and clinical clients. It is well documented that the ability to express high power outputs is a primary contributor to success in sporting activities (6, 55). For example, Carlock et al. (10) found that an athlete's weightlifting ability is highly correlated with the ability to express high power outputs during jumping tasks. Similarly, among American football and soccer players, performance capacity is strongly related to power-generating capacity as indicated by jumping performance and sprinting performance (6, 14, 15, 55). Because of these relationships, power output is one of the most commonly tested performance characteristics in sport science and athletic settings (55). Thus, it is important that students in sport or exercise science understand the various methods for assessing power.

Anaerobic fitness may also be related to the ability to undertake activities of daily living. People whose power-generating capacity is compromised are often hindered in their ability to perform simple tasks of daily life. If, however, they undertake activities to improve their power output, they can markedly improve their performance in these tasks. The importance of power-generating capacity in activities of daily living means that exercise physiology students must understand the types of testing available for evaluating a client's power output, as well as how these tests are implemented.

Generally, **power** can be defined as the rate of performing work (36), whereas **anaerobic power** is the rate of performing work under conditions in which aerobic metabolism offers little contribution (42). Tests of anaerobic power generally evaluate activities in which the primary energy supplier is the phosphagen or ATP-PC system, whereas tests of **anaerobic capacity** examine activities in which the pri-

mary energy supplier is the glycolytic system. Typically, anaerobic power tests last less than 10 s, whereas anaerobic capacity tests last longer than 10 s but no longer than 2 min.

When evaluating anaerobic capacity tests, you can often calculate a peak power, mean power, or fatigue rate. **Peak power** is the highest power output attained during a specific test, whereas mean power is an average collected over the duration of the test. Sometimes **mean power** is referred to as *anaerobic capacity* (29) or as **power endurance**. **Fatigue rate** is the degree of power drop-off during the test. It is generally calculated during tests based on power endurance or anaerobic capacity as a percentage of peak power output (29). Depending on the systems and techniques used, anaerobic capacity and power tests can also involve calculation of other variables, such as the time to peak power, the rate of power development, and the velocity of movement.

Numerous anaerobic power and capacity tests can be conducted. In the following laboratory activities, we discuss five tests, five of which examine anaerobic peak power output and two of which examine both peak anaerobic power and anaerobic capacity (table 13.1). We also discuss a test that estimates **horizontal power**.

SPRINTING PERFORMANCE TESTS FOR ESTIMATING HORIZONTAL POWER

Sprinting performance is commonly measured in many athletic settings. Successful sprinting depends largely on the ability of the body to exert vertical forces, the ability of the bioenergetic systems to meet the metabolic demands of the activity, and the muscle-fiber-type profile of the individual (56).

Horizontal sprinting activities are often used to evaluate horizontal power-generating capac-

Table 13.1 Classification of Anaerobic Power Tests

Test	VARIABLES QUANTIFIED BY TEST	
	Anaerobic peak power	Anaerobic capacity
Countermovement vertical jump test	Yes	
Static vertical jump test	Yes	
Margaria-Kalamen test	Yes	
Bosco 60 s continuous jump test	Yes	Yes
Wingate anaerobic cycle test (WAnT)	Yes	Yes

ity, but they do not involve measurement of a power output (1). For sprinting activities, the only knowns are the athlete's body mass, the horizontal distance covered, and the time used to cover that distance. Therefore, these variables are used to estimate horizontal power according to the following formula:

$$\text{Horizontal power (kg} \cdot \text{m} \cdot \text{s}^{-1}) = \text{force (kg)} \times \text{average velocity (m} \cdot \text{s}^{-1})$$

This equation can also be represented as follows:

$$\text{Horizontal power (N} \cdot \text{m} \cdot \text{s}^{-1}) = \text{force (N)} \times \text{average velocity (m} \cdot \text{s}^{-1})$$

In both equations, force is represented by the individual's body mass, and average velocity is calculated by dividing the known distance by the time it took to cover that distance. For example, a 100 kg athlete who covered a distance of 200 m in 22 s would achieve a horizontal power of 909 kg $\cdot$ m $\cdot$ s^{-1}. This value could be calculated as follows:

$$\text{Horizontal power (kg} \cdot \text{m} \cdot \text{s}^{-1}) = 100 \text{ kg} \times (200 \text{ m} / 22 \text{ s})$$

$$= 100 \text{ kg} \times 9.09 \text{ m} \cdot \text{s}^{-1}$$

$$= 909 \text{ kg} \cdot \text{m} \cdot \text{s}^{-1}$$

Thus, if two people of similar mass cover the same distance, the faster person has a higher power output.

Horizontal power is therefore usually tested with sprinting activities such as the 30 yd (27.4 m), 40 yd (36.6 m), and 60 yd (54.9 m) sprints (25). When performing sprint testing, you can use either a stopwatch or an electronic timer to quantify the athlete's sprint time. Stopwatches are generally well correlated with electronic timers ($r = .98$, $r^2 = .95$), but there is usually a 0.24 s error when using a stopwatch, which results in a faster running time. Stopwatch times can be converted to electronic-equivalent times using the following equation:

$$\text{Electronic timer time (s)} = 1.0113 \times \text{handheld stopwatch time (s)} + 0.2252$$

This difference in the results achieved by the two timing methods makes it important to know which method was used when comparing results with norms presented in the literature (25).

JUMPING PERFORMANCE TESTS FOR DETERMINING VERTICAL POWER

Vertical power or jumping performance is one of the most popular field tests used to assess anaerobic power in athletic populations (25). The ability to express high vertical power outputs or jump heights has been correlated with a variety of sporting activities, including competitive weightlifting (10, 17), track cycling sprint performance (55), and sprinting performance (27). Jumping ability has also been shown to differentiate between level of play and playing abilities in American football (14) and soccer players (15). Because of the simplicity of jumping tests, they are commonly included in field-based performance test batteries (16, 20, 54).

Vertical jumps for testing vertical power output can be of two types: countermovement and static jumps. The countermovement vertical jump requires the athlete to start in a standing position, drop into a squat position, and then immediately jump as high as possible (figure 13.1). The static vertical jump, or squat jump (7), requires the athlete to squat with the tops of the thighs parallel to the floor (with a knee angle of approximately 90°), hold this position during a countdown ("3, 2, 1, jump"), and then jump as high and as fast as possible (18) (figure 13.2). The static vertical jump is often the preferred testing method because countermovement jump performance can be affected by the subject's skill level (52). However, both types of jumps are commonly used in testing batteries, in which case the countermovement jump test precedes the static jump test. Regardless of which jump type you use, the goal of the testing process is to determine the vertical jump displacement that the individual can achieve. This can be done by means of various methods (35).

Figure 13.1 Countermovement vertical jump positions: *(a)* start position, *(b)* dip, *(c)* maximum height, and *(d)* landing.

Figure 13.2 Static vertical jump positions: *(a)* start position, *(b)* takeoff position, *(c)* maximum height, and *(d)* landing.

Force Plate

The **force plate** is considered by many to be the gold standard when evaluating vertical jumping and power outputs (26). The vertical displacement achieved during a vertical jump performed on a force plate is determined by quantifying flight time (i.e., the time between takeoff from the force plate and landing on it during the jump)—the higher the athlete jumps, the longer the flight time. Therefore, flight time and gravity can be used to estimate the vertical displacement by means of the following equation (8):

$$\text{Vertical displacement (m)} = \frac{9.81\,(\text{m} \cdot \text{s}^{-2}) \times \text{flight time (s)} \times \text{flight time (s)}}{8}$$

Using a force plate to determine vertical displacement has been shown to be highly reliable (ICC = .94-.99) (19, 32, 37). In examining the power output generated during a jumping activity, you can use two methods when working with force plates.

The first method involves measuring the force generated during a vertical jump and then deriving the velocity of movement achieved during the jump. The force generated during the vertical jump is directly assessed by measuring the ground reaction forces created during the movement (11). The force at each time point during the movement is then divided by the system mass in order to determine acceleration. The acceleration due to gravity is then subtracted from the calculated acceleration in order to represent the acceleration produced by the individual when calculating the velocity of movement. The velocity of movement is then determined by the product of acceleration and time data at each time point (11). The derived velocity is then multiplied by the quantified force in order to calculate the power output achieved in the jumping activity. Generally, this method for quantifying vertical jump power output has been shown to be highly reliable (ICC = .93-.98) (32, 33).

The second method for estimating power output uses the force plate to measure the flight time achieved during the vertical jump in order to calculate the vertical displacement achieved during the jump (8). The reliability or accuracy of the test can be compromised if the subject tries to extend the flight time by tucking the legs when jumping or landing. Therefore, it is important to monitor the jump and landing to make sure the landing is performed properly. If the jump is properly performed, the vertical displacement determined can then be placed into one of the power estimation equations presented in the section of this laboratory that presents formulas for estimating vertical power.

Switch Mat

Another method for evaluating vertical jump displacement uses a **switch mat** (35). Switch mats (which typically measure 0.6858 by 0.6858 m) contain embedded microswitches that measure the time interval between takeoff and landing and quantify it as flight time. As with the force plate, flight time can then be used to determine vertical displacement. To use a switch mat, the subject stands on the mat and performs either a countermovement or a static vertical jump (figure 13.3). Flight time is then determined and placed into the previously presented equation for determining vertical displacement.

Vertical displacement determined with a switch mat has been reported to be highly reliable (ICC = .96-.98) (10). As with the force plate, you must carefully monitor the jumping and landing technique to ensure that accurate displacements are estimated. Once the vertical displacement is derived, it is then placed into one of the power estimation equations presented later in this laboratory (Formulas for Estimating Vertical Power). Normative data for flight time and leg power during jumping tasks can be found in table 13.2.

Table 13.2 Flight Time and Leg Power Normative Data

Percentile	Classification	FEMALES		MALES	
		Flight time (ms)	Power (W)	Flight time (ms)	Power (W)
>81	Excellent	>519	>983	>668	>1,393
61-80	Good	506-519	770-983	634-668	1,283-1,393
41-60	Average	487-505	726-769	606-633	1,118-1282
21-40	Poor	456-486	661-725	556-605	1,027-1,117
<20	Very poor	<456	<661	<556	<1,027
Mean		487	818	607	1,229
SD		50	232	43	261

Adapted, by permission, from T. Skinner, R.U. Newton, and G.G. Haff, 2014, Neuromuscular strength, power, and strength endurance. In *ESSA's student manual for health, exercise, and sport assessment*, edited by J.S. Coombes and T. Skinner (Australia: Elsevier), 160-161.

Figure 13.3 Countermovement vertical jump test performed on a switch mat: *(a)* start position, *(b)* dip, *(c)* maximum height, and *(d)* landing.

Reactive Strength Index

The **reactive strength index (RSI)** is a useful measure for athletes in sports that require a rapid landing followed by a jump or change of direction (53). The most common method for evaluating the RSI is the depth jump, where the athlete steps off a box, lands on a force plate or switch mat, and then jumps up for a maximal height (figure 13.4).

When performed in this manner, the ability to land from a known height and rapidly perform a stretch–shortening cycle to transition from an eccentric to a concentric muscle action is evaluated (48). The testing protocol was first presented by the Australian Institute of Sport in the mid-1990s (58) and has become a common performance test used by sport scientists and strength scientists. The test determines the RSI over a series of box heights ranging from 0.3 m to 0.6 m to quantify the athlete's reactive strength

Figure 13.4 Box jump method for RSI testing: *(a)* start position, *(b)* contact on mat, and *(c)* maximal jump.

over a series of stretch loads. The RSI is then calculated as follows:

RSI = jump height (m) / contact time (s)

Using a range of drop jump heights from 0.3 m to 0.6 m has been found to be highly reliable (ICC = .96-.99) and precise (CV = 2.1%-3.1%) (41).

The results of the RSI test can be used to determine which box heights are best suited for the optimization of performance. Specifically, the box height that elicits the highest RSI score is the best box height for performing drop jumps (48) (figure 13.5).

Research by Beattie et al. (3) demonstrates that stronger athletes perform markedly better on the series of box heights used in the standard RSI test when compared with weaker athletes. This is partially because stronger athletes are better able to tolerate eccentric stretch loads.

Jump and Reach

The jump-and-reach test is the most basic method for determining vertical jump displace-

ment and is generally based on the original Sargent jump test (35). This method measures the difference between the standing reach height of the subject and the height that the subject can jump and touch. Standing reach is determined by having the subject place the feet together with the dominant arm near a wall (1). The subject then reaches as high as possible so that the palm is against the marking scale. The highest point achieved is the standing reach height. Next, the subject jumps as high as possible and reaches for the highest point possible. The vertical jump displacement is then measured as the difference between the standing reach height and the maximal reach height achieved during the jump (35).

Vertical displacement (cm) = jump height (cm) − reach height (cm)

An alternative jump-and-reach method uses a commercial apparatus such as the Vertec, which allows you to measure the reach height and jump height without needing a wall. The device contains a telescopic metal pole and plastic swivel

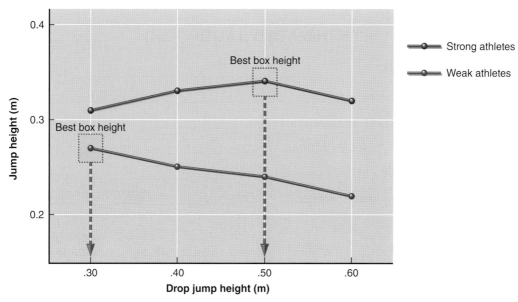

Figure 13.5   Using the drop jump to determine best box height.

Based on Beattie et al. 2017.

vanes that are separated by 0.05 in. (1.3 mm) intervals (35). As with the classic vertical jump, you first determine reach height, in this case by having the subject reach as high as possible with the dominant hand and push away plastic swivel vanes while standing flat-footed under the Vertec (figure 13.6*a*). Because the Vertec can be raised to specific known distances and each vane corresponds to 0.05 in. increments, the standing height is easily obtained. Depending on the athlete's jumping ability, the Vertec can then be raised. At this point, the subject performs a two-footed, nonstep jump in an attempt to displace as many of the horizontal vanes as possible (figure 13.6*b*). The maximal height is determined as the highest displacement achieved. The vertical jump displacement is then calculated by means of the same formula used with the classic jump-and-reach method (35).

Though considered very reliable, the jump-and-reach method does have several limitations. Specifically, it gives little insight into the force and velocity aspects of the jumping movements. In addition, research suggests that this method underestimates vertical displacement as compared with video analysis and switch mats (38).

FORMULAS FOR ESTIMATING VERTICAL POWER

Though the classical vertical jump test is important, it does have some drawbacks; specifically, it yields only a vertical jump displacement and does not truly yield a power output (25). To compensate for this limitation, numerous methods can be used for estimating the power output achieved during a vertical jump. One oft-used method is the Lewis formula (25):

$$\text{Power (W)} = \sqrt{4.9} \times$$
$$\text{weight (kg)} \times \sqrt{\text{jump height (m)}}$$

However, even though the Lewis formula is popular with coaches, physical educators, and sports scientists, its power estimate does not

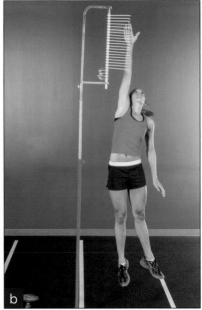

Figure 13.6 Positions for the standing jump-and-reach test using the Vertec: *(a)* establishment of reach height and *(b)* the jump portion of the test.

accurately represent the values achieved for peak or average power output on a force plate (21). To address these inaccuracies, Harman et al. (21) developed two formulas to estimate peak and average power output based on force plate data and regression analysis:

$$\text{Peak power (W)} = 61.9 \times \text{jump height (cm)} + 36 \times \text{body mass (kg)} + 1{,}822$$

$$\text{Average power (W)} = 21.2 \times \text{jump height (cm)} + 23 \times \text{body mass (kg)} - 1{,}393$$

Though these two formulas do improve on the Lewis equation, they were generated with a small sample size (*n* = 17) and a relatively homogenous group. As a result, their validity has been challenged in the scientific literature. Sayers et al. (52) have questioned the validity of the Harman peak and average power equations because they were generated using a static vertical jump test. In contrast, a person doing a countermovement vertical jump test engages the stretch reflex and thus can generate significantly more force (4); specifically, the countermovement vertical jump

can result in a 10% to 23% higher vertical displacement compared with a static vertical jump (52). To address these issues, Sayers et al. (52) used a force plate system to test both countermovement and static vertical jump performance of a large heterogeneous sample (*n* = 108). After performing regression analyses on these data, they generated the following equation:

$$\text{Peak power (W)} = 60.7 \times \text{jump height (cm)} + 45.3 \times \text{body mass (kg)} - 2{,}055$$

This equation was found to be a valid method for estimating peak power output from a static or a countermovement vertical jump, and it has become increasingly popular in research and applied uses. It has also been consistently reported to be highly reliable (ICC = .99) (10, 19).

An additional method for evaluating both peak and average power during a vertical jump has been reported by Johnson and Bahamonde (30). Their equation differs from the Sayers equations in that they include the jumper's height:

Peak power (W) = 78.5 × jump height (cm) + 60.6 × body mass (kg) − 15.3 × height (cm) − 1,308

Average power (W) = 41.4 × jump height (cm) + 31.2 × body mass (kg) − 13.9 × height (cm) + 431

BOSCO TEST FOR ESTIMATING POWER ENDURANCE

Power endurance can be defined as the ability to repetitively perform high-power movements such as jumping. One of the most popular methods for estimating power endurance is the Bosco test (8), which is a repetitive jumping test that lasts between 15 and 60 s and requires the athlete to jump as high and as rapidly as possible (9).

Power output is calculated by determining the number of jumps completed and the sum of the flight times recorded during the work time (8). These data can then be placed into the following equation to calculate mechanical power:

$$\text{Mechanical power (W} \cdot \text{kg}^{-1}) = \frac{g^2 \times T_f \times T_t}{4 \times n \times (T_t - T_f)}$$

where g = 9.81 m · s^{-2}, T_f = total flight time, T_t = total time, and n = number of jumps. The base equation can accommodate a variety of testing lengths, but the most common testing duration is 60 s, in which case the formula would be rewritten as follows:

$$\text{Mechanical power (W} \cdot \text{kg}^{-1}) = \frac{g^2 \times T_f \times 60}{4 \times n \times (60 - T_f)}$$

The mechanical power value calculated by the Bosco test represents the subject's power endurance because it represents an average power output.

Regardless of the test duration, this test should be performed on either a force plate or a switch mat to collect the data required for quantifying power endurance.

DETERMINING THE ECCENTRIC UTILIZATION RATIO

The ability to use the **stretch–shortening cycle (SSC)** is important for success in numerous sporting activities. The SSC involves performing an eccentric muscle action immediately followed by the concentric muscle action. The muscle is activated, stretched, and then immediately shortened, which augments the force and power generated as compared with a concentric-only muscle action. Using the SSC during a countermovement jump results in a higher power output as compared with a static vertical jump, which involves only a concentric muscle action. Quantifying the effect of training interventions on a person's ability to engage the SSC is of particular importance to sport scientists and strength and conditioning professionals. One method involves quantifying the **eccentric utilization ratio (EUR)**, which can be measured to gain insight into the contribution of the SSC to the athlete's performance.

To calculate the EUR, the individual must perform a static vertical jump test and a countermovement vertical jump test. The results can be used to calculate the EUR with the following equation:

EUR = countermovement vertical jump / static vertical jump

This formula can then be used with countermovement and static vertical jump displacements or power outputs. McGuigan et al. (44) suggest that the EUR is sensitive to training stressors and can be used to monitor training outcomes across a periodized training plan. As a monitoring tool, a higher EUR value indicates greater reliance on the SSC, whereas a lower value indicates less reliance on the SSC. For example, McGuigan et al. (44) present data showing that field hockey players have a higher EUR (1.26) value in the preseason than in the off-season (1.05). These findings seem logical because preseason training generally targets the development of power output, whereas off-season training generally targets strength development. Because higher power outputs are associated with use of the SSC, it is clear that when training targets power development, improvements in the EUR should be noticed. The influence of training on the EUR is clearly noted by the fact that endurance athletes consistently express the lowest EURs because their training typically does not target the maximization of power output and can negatively affect Type II fiber content.

WINGATE ANAEROBIC TEST FOR DETERMINING ANAEROBIC CYCLING POWER

The most popular anaerobic cycling power test is the **Wingate anaerobic test (WAnT)**, which was developed at the Wingate Institute in Israel (29). The classic WAnT is performed for 30 s against 0.075 kp per kg of body mass (kp · kg^{-1}) and is designed to test anaerobic performance capacity (29). The WAnT appears ideally suited for this usage because 60% to 85% of energy supply comes from the ATP-PC and glycolytic energy systems during the test (5, 13, 45). The first 3 to 15 s of the test appear to tax the ATP-PC system, and the glycolytic systems are used for the remainder of the test (59).

This test can be used to determine peak anaerobic power, lowest anaerobic power output, mean anaerobic power, total work, and rate of fatigue during an all-out cycling bout performed against a set resistance for a predetermined duration. **Peak anaerobic power** is the highest mechanical power generated during the test. It typically occurs in the first 5 s (59) and is usually calculated as the average power over any 3 or 5 s time period by dividing the amount of work completed by time:

$$\text{Power} = \frac{\text{work}}{\text{time}} = \frac{\text{force} \times \text{displacement}}{\text{time}}$$

If working with 5 s time periods, the maximum number of revolutions completed in one of the six

5 s intervals is multiplied by the distance covered by the flywheel during one revolution and then divided by 5 s. When using a Monark cycle, the flywheel is 1.62 m in circumference and makes 3.7 rotations per pedal revolution, thus producing a total distance of 6 m per revolution (1).

$$\text{Power (W)} = [\text{force (N)} \times (\text{maximum revolutions} \times 6\text{ m})] / 5\text{ s}$$

It is often important to examine relative power outputs, and peak power output can be calculated relative to body mass (W · kg^{-1}) with the following equation (29):

$$\text{Power (W · kg}^{-1}) = \text{power (W)} / \text{body mass (kg)}$$

Power can also be calculated relative to lean body mass (W · kgLBM^{-1}) with the following equation (29):

$$\text{Power (W · kg}^{-1}) = \text{power (W)} / \text{lean body mass (kg)}$$

The lowest anaerobic power is the lowest power output achieved during any individual 3 or 5 s interval during the 30 s test. This value typically occurs during the last 3 to 5 s of the test and is used in calculating the fatigue index (59).

Total work can be calculated if you know the total number of revolutions completed at the end of the 30 s cycle test and the distance covered by the flywheel per revolution (again, 6 m when using a Monark cycle ergometer) (1).

$$\text{Work (J)} = \text{force setting (N)} \times (\text{rev in 30 s} \times \text{distance covered by flywheel per revolution})$$

Therefore, the following equation can be used to calculate total work when using a Monark cycle ergometer:

$$\text{Work (J)} = \text{force setting (N)} \times (\text{rev in 30 s} \times 6\text{ m})$$

Total work can also be normalized to body mass by dividing the total work by the individual's body mass.

$$\text{Work (J · kg}^{-1}) = \frac{\text{work (J)}}{\text{body mass (kg)}} = \frac{\text{force setting (N)} \times (\text{rev in 30 s} \times 6\text{ m})}{\text{body mass (kg)}}$$

Mean anaerobic power is the average power maintained during the 30 s test (59). It can be calculated by averaging the values obtained during the ten 3 s or six 5 s segments of the test (29) or by dividing the total amount of work by the duration of the test (1). This value can be calculated with the following equation:

$$\text{Mean power (W)} = \frac{\text{work}}{\text{time}} = \frac{\text{force setting (N)} \times (\text{rev in 30 s} \times 6 \text{ m})}{30 \text{ s}}$$

Mean power output can also be normalized to body mass by dividing the total work by the individual's body mass.

$$\text{Mean power (W} \cdot \text{kg}^{-1}) = \frac{\text{mean power (W)}}{\text{body mass (kg)}}$$

Rate of fatigue is also known as the *fatigue index* and is represented as the degree of power drop-off during the test (29). It is generally calculated with the following equation:

$$\text{Fatigue index (\%)} = \frac{(\text{peak power} - \text{lowest power})}{\text{peak power}} \times 100$$

Typically, a fatigue rate of $\geq 40\%$ can be found from the first 5 s (peak power) to the last 5 s (lowest power) (1, 59).

When working with the WAnT, two considerations are the duration of the test and the amount of resistance applied. The duration of the WAnT is typically 30 s, although shorter (e.g., 5 s) tests (50) have been used to examine only peak power output.

Resistance in the WAnT is generally set at 0.075 kp $\cdot$ kg^{-1} body mass for children, older adults, and sedentary people (1, 29). Some researchers have also used this resistance with athletes (53), but it is generally recommended that athletes use between 0.090 and 0.100 kp $\cdot$ kg^{-1} body mass (table 13.3) (29). Here is the basic method for determining resistance load or force with the WAnT:

$$\text{Force (N)} = \text{body mass (kg)} \times 9.81 \text{ m} \cdot \text{s}^{-2} \times \text{resistance factor}$$

Therefore, when working with a sedentary adult, you would use the following equation to determine the resistance load for the WAnT:

$$\text{Force (N)} = \text{body mass (kg)} \times 9.81 \text{ m} \cdot \text{s}^{-2} \times 0.075$$

When working with an athlete, you would use this equation:

$$\text{Force (N)} = \text{body mass (kg)} \times 9.81 \text{ m} \cdot \text{s}^{-2} \times 0.100$$

Finally, the WAnT has been consistently shown to be highly reliable (with reliability

Table 13.3 Optimal Loads for the WAnT

Subject		Force load (kp) / body mass (kg)	Reference
Adult male	Sedentary	0.075	Inbar et al. (29)
	Active	0.098	Evans and Quinney (13)
	Athlete	0.098	Evans and Quinney (13)
	Physical education student	0.087	Dotan and Bar-Or (12)
	Recreational	0.085	Vargas et al. (57)
	Track athlete	0.100	Kirksey et al. (34)
Junior male	Wrestler	0.090	Mirzaei et al. (46)
Adult female	Sedentary	0.075	Inbar et al. (29)
	Physical education student	0.085	Dotan and Bar-Or (12)
	Track athlete	0.100	Kirksey et al. (34)

coefficients ranging between .89 and .96) for both peak and mean anaerobic power (29). This high reliability demonstrates the test's repeatability and ensures that any changes in performance are indeed related to actual changes rather than testing error. In addition, a large body of normative data concerning the results of the WAnT test for various populations is presented in the scientific literature, which allows for comparisons between testing populations; see Inbar et al. (29) for detailed normative data.

MARGARIA-KALAMEN STAIR-CLIMB TEST FOR DETERMINING ANAEROBIC POWER

One of the more popular anaerobic stepping tests performed in both athletic and clinical settings is the Margaria stair-climb test (40), which was modified by Kalamen (31). The modern version of the **Margaria-Kalamen test** is designed to test anaerobic power; specifically, because of its relatively short duration (usually less than 5 s), it tests the contribution of the phosphagen system (40).

To perform this test, you need a staircase with a 6 m run-up and at least nine stairs, each of which is between 174 and 175 cm in height (54). It is generally recommended that you place an electronic switch mat or photoelectric cell on the third and ninth steps to ensure accurate timing (49), but careful timing with a stopwatch can provide data that are somewhat less accurate yet still reasonable (54). The test requires the subject to move up the stairs as quickly as possible by taking three steps at a time (i.e., stepping on the third, sixth, and ninth steps). A graphic summary of the testing setup is presented in figure 13.7.

You can calculate power from this test with the following equation:

$$\text{Power}(\text{kg} \cdot \text{m} \cdot \text{s}^{-1}) = \frac{\text{weight (kg)} \times \text{distance (m)}}{\text{time (s)}}$$

where weight is the subject's body mass in kilograms, distance is the vertical height between step 3 and step 9, and time is the number of seconds it takes the subject to move between step 3 and step 9. Therefore, based on the setup shown in figure 13.7, the equation could be modified as follows:

$$\text{Power}(\text{kg} \cdot \text{m} \cdot \text{s}^{-1}) = \frac{\text{weight (kg)} \times 1.05\,\text{m}}{\text{time (s)}}$$

If the power value is to be converted to watts, the power value in kg · m · s⁻¹ is multiplied by 9.807, which is equivalent to the normal acceleration due to gravity.

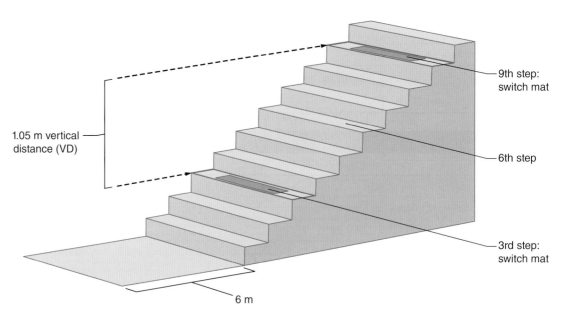

1.05 m vertical distance (VD)

9th step: switch mat

6th step

3rd step: switch mat

6 m

Figure 13.7 Setup for the Margaria-Kalamen stair-climb test.

The Margaria-Kalamen test is considered very reliable, exhibiting a test–retest reliability of $r = .85$ and a CV of <4% (39). Nonetheless, it should be administered with some caution; people who are less experienced or untrained may have difficulty with the every-third-step foot placement required by the test, thus increasing the risk of injury or inaccurate results.

Go to the web study guide to access electronic versions of individual and group data sheets, the question sets from each laboratory activity, and case studies for each laboratory activity.

Sprinting Performance

Equipment

- Track or other area affording 80 to 100 m that can be used for sprinting
- Four stopwatches or an electronic timing gate system
- Calculator
- Individual data sheet
- Group data sheet
- Microsoft Excel or equivalent spreadsheet program

Testing Setup

The sprint test requires a track or other area that offers 80 to 100 m in which the following distances can be marked: 9.14 m (10 yd), 36.6 m (40 yd), 45.7 m (50 yd), and 54.9 m (60 yd). Four timers are used to time each of the four distances. Position the timers such that they can see both the start and finish of the sprint. A summary of the testing area setup is shown in figure 13.8.

Before beginning a maximal sprint test, have the subject complete a structured warm-up that includes 5 min of general warm-up activities (e.g., jogging, cycling, rope jumping) followed by 5 min of dynamic stretching (e.g., leg swings, high knees, walking knee tucks, walking lunges, butt kicks, inchworms, power skips). Next, have the subject perform moderate-speed runs for the 54.9 m (60 yd) distance; specifically, the subject should perform a series of moderate-speed sprints that get progressively faster. The subject should also practice the start technique. It is recommended that the subject *not* use starting blocks or brace against another person's foot because of the technical skill required to use this technique and the inability to replicate positions consistently. The subject should be in a standing position with a low center of gravity and a slight forward lean (1). Once the warm-up is completed, the testing session can begin.

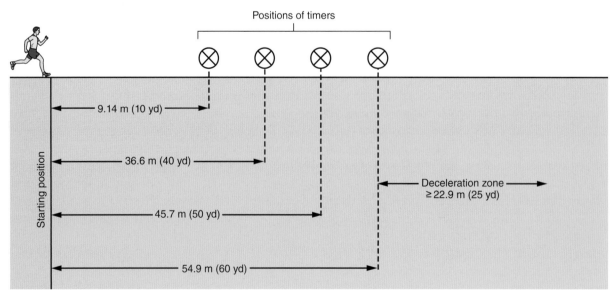

Figure 13.8 Setup for the sprint test.

Maximal Sprint Test

Step 1: Set up the sprint area by placing marks at 9.14 m (10 yd), 36.6 m (40 yd), 45.7 m (50 yd), and 54.9 m (60 yd). At the end of the sprint area, include a deceleration zone of at least 22.9 m (25 yd) (see figure 13.8).

Step 2: Gather basic data (e.g., age, height, weight) for the individual data sheet as described in laboratory activity 1.1. The subject should put his or her shoes back on before continuing the test.

Step 3: Have the subject perform a structured warm-up designed to prepare for a maximal sprint bout (table 13.4).

Step 4: Direct the four timers to stand at the locations noted in figure 13.8. Make sure they can see both the start and finish of the sprint.

Step 5: Instruct the timers to start their stopwatches at the subject's first movement. Tell the subject that the objective of the test is to cover the distance as quickly as possible and that the timers will start their stopwatches upon the subject's first movement.

Step 6: Have the subject move to the starting line and assume the starting position.

Step 7: Before you start the test, have the subject and all timers acknowledge that they are ready.

Step 8: Encourage the subject to run as fast as possible through the 9.14 m, 36.6 m, 45.7 m, and 54.9 m lines and then decelerate in the designated zone.

Step 9: The timers should stop their stopwatch when the subject passes the individual finish line at which the timer is standing—9.14 m, 36.6 m, 45.7 m, or 54.9 m. For example, as the sprinter passes the 9.14 m mark, the timer standing at this point should stop the watch to mark the sprinter's 9.14 m time.

Step 10: Record the time achieved for each distance to the nearest tenth of a second and note it in the appropriate location on the individual data sheet.

Step 11: Have subject rest and recover for 3 min before initiating the next sprint. Repeat steps 4 through 10.

Step 12: After completing the test, have the subject perform a 5 to 10 min cool-down.

Step 13: Calculate the results of the test. Average trials 1 and 2 for each distance covered and record the results in the appropriate locations on the individual and group data sheets.

Step 14: Calculate the velocity of each sprint distance with the following formula:

$$\text{Velocity (m} \cdot \text{s}^{-1}) = \text{distance (m) / time (s)}$$

Record the results of this calculation on the individual data sheet.

Step 15: Calculate the horizontal power output for each distance with the following equation:

$$\text{Power (N} \cdot \text{m} \cdot \text{s}^{-1}) = \text{force} \times \text{velocity} = \text{body mass (N)} \times \text{velocity (m} \cdot \text{s}^{-1})$$

Record the results of this calculation on the individual and group data sheets.

Step 16: Use the following formula to calculate the acceleration between these marks: 0 and 9.14 m, 0 and 36.6 m, 0 and 45.7 m, 0 and 54.9 m, 9.14 m and 36.6 m, 36.6 m and 45.7 m, and 45.7 m and 54.9 m.

$$\text{Acceleration } (m \cdot s^{-2}) = \frac{\text{final velocity } (m \cdot s^{-1}) - \text{initial velocity } (m \cdot s^{-1})}{\text{time during interval (s)}}$$

Record the results of this calculation on the individual and group data sheets. For example, if the sprinter ran 9.14 m in 2 s, then the final velocity would be 4.57 m · s⁻¹. Because the subject is starting from a standing position, the initial velocity would be equal to 0 m · s⁻¹. Therefore, the acceleration would be calculated as follows:

$$\text{Acceleration } (m \cdot s^{-2}) = \frac{4.57 \ m \cdot s^{-1} - 0 \ m \cdot s^{-1}}{2 \ s} = 2.285 \ m \cdot s^{-2}$$

Thus, the subject is accelerating at a rate of 2.285 m · s⁻².

Step 17: Compare the results of this test with those presented in table 13.5.

Step 18: Use Excel or an equivalent spreadsheet program to graph the individual and class velocities achieved at each distance.

Table 13.4 Warm-Up for Sprints

Time (min:s)	Activity	
0:00-5:00	General warm-up	Activities such as jogging, cycling, and rope jumping
5:00-10:00	Dynamic stretching warm-up	Activities such as leg swings, high knees, walking knee tucks, walking lunges, butt kicks, inchworms, and power skips
10:00-15:00	Sprint-specific warm-up	Series of moderate-speed sprints that get progressively faster (while practicing correct starting position)
15:00-17:00	Recovery	Active rest in which the subject prepares for the all-out sprint test

Table 13.5 Percentile Ranks for the 36.6 m (40 yd) Sprint

AGE (y)	12-13		14-15		16-18	
Percentile	Male	Female	Male	Female	Male	Female
90	5.41	5.79	5.02	5.36	4.76	4.93
80	5.63	6.14	5.15	5.68	4.85	5.22
70	5.77	6.49	5.24	6.01	4.90	5.52
60	5.84	6.84	5.32	6.33	4.98	5.82
50	5.97	7.19	5.46	6.65	5.10	6.11
40	6.08	7.54	5.54	6.97	5.13	6.41
30	6.25	7.89	5.78	7.30	5.21	6.71
20	6.32	8.24	6.02	7.62	5.30	7.00
10	6.64	8.59	6.08	7.95	5.46	7.31
Mean	5.99	7.19	5.51	6.65	5.08	6.11

Adapted from Hoffman 2002; Housh et al. 2009.

Question Set 13.1

1. Based on your time, what energy systems contributed to your sprinting performance?

2. If the sprint were extended to 400 m, how would the contribution of the various energy systems change?

3. At what interval (0-9.14 m, 9.14-36.6 m, 36.6-45.7 m, or 45.7-54.9 m) was the highest velocity achieved? How do your individual data compare with the class data?

4. At what point during the sprint was the highest acceleration achieved? How do your individual data compare with the class data?

5. How do the class data for the 36.6 m distance compare with the norm table (see table 13.5) for sprint performance?

6. Were there any power output differences between intervals? If so, what may have contributed to them?

7. Are there any sex differences in sprint times and power outputs? Physiologically, how could these differences be explained?

8. Explain in detail the relationship between muscle fiber type and sprinting performance. (Hint: What fiber type would be important for sprinting? Why?)

LABORATORY ACTIVITY 13.1

INDIVIDUAL DATA SHEET

Name or ID number: _____ Date: _____

Tester: _____ Time: _____

Sex: M / F (circle one)   Age: _____ y   **Location of Testing**   **Footwear**

Height: _____ in. _____ m   ❑ Outdoor Field   ❑ Jogging shoe

Weight: _____ lb _____ kg   ❑ Indoor Field   ❑ Walking shoe

Temperature: _____ °F _____ °C   ❑ Indoor Track   ❑ Tennis shoe

Barometric pressure: _____ mmHg   ❑ Outdoor Track   ❑ Basketball shoe

Relative humidity: _____ %   ❑ Gym   ❑ Running Shoe

❑ Other _____   ❑ Cross Training Shoe

❑ CrossFit Shoe

Sprint Times

❑ Other _____

Round the sprint times to the nearest 0.1 s.

9.14 m SPRINT			
Trial	Distance (m)	Time (s)	Velocity (m · s⁻¹)
1	9.14		
2	9.14		
Mean	——		

36.6 m SPRINT			
Trial	Distance (m)	Time (s)	Velocity (m · s⁻¹)
1	36.6		
2	36.6		
Mean	——		

45.7 m SPRINT			
Trial	Distance (m)	Time (s)	Velocity (m · s⁻¹)
1	45.7		
2	45.7		
Mean	——		

54.9 m SPRINT			
Trial	Distance (m)	Time (s)	Velocity (m · s⁻¹)
1	54.9		
2	54.9		
Mean	——		

Calculation of Horizontal Power

Power (N · m · s⁻¹) = force × velocity = body mass (N) × velocity (m · s⁻¹)

$$\text{Power (N} \cdot \text{m} \cdot \text{s}^{-1}) = \text{force} \times \text{velocity} = \text{body mass (N)} \times \text{velocity (m} \cdot \text{s}^{-1})$$

Distance (m)	Body mass (N)		Velocity (m · s⁻¹)		Power (N · m · s⁻¹)
9.14		×		=	
36.6		×		=	
45.7		×		=	
54.9		×		=	

To convert body mass to newtons, multiply body mass in kg × 9.81 m · s⁻².

Calculation of Acceleration

$$\text{Acceleration (m} \cdot \text{s}^{-2}) = \frac{\text{final velocity (m} \cdot \text{s}^{-1}) - \text{initial velocity (m} \cdot \text{s}^{-1})}{\text{time during interval (s)}}$$

Intervals	Acceleration calculation	Acceleration (m · s⁻²)
Acceleration from 0-9.14 m	_____ (m · s⁻¹) − _____ (m · s⁻¹) / _____ (s)	
Acceleration from 9.14-36.6 m	_____ (m · s⁻¹) − _____ (m · s⁻¹) / _____ (s)	
Acceleration from 36.6 m-45.7 m	_____ (m · s⁻¹) − _____ (m · s⁻¹) / _____ (s)	
Acceleration from 45.7-54.9 m	_____ (m · s⁻¹) − _____ (m · s⁻¹) / _____ (s)	
Acceleration from 0-36.6 m	_____ (m · s⁻¹) − _____ (m · s⁻¹) / _____ (s)	
Acceleration from 0-45.7 m	_____ (m · s⁻¹) − _____ (m · s⁻¹) / _____ (s)	
Acceleration from 0-54.9 m	_____ (m · s⁻¹) − _____ (m · s⁻¹) / _____ (s)	

Jumping Performance

Equipment

- Area in which to perform a dynamic warm-up
- Vertec or wall-mounted board for marking vertical displacements
- Calculator
- Individual data sheet
- Group data sheet
- Microsoft Excel or equivalent spreadsheet program

Warm-Up

This test requires an area with sufficient ceiling clearance and with space for jumpers to perform an appropriate warm-up. As with any physical test, it is essential to have the subject perform a proper warm-up in order to maximize performance. Begin the general warm-up with activities such as jogging, cycling, or rope jumping. Then have the subject move on to a dynamic stretching routine consisting of activities that target the lower body (e.g., body-weight squats, squat thrusts, walking lunges, butt kicks, high knees, walking knee tucks). Next, have the subject complete a warm-up tailored to the vertical jump, practicing both the countermovement (see figure 13.1) and the static (see figure 13.2) vertical jump tests. A summary of this warm-up procedure can be found in table 13.6.

Table 13.6 Warm-Up for Vertical Jump

Time (min:s)	Activity	
0:00-5:00	General warm-up	Activities such as jogging, cycling, and rope jumping
5:00-10:00	Dynamic stretching warm-up	Activities such as leg swings, high knees, walking knee tucks, walking lunges, butt kicks, inchworms, and power skips
10:00-15:00	Warm-up specific to the vertical jump	Series of static and countermovement jumps that allow the subject to practice jumping positions and motions
15:00-17:00	Recovery	Active rest in which the subject prepares for the all-out jumping tasks

Countermovement Vertical Jump Test

Step 1: Set up the testing area by positioning the Vertec with plenty of room around it. If you are using a wall-mounted jump board, make sure it is clearly marked and easily accessible.

Step 2: Gather basic data (e.g., age, height, weight) for the individual data sheet as described in laboratory activity 1.1. Have the subject put his or her shoes back on.

Step 3: Direct the subject through a structured warm-up to prepare for the vertical jump test (see table 13.6).

Step 4: Explain to the subject the position used to establish reach height. Remember that reach height is measured by having the subject stand with the feet together and the dominant side near the wall or Vertec. In this position, have the subject reach as high as possible with the dominant hand while the feet remain flat on the floor. If using a wall-mounted measuring device, the subject should keep the palm flat against the wall. Record the highest reach point. If using a Vertec, the subject should reach as high as possible, moving as many vanes as possible until the highest displacement is achieved. Record the highest value as the reach height in the appropriate location on the individual data sheet.

Step 5: After establishing the reach height, explain the procedures associated with the countermovement vertical jump test. Demonstrate the jumping process.

Step 6: Position the subject so that the feet are approximately shoulder-width apart, and instruct the subject to remain in this position (see figure 13.1*a*) during the jump. After achieving this position, instruct the subject to perform a countermovement, dipping the hips and legs while swinging the arms prior to pushing off the ground (see figure 13.1*b*). While in the air, the subject should reach for the highest possible point on the wall or hit the highest possible vanes of the Vertec. Upon landing, the subject should bend the knees to absorb landing forces (see figure 13.1*d*).

Step 7: Have the subject perform a total of three countermovement vertical jumps separated by 2 min each. Record the vertical jump height achieved during each jump on the individual data sheet.

Step 8: Subtract the reach height from the highest vertical jump height. Convert this measure to centimeters (if it is initially calculated in inches), and average the results in the appropriate location on the individual data sheet. Transfer these results to the group data sheet.

Static Vertical Jump Test

Step 1: After completing the countermovement vertical jump test, explain and demonstrate the static vertical jump test to the subject. As with the countermovement test, have the subject assume a position in which the feet are approximately shoulder-width apart (see figure 13.2*a*). After achieving this position, instruct the subject to squat so that the tops of the thighs are parallel to the floor (see figure 13.2*b*). Have the subject hold this squat position for a count of three. Then encourage the subject to rapidly extend the knees, hips, and ankles while swinging the arms up in order to touch the highest possible point on the wall or move the highest possible vane on the Vertec (see figure 13.2*c*). Encourage the subject to bend the knees upon landing to absorb landing forces (see figure 13.2*d*).

Step 2: Have the subject perform three static vertical jumps separated by 2 min each. Record the vertical jump height achieved during each jump on the individual data sheet.

Step 3: Subtract the reach height from the highest vertical jump height. Convert this measure to centimeters (if it is initially calculated in inches), and average the results in the appropriate location on the individual data sheet. Transfer the average jump height to the group data sheet.

Calculations

Step 1: Calculate the power outputs for both the countermovement and static vertical jump tests using the equations presented on the individual data sheet.

Step 2: Calculate the EUR by dividing the vertical heights and power outputs calculated from the countermovement vertical jump test by the results of the static jump test.

Question Set 13.2

1. Was there any difference between the highest vertical jump heights for the countermovement and static vertical jump tests? If so, what neuromuscular characteristics might explain these findings?

2. When you examine the calculated power outputs, how does the Lewis equation compare with the Johnson average and the Harman average power equations? With the Johnson and Harman peak power equations? The Sayers equation? If differences exist between the results for the various formulas, explain what may contribute to these differences.

3. How do the Johnson and Harman equation results compare?

4. Were there any sex differences in either the static or the countermovement vertical jump power outputs? If so, what neuromuscular characteristics might explain these differences?

5. When you look at the EUR, how do your results compare with those for the class for vertical jump height? For power output?

6. Explain in detail what neuromuscular characteristics can contribute to a high EUR and how these characteristics might be affected by training.

LABORATORY ACTIVITY 13.2

INDIVIDUAL DATA SHEET

Name or ID number: _____ Date: _____

Tester: _____ Time: _____

Sex: M / F (circle one)   Age: _____ y

Height: _____ in. _____ m

Weight: _____ lb _____ kg

Temperature: _____°F _____°C

Barometric pressure: _____ mmHg

Relative humidity: _____ %

Location of Testing

❏ Outdoor Field
❏ Indoor Field
❏ Indoor Track
❏ Outdoor Track
❏ Gym
❏ Other _____

Footwear

❏ Jogging shoe
❏ Walking shoe
❏ Tennis shoe
❏ Basketball shoe
❏ Running Shoe
❏ Cross Training Shoe
❏ CrossFit Shoe
❏ Other _____

Countermovement Jump Test Results

Reach height = _____ in. = _____ in. / 0.3937 = _____ cm

Trial	VERTICAL DISPLACEMENT (in.)				
	Jump height (in.)	−	Reach height (in.)	=	Jump height
1		−		=	
2		−		=	
3		−		=	
Mean				=	

Trial	VERTICAL DISPLACEMENT (cm)				
	Jump height (cm)	−	Reach height (cm)	=	Jump height
1		−		=	
2		−		=	
3		−		=	
Mean				=	

Countermovement Vertical Jump Power Calculations

When working with the power equations, insert the mean displacements into the following equation.

Lewis Power

$$\text{Power}(W) = \sqrt{4.9} \times \underset{\text{weight(kg)}}{\underline{\hspace{2cm}}} \times \sqrt{\underset{\text{jump height(m)}}{\underline{\hspace{2cm}}}} = \underline{\hspace{2cm}} \; W$$

Sayers Peak Power

$$\text{Peak power}(W) = 60.7 \times \underset{\text{jump height(cm)}}{\underline{\hspace{2cm}}} + 45.3 \times \underset{\text{body mass (kg)}}{\underline{\hspace{2cm}}} - 2{,}055 = \underline{\hspace{2cm}} \; W$$

Harman Peak Power

$$\text{Peak power (W)} = 61.9 \times \underset{\text{jump height (cm)}}{\underline{\hspace{2cm}}} + 36 \times \underset{\text{body mass (kg)}}{\underline{\hspace{2cm}}} + 1{,}822 = \underline{\hspace{2cm}} \; W$$

Harman Average Power

$$\text{Average power (W)} = 21.2 \times \underset{\text{jump height (cm)}}{\underline{\hspace{2cm}}} + 23 \times \underset{\text{body mass (kg)}}{\underline{\hspace{2cm}}} - 1{,}393 = \underline{\hspace{2cm}} \; W$$

Johnson Peak Power

$$\text{Peak power}(W) = 78.5 \times \underset{\text{jump height (cm)}}{\underline{\hspace{2cm}}} + 60.6 \times \underset{\text{body mass (kg)}}{\underline{\hspace{2cm}}} - 15.3 \times \underset{\text{height (cm)}}{\underline{\hspace{2cm}}} - 1{,}308 = \underline{\hspace{2cm}} \; W$$

Johnson Average Power

$$\text{Average power (W)} = 41.4 \times \underset{\text{jump height (cm)}}{\underline{\hspace{2cm}}} + 31.2 \times \underset{\text{body mass (kg)}}{\underline{\hspace{2cm}}} - 13.9 \times \underset{\text{height (cm)}}{\underline{\hspace{2cm}}} + 431 = \underline{\hspace{2cm}} \; W$$

Static Jump Test Results

Reach height = _____ in. = _____ in. / 0.3937 = _____ cm

	VERTICAL DISPLACEMENT (in.)				
Trial	Jump height (in.)	–	Reach height (in.)	=	Jump height
1		–		=	
2		–		=	
3		–		=	
Mean				=	

	VERTICAL DISPLACEMENT (cm)				
Trial	Jump height (cm)	–	Reach height (cm)	=	Jump height
1		–		=	
2		–		=	
3		–		=	
Mean				=	

Static Vertical Jump Power Calculations

When working with the power equations, insert the mean displacements into the following equation.

Lewis Power

$$\text{Power (W)} = \sqrt{4.9} \times \underbrace{}_{\text{weight (kg)}} \times \sqrt{\underbrace{}_{\text{jump height (m)}}} = \underline{} \text{ W}$$

Sayers Peak Power

$$\text{Peak power (W)} = 60.7 \times \underbrace{}_{\text{jump height (cm)}} + 45.3 \times \underbrace{}_{\text{body mass (kg)}} - 2{,}055 = \underline{} \text{ W}$$

Harman Peak Power

$$\text{Peak power (W)} = 61.9 \times \underbrace{}_{\text{jump height (cm)}} + 36 \times \underbrace{}_{\text{body mass (kg)}} + 1{,}822 = \underline{} \text{ W}$$

Harman Average Power

$$\text{Average power (W)} = 21.2 \times \underbrace{}_{\text{jump height (cm)}} + 23 \times \underbrace{}_{\text{body mass (kg)}} - 1{,}393 = \underline{} \text{ W}$$

Johnson Peak Power

$$\text{Peak power (W)} = 78.5 \times \underbrace{}_{\text{jump height (cm)}} + 60.6 \times \underbrace{}_{\text{body mass (kg)}} - 15.3 \times \underbrace{}_{\text{height (cm)}} - 1{,}308 = \underline{} \text{ W}$$

Johnson Average Power

$$\text{Average power (W)} = 41.4 \times \underbrace{}_{\text{jump height (cm)}} + 31.2 \times \underbrace{}_{\text{body mass (kg)}} - 13.9 \times \underbrace{}_{\text{height (cm)}} + 431 = \underline{} \text{ W}$$

Eccentric Utilization Ratio (EUR) Calculation

To calculate the EUR, use the following:

Vertical Displacement

$$\underbrace{}_{\text{countermovement jump (cm)}} / \underbrace{}_{\text{static jump (cm)}} = \underbrace{}_{\text{eccentric utilization ratio}}$$

Lewis

$$\underbrace{}_{\text{countermovement jump (W)}} / \underbrace{}_{\text{static jump (W)}} = \underbrace{}_{\text{eccentric utilization ratio}}$$

Harman Peak Power

$$\underbrace{}_{\text{countermovement jump (W)}} / \underbrace{}_{\text{static jump (W)}} = \underbrace{}_{\text{eccentric utilization ratio}}$$

Harman Average Power

$$\underbrace{}_{\text{countermovement jump (W)}} / \underbrace{}_{\text{static jump (W)}} = \underbrace{}_{\text{eccentric utilization ratio}}$$

Sayers Peak Power

$$\frac{\text{\underline{\hspace{4cm}}}}{\text{countermovement jump (W)}} \ / \ \frac{\text{\underline{\hspace{4cm}}}}{\text{static jump (W)}} = \frac{\text{\underline{\hspace{4cm}}}}{\text{eccentric utilization ratio}}$$

Johnson Average Power

$$\frac{\text{\underline{\hspace{4cm}}}}{\text{countermovement jump (W)}} \ / \ \frac{\text{\underline{\hspace{4cm}}}}{\text{static jump (W)}} = \frac{\text{\underline{\hspace{4cm}}}}{\text{eccentric utilization ratio}}$$

Johnson Peak Power

$$\frac{\text{\underline{\hspace{4cm}}}}{\text{countermovement jump (W)}} \ / \ \frac{\text{\underline{\hspace{4cm}}}}{\text{static jump (W)}} = \frac{\text{\underline{\hspace{4cm}}}}{\text{eccentric utilization ratio}}$$

Jumping Performance With a Switch Mat

Equipment

- Area in which to perform a dynamic warm-up
- Boxes (20 cm, 30 cm, 40 cm, 50 cm, and 60 cm)
- Switch mat
- Calculator
- Individual data sheet
- Group data sheet
- Microsoft Excel or equivalent spreadsheet program

Warm-Up

This test requires an area with sufficient ceiling clearance and with space for the subject to perform an appropriate warm-up. As with any physical test, it is essential to have the subject perform a proper warm-up in order to maximize performance. Begin the general warm-up with activities such as jogging, cycling, or rope jumping. Then have the subject move on to a dynamic stretching routine consisting of activities that target the lower body (e.g., body-weight squats, squat thrusts, walking lunges, butt kicks, high knees, walking knee tucks). Next, have the subject complete a warm-up tailored to the vertical jump, practicing both the countermovement (see figure 13.1) and static (see figure 13.2) vertical jump tests on the switch mat. A summary of this warm-up procedure can be found in table 13.6.

Countermovement Vertical Jump Test

Step 1:   Prepare the testing area by positioning the switch mat with plenty of room around it and appropriate ceiling clearance above it.

Step 2:   Gather basic data (e.g., age, height, weight) for the individual data sheet as described in laboratory activity 1.1. Have the subject put his or her shoes back on.

Step 3:   Direct the subject to perform a structured warm-up to prepare for the vertical jump test (table 13.6).

Step 4:   Explain the protocol for the countermovement jump and demonstrate the technique used in performing this type of jump.

Step 5:   Position the subject so that the feet are approximately shoulder-width apart (see figure 13.3a). Instruct the subject to dip the hips and legs while keeping the hands at the hips and then rapidly extend the legs and hips to push off the ground (see figure 13.3b). Encourage the subject to extend the body as much as possible while in the air (see figure 13.3c). Instruct the subject to bend the knees when landing to absorb landing forces (see figure 13.3d).

Step 6:   Have the subject perform a total of three countermovement vertical jumps separated by 2 min each. Record the flight time achieved during each jump on the individual data sheet.

Step 7:   Calculate the mean jump height for the three static vertical jumps preformed and place this value in the appropriate location on the individual and group data sheets.

Static Vertical Jump Test

Step 1: Explain the protocol for the static vertical jump and demonstrate the appropriate technique for performing this type of jump.

Step 2: Have the subject assume a position in which the feet are shoulder-width apart and the hands are on the hips (see figure 13.2*a*).

Step 3: Instruct the subject to squat to a position in which the tops of the thighs are parallel to the ground (see figure 13.2*b*) while keeping the hands on the hips.

Step 4: Give a countdown: "*3, 2, 1, jump.*" Encourage the subject to rapidly extend the knees, hips, and ankles to jump off the switch mat (see figure 13.2*c*). Encourage the subject to bend the knees upon landing to absorb landing forces (see figure 13.2*d*).

Step 5: Have the subject perform a total of three static vertical jumps separated by 2 min each. Record the flight time achieved during each jump on the individual data sheet.

Step 6: Calculate the mean jump height for the three static vertical jumps preformed and place this value in the appropriate location on the individual and group data sheets.

Calculations

Step 1: For each jump performed in this lab, convert the flight time to vertical displacement with the following equation:

$$\text{Vertical displacement (m)} = \frac{9.81(\text{m} \cdot \text{s}^{-2}) \times \text{flight time (s)} \times \text{flight time (s)}}{8}$$

Step 2: Use the various formulas presented on the individual data sheet for this lab to calculate the countermovement and static vertical jump power outputs. Record these values in the appropriate locations on the individual and group data sheets.

Step 3: Calculate the EUR for the vertical displacement and the power outputs achieved during the countermovement and static vertical jumps, and record it in the appropriate location on the individual and group data sheets.

Step 4: Compare the EUR with the norms presented in table 13.7.

Table 13.7 Eccentric Utilization Ratio (EUR) Norms

Sport	Men	Women
Field hockey		1.02 ± 0.13
Rugby union	1.13 ± 0.14	
Soccer	1.14 ± 0.15	1.17 ± 0.16
Softball	1.03 ± 0.09	1.04 ± 0.13
Athletics: Throwing	1.11 ± 0.10	1.05 ± 0.06
Athletics: Jumping and sprinting	1.11 ± 0.10	1.11 ± 0.10
Untrained	1.17 ± 0.17	

Data from McGuigan et al. 2006; Hawkins et al. 2009; unpublished data.

RSI

Step 1: Prepare the testing area by positioning the 20 cm (7.9 in.) box with a switch mat placed at least 0.2 m in front of the box. Ensure appropriate ceiling clearance above the box so that athlete does not hit the ceiling.

Step 2: Give the athlete these instructions: Step forward off the box without stepping down or jumping downward or upward. Upon contact with the ground, minimize contact time with the ground and jump as high as possible (see figure 13.4, *a-c*).

Step 3: Have the athlete stand on top of the box to initiate the test.

Step 4: The athlete places the hands on the hips, steps off the box, and minimizes contact with the ground while jumping as high as possible.

Step 5: Record the jump height and contact time from the switch match on the individual data sheet

Step 6: Repeat steps 3 through 5.

Step 7: Switch the box with a 30 cm (11.8 in.) box and repeat steps 3 through 5 a total of two times.

Step 8: Switch the box with a 40 cm (15.7 in.) box and repeat steps 3 through 5 a total of two times.

Step 9: Switch the box with a 50 cm (19.7 in.) box and repeat steps 3 through 5 a total of two times.

Step 10: Switch the box with a 60 cm (23.6 in.) box and repeat steps 3 through 5 a total of two times.

Step 11: Calculate the RSI for each box and record them in the appropriate space on the individual data sheet. Calculate the mean for the two trials performed at each box height, and record them on the individual and group data sheets.

Step 12: Determine which box height achieves the best RSI results and record it on the individual data sheet.

Question Set 13.3

1. Was there any difference between the highest vertical jump displacement for the countermovement and static vertical jump tests? Is this expected? Why or why not? If there is a difference, can it be explained in neuromuscular terms? How?

2. When you examine the calculated power outputs, how does the Lewis equation compare with the Johnson average and the Harman average power equations? With the Johnson and Harman peak power equations? The Sayers equation? What might explain these differences?

3. How do the Johnson and Harman equation results compare?

4. Were there any sex differences in either the static or the countermovement vertical jump power outputs?

5. When you look at the EUR, how do your results compare with those for the class for vertical displacement? For power output?

6. Explain procedural issues that can increase the reliability of testing on a switch mat. Be specific.

7. What are some pros and cons of using switch mats in a testing situation?

8. How do the heights achieved in the countermovement and static vertical jump compare with the heights achieved with the various box jumps used in the RSI test?

LABORATORY ACTIVITY 13.3

INDIVIDUAL DATA SHEET

Name or ID number: _____ Date: _____

Tester: _____ Time: _____

Sex: M / F (circle one)   Age: _____ y

Height: _____ in. _____ m

Weight: _____ lb _____ kg

Temperature: _____°F _____ °C

Barometric pressure: _____ mmHg

Relative humidity: _____ %

Location of Testing

❏ Outdoor Field
❏ Indoor Field
❏ Indoor Track
❏ Outdoor Track
❏ Gym
❏ Other _____

Footwear

❏ Jogging shoe
❏ Walking shoe
❏ Tennis shoe
❏ Basketball shoe
❏ Running Shoe
❏ Cross Training Shoe
❏ CrossFit Shoe
❏ Other _____

Countermovement Jump Test Results

To determine the vertical displacement, use the following equation:

$$\text{Vertical displacement (m)} = \frac{9.81 (\text{m} \cdot \text{s}^{-2}) \times \text{flight time (s)} \times \text{flight time (s)}}{8}$$

Trial	Flight time (s)	Calculation	=	Displacement (m)
1			=	
2			=	
3			=	
Mean			=	

Countermovement Vertical Jump Power Calculations

When working with the power equations, insert the mean displacements into the following equation.

Lewis Power

$$\text{Power (W)} = \sqrt{4.9} \times \underset{\text{weight (kg)}}{\underline{\hspace{1.5cm}}} \times \sqrt{\underset{\text{jump height (m)}}{\underline{\hspace{1.5cm}}}} = \underline{\hspace{1.5cm}} \text{W}$$

Sayers Peak Power

$$\text{Peak power (W)} = 60.7 \times \underset{\text{jump height (cm)}}{\underline{\hspace{1.5cm}}} + 45.3 \times \underset{\text{body mass (kg)}}{\underline{\hspace{1.5cm}}} - 2{,}055 = \underline{\hspace{1.5cm}} \text{W}$$

Harman Peak Power

$$\text{Peak power (W)} = 61.9 \times \underset{\text{jump height (cm)}}{\underline{\hspace{1.5cm}}} + 36 \times \underset{\text{body mass (kg)}}{\underline{\hspace{1.5cm}}} + 1{,}822 = \underline{\hspace{1.5cm}} \text{W}$$

Harman Average Power

$$\text{Average power (W)} = 21.2 \times \underset{\text{jump height (cm)}}{\underline{\hspace{1.5cm}}} + 23 \times \underset{\text{body mass (kg)}}{\underline{\hspace{1.5cm}}} - 1{,}393 = \underline{\hspace{1.5cm}} \text{W}$$

Johnson Peak Power

$$\text{Peak power (W)} = 78.5 \times \underline{\hspace{3cm}}_{\text{jump height (cm)}} + 60.6 \times \underline{\hspace{3cm}}_{\text{body mass (kg)}} - 15.3 \times \underline{\hspace{2cm}}_{\text{height (cm)}} - 1{,}308 = \underline{\hspace{1.5cm}} \text{ W}$$

Johnson Average Power

$$\text{Average power (W)} = 41.4 \times \underline{\hspace{3cm}}_{\text{jump height (cm)}} + 31.2 \times \underline{\hspace{3cm}}_{\text{body mass (kg)}} - 13.9 \times \underline{\hspace{2cm}}_{\text{height (cm)}} + 431 = \underline{\hspace{1.5cm}} \text{ W}$$

Static Jump Test Results

To determine the vertical displacement, use the following equation:

$$\text{Vertical displacement (m)} = \frac{9.81\,(\text{m}\cdot\text{s}^{-2}) \times \text{flight time (s)} \times \text{flight time (s)}}{8}$$

Trial	Flight time (s)	Calculation	=	Displacement (m)
1			=	
2			=	
3			=	
Mean			=	

Static Vertical Jump Power Calculations

When working with the power equations, insert the mean displacements into the following equation.

Lewis Power

$$\text{Power (W)} = \sqrt{4.9} \times \underline{\hspace{2cm}}_{\text{weight (kg)}} \times \sqrt{\underline{\hspace{2cm}}_{\text{jump height (m)}}} = \underline{\hspace{2cm}} \text{ W}$$

Sayers Peak Power

$$\text{Peak power (W)} = 60.7 \times \underline{\hspace{2cm}}_{\text{jump height (cm)}} + 45.3 \times \underline{\hspace{2cm}}_{\text{body mass (kg)}} - 2{,}055 = \underline{\hspace{2cm}} \text{ W}$$

Harman Peak Power

$$\text{Peak power (W)} = 61.9 \times \underline{\hspace{3cm}}_{\text{jump height (cm)}} + 36 \times \underline{\hspace{3cm}}_{\text{body mass (kg)}} + 1{,}822 = \underline{\hspace{1.5cm}} \text{ W}$$

Harman Average Power

$$\text{Average power (W)} = 21.2 \times \underline{\hspace{3cm}}_{\text{jump height (cm)}} + 23 \times \underline{\hspace{3cm}}_{\text{body mass (kg)}} - 1{,}393 = \underline{\hspace{1.5cm}} \text{ W}$$

Johnson Peak Power

$$\text{Peak power (W)} = 78.5 \times \underline{\hspace{3cm}}_{\text{jump height (cm)}} + 60.6 \times \underline{\hspace{3cm}}_{\text{body mass (kg)}} - 15.3 \times \underline{\hspace{2cm}}_{\text{height (cm)}} - 1{,}308 = \underline{\hspace{1.5cm}} \text{ W}$$

Johnson Average Power

$$\text{Average power (W)} = 41.4 \times \underline{\hspace{3cm}}_{\text{jump height (cm)}} + 31.2 \times \underline{\hspace{3cm}}_{\text{body mass (kg)}} - 13.9 \times \underline{\hspace{2cm}}_{\text{height (cm)}} + 431 = \underline{\hspace{1.5cm}} \text{ W}$$

Eccentric Utilization Ratio (EUR) Calculation

To calculate the EUR, use the following:

Vertical Displacement

$$\frac{\rule{3cm}{0.4pt}}{\text{countermovement jump (cm)}} / \frac{\rule{3cm}{0.4pt}}{\text{static jump (cm)}} = \frac{\rule{3cm}{0.4pt}}{\text{eccentric utilization ratio}}$$

Lewis

$$\frac{\rule{3cm}{0.4pt}}{\text{countermovement jump (W)}} / \frac{\rule{3cm}{0.4pt}}{\text{static jump (W)}} = \frac{\rule{3cm}{0.4pt}}{\text{eccentric utilization ratio}}$$

Harman Peak Power

$$\frac{\rule{3cm}{0.4pt}}{\text{countermovement jump (W)}} / \frac{\rule{3cm}{0.4pt}}{\text{static jump (W)}} = \frac{\rule{3cm}{0.4pt}}{\text{eccentric utilization ratio}}$$

Harman Average Power

$$\frac{\rule{3cm}{0.4pt}}{\text{countermovement jump (W)}} / \frac{\rule{3cm}{0.4pt}}{\text{static jump (W)}} = \frac{\rule{3cm}{0.4pt}}{\text{eccentric utilization ratio}}$$

Sayers Peak Power

$$\frac{\rule{3cm}{0.4pt}}{\text{countermovement jump (W)}} / \frac{\rule{3cm}{0.4pt}}{\text{static jump (W)}} = \frac{\rule{3cm}{0.4pt}}{\text{eccentric utilization ratio}}$$

Johnson Average Power

$$\frac{\rule{3cm}{0.4pt}}{\text{countermovement jump (W)}} / \frac{\rule{3cm}{0.4pt}}{\text{static jump (W)}} = \frac{\rule{3cm}{0.4pt}}{\text{eccentric utilization ratio}}$$

Johnson Peak Power

$$\frac{\rule{3cm}{0.4pt}}{\text{countermovement jump (W)}} / \frac{\rule{3cm}{0.4pt}}{\text{static jump (W)}} = \frac{\rule{3cm}{0.4pt}}{\text{eccentric utilization ratio}}$$

Reactive Strength Index (RSI)

To determine the RSI, use the following equation:

$$\text{RSI} = \text{jump height (m)} / \text{contact time (s)}$$

Box height (cm)	Jump #	Jump height (m)	Contact time (s)	Calculation	RSI score
20	1				
	2				
	Mean				
30	1				
	2				
	Mean				
40	1				
	2				
	Mean				
50	1				
	2				
	Mean				
60	1				
	2				
	Mean				

Which box height is the best for this athlete?

_____

Power Endurance

Equipment

- Area in which to perform a dynamic warm-up
- Switch mat
- Calculator
- Individual data sheet
- Group data sheet
- Microsoft Excel or equivalent spreadsheet program

Warm-Up

This test can be performed only on a force platform or switch mat. As with other vertical jump tests, this test requires an area with sufficient space and ceiling clearance for safety. In addition, it is essential to have the subject perform a jump-specific warm-up to prepare for maximal repetitive jumping in order to ensure the accuracy of the testing protocol. Begin with a general warm-up consisting of activities such as jogging, cycling, or rope jumping. Then have the subject move on to a dynamic stretching routine consisting of activities that target the lower body (e.g., body-weight squats, squat thrusts, walking lunges, butt kicks, high knees, walking knee tucks). Next, have the subject complete a warm-up tailored to the vertical jump, practicing a series of countermovement vertical jumps (see figure 13.1) with hands on hips. A summary of this warm-up procedure can be found in table 13.8.

Table 13.8 Warm-Up for Multiple Vertical Jumps

Time (min:s)	Activity	
0:00-5:00	General warm-up	Activities such as jogging, cycling, and rope jumping
5:00-10:00	Dynamic stretching warm-up	Activities such as leg swings, high knees, walking knee tucks, walking lunges, butt kicks, inchworms, and power skips
10:00-15:00	Warm-up specific to vertical jump	Series of countermovement jumps that allow the subject to practice jumping positions and motions
15:00-17:00	Recovery	Active rest in which the subject prepares for the all-out jumping tasks

Power Endurance Jump Test

Step 1: Prep the test area by setting up the switch mat or force plate according to the manufacturer's standards.

Step 2: Gather basic data (e.g., age, height, weight) for the individual data sheet as described in laboratory activity 1.1. Have the subject put his or her shoes back on.

Step 3: Direct the subject through a structured warm-up to prepare for multiple jumping tasks (see table 13.8).

Step 4: Explain the power endurance test to the subject. The subject is to perform as many countermovement vertical jumps as possible in the 1 min time frame. Make it clear that the objective includes jumping as high as possible on each jump.

Step 5: Explain that the subject should bend the knees to about 90° during each contact phase of the jumping.

Step 6: Initiate the test by having the subject stand on the force platform or switch mat with the hands on the hips. Give a countdown: "*3, 2, 1, jump.*" When you say "*jump,*" the subject should jump as high as possible and continue doing so until 1 min has elapsed.

Step 7: Count the number of jumps completed by the subject in 15 s, 30 s, and 60 s, recording these values in the appropriate locations on the individual and group data sheets.

Step 8: Summate and record the flight times achieved during the 60 s test.

Step 9: Calculate the individual's mechanical power over 15 s with the following equation:

$$\text{Mechanical power } (\text{W} \cdot \text{kg}^{-1}) = \frac{9.81^2 \times T_f \times 15}{4 \times n \times (15 - T_f)}$$

where T_f = total flight time and n = number of jumps. Record the resulting information on the individual and group data sheets.

Step 10: Calculate the individual's mechanical power over 30 s with the following equation:

$$\text{Mechanical power } (\text{W} \cdot \text{kg}^{-1}) = \frac{9.81^2 \times T_f \times 30}{4 \times n \times (30 - T_f)}$$

where T_f = total flight time and n = number of jumps. Record the resulting information on the individual and group data sheets.

Step 11: Calculate the individual's mechanical power over 60 s with the following equation:

$$\text{Mechanical power } (\text{W} \cdot \text{kg}^{-1}) = \frac{9.81^2 \times T_f \times 60}{4 \times n \times (60 - T_f)}$$

where T_f = total flight time and n = number of jumps. Record the resulting information on the individual and group data sheets.

Question Set 13.4

1. How does your mechanical power output compare with that of your classmates? With the norms presented in table 13.9 and table 13.10?

2. How do your mechanical power outputs for the 15 s, 30 s, and 60 s intervals compare with each other? Is this what you would expect? Give a physiological explanation for this finding.

3. Using Excel or an equivalent spreadsheet program, graph the mechanical power outputs for yourself and for the class at 15 s, 30 s, and 60 s. (Hint: Use time as the *x*-axis and power as the *y*-axis.)

4. What bioenergetic systems are targeted in this test?

Table 13.9 Norms for Power Endurance Jump Tests Lasting 60 s

Classification	Number of jumps in 60 s	Power output (W · kg⁻¹)
Male basketball players	56.8 ± 4.3	19.8 ± 2.2
Male volleyball players	50.8 ± 2.7	19.6 ± 2.6
Adolescent males	63.2 ± 5.8	22.2 ± 1.8
Female athletes		12.2 ± 2.4
Male athletes		17.8 ± 2.7

Data from Bosco et al. 1983; Hespanhol et al. 2007; Sands et al. 2004.

Table 13.10 Percentile Ranks for Power Endurance Jump Tests Lasting 60 s

Percentile rank	MALE Absolute (W)	MALE Relative (W · kg⁻¹)	FEMALE Absolute (W)	FEMALE Relative (W · kg⁻¹)
95	2,385	29.85	961	15.32
90	1,556	19.90	885	13.46
85	1,481	18.80	848	13.34
80	1,464	17.80	810	13.26
75	1,395	17.35	746	12.80
70	1,367	17.30	740	12.52
65	1,309	16.35	730	11.94
60	1,267	16.10	723	11.80
55	1,249	16.05	705	11.60
50	1,223	15.90	703	11.60
45	1,203	15.55	698	11.42
40	1,172	15.30	667	11.04
35	1,140	15.30	639	10.76
30	1,120	15.10	623	10.08
25	1,101	14.70	619	9.70
20	1,083	14.70	594	9.52
15	1,060	14.45	583	9.40
10	986	14.10	547	9.20
5	922	12.50	470	8.48
Mean	1,289	16.69	700	11.47

Reprinted, by permission, from P. Maud and C. Foster, 2005, *Physiological assessment of human fitness*, 2nd ed. (Champaign, IL: Human Kinetics), 229.

LABORATORY ACTIVITY 13.4

INDIVIDUAL DATA SHEET

Name or ID number: _____ Date: _____

Tester: _____ Time: _____

Sex: M / F (circle one)   Age: _____ y

Height: _____ in. _____ m

Weight: _____ lb _____ kg

Temperature: _____ °F _____ °C

Barometric pressure: _____ mmHg

Relative humidity: _____ %

Location of Testing
❏ Outdoor Field
❏ Indoor Field
❏ Indoor Track
❏ Outdoor Track
❏ Gym
❏ Other _____

Footwear
❏ Jogging shoe
❏ Walking shoe
❏ Tennis shoe
❏ Basketball shoe
❏ Running Shoe
❏ Cross Training Shoe
❏ CrossFit Shoe
❏ Other _____

Calculation of Mechanical Power

	15 s interval	30 s interval	60 s interval
Number of jumps completed (n)			
Total flight time (T_f)			

When calculating mechanical power, use the following equations:

15 s $$\text{Mechanical power (W} \cdot \text{kg}^{-1}) = \frac{9.81^2 \times T_f \times 15}{4 \times n \times (15 - T_f)}$$

30 s $$\text{Mechanical power (W} \cdot \text{kg}^{-1}) = \frac{9.81^2 \times T_f \times 30}{4 \times n \times (30 - T_f)}$$

60 s $$\text{Mechanical power (W} \cdot \text{kg}^{-1}) = \frac{9.81^2 \times T_f \times 60}{4 \times n \times (60 - T_f)}$$

Time interval (s)	Calculation	Mechanical power (W · kg⁻¹)
15		
30		
60		

Anaerobic Cycling Power

Equipment

- Monark cycle or equivalent ergometer equipped to perform a WAnT (e.g., Monark 814E, 824E, or 834E, all of which contain a peg or basket upon which the external load can be applied)
- Revolutions counter that calculates the number of pedal revolutions performed in the test
- Timing apparatus (e.g., laboratory set-timer, stopwatch, or timer attached to the cycle ergometer)
- Calibrated platform-beam scale or electronic digital scale
- Calculator
- Individual data sheet
- Group data sheet
- Microsoft Excel or equivalent spreadsheet program

This laboratory activity is supported by a virtual lab experience in the web study guide.

www

Laboratory Orientation

The WAnT should be performed with a cycle ergometer that can immediately apply an accurate constant force throughout the test. It is essential to select an appropriate cycle ergometer in order to maximize the accuracy of the test. Typically, testers choose a Monark cycle or equivalent in which a basket device applies the resistance load (figure 13.9). The test should be performed in a well-ventilated area.

Figure 13.9   Monark cycle ergometer.

The testing protocol is presented in table 13.11, which outlines four phases: warm-up, recovery, 30 s WAnT, and cool-down.

- *Warm-up:* As with all testing protocols, a proper warm-up is essential to minimizing injury risk and maximizing performance. The WAnT includes a standardized warm-up in which the subject cycles at about 60 to 70 rev · min^{-1} against 20% of the calculated resistance load for a total of 4 to 5 min. At the end of each minute, the subject performs an all-out sprint lasting 4 to 6 s. The load increases with each minute until the last 4 to 6 s sprint is performed with the target load.

- *Recovery:* After completing the warm-up, the subject should take 2 to 5 min to recover. This time is generally spent cycling against minimal resistance.

- *30 s WAnT:* The lead-in to the WAnT consists of two phases that allow the subject to progressively increase cycling rev · min^{-1}. The first phase, lasting 5 to 10 s, requires the subject to pedal at 20 to 50 rev · min^{-1} against one-third of the prescribed resistance. The second phase requires the subject to increase the pedaling rate to maximum, which generally takes 2 to 5 s. The prescribed load is then applied while the subject cycles at the highest possible rev · min^{-1} until the end of the 30 s.

- *Cool-down:* After completing the WAnT, it is essential that the subject perform a cool-down lasting 2 to 5 min. During this time, the subject should pedal at 25 to 100 W to facilitate recovery from the testing bout.

Thus, the basic structure of the WAnT is standardized, but it can be modified depending on the individual needs of the subject. Some typical modifications include lengthening the duration to 60 s or reducing the duration to 5, 10, or 15 s. One can also modify the resistance used during the test.

If the test is not performed with a computerized cycle ergometer, it requires four testers: timer, force setter, counter, and recorder. The timer initiates the test by yelling *"Go,"* maintains control of the testing clock, shouts out the time at 5 s

Table 13.11   WAnT Procedure

Step	Phase	Duration	Activity
1	Warm-up	4-5 min	Cycle at a comfortable pace (about 60-70 rev · min^{-1}) against a resistance equal to 20% of the calculated load for the subsequent test. At the end of each minute, perform a 4-6 s sprint; resistance should increase with each sprint until the last sprint is performed against the prescribed load.
2	Recovery	2-5 min	Rest or cycle slowly against a minimal resistance.
3	Wingate test	5-10 s	Pedal against one-third of the prescribed resistance at 20-50 rev · min^{-1}.
		2-5 s	Increase pedaling rate to maximum; once maximal pedaling rate is achieved, the prescribed resistance should be applied.
		30 s	Cycle at the highest possible rev · min^{-1} against the prescribed resistance.
4	Cool-down	2-5 min	Pedal at a low to moderate power level (25-100 W) for 2-5 min. If repeated tests are performed, the cool-down should be extended.

Adapted, by permission, from SHAPE America, 1989, "Norms for the Wingate Anaerobic Test with comparison to another similar test," *Research Quarterly for Exercise and Sport* Vol. 60(2): 144-151.

intervals during the test, and yells *"Stop"* when the 30 s time limit is reached (2). The resistance setter establishes the prescribed load, makes sure the load is applied at the right time and is maintained, and lowers the load while the subject performs the cool-down.

The counter's main job is to quantify the number of complete pedal revolutions during each 5 s interval. A complete pedal revolution involves full rotation of the pedal from its originating position. The counter relays the number of pedal revolutions for each period to the recorder, who writes the number of revolutions on an individual data sheet throughout the duration of the test. Though this manual process is functional, computers make the testing process much simpler (47), more accurate (thanks to automatic counts of the number of revolutions), and doable with fewer testers.

30 s WAnT

Step 1: Set up the cycle ergometer and check to see if it is in working order.

Step 2: Gather basic data (e.g., age, height, weight) for the individual data sheet as described in laboratory activity 1.1. Have the subject put on his or her shoes.

Step 3: Calculate the prescribed load based on the subject's training status (see table 13.3). Record this load on the individual data sheet.

Step 4: Fit the cycle ergometer to the subject so that while the subject is seated on the bike, the extended leg has a slight bend (5°-15°) (see figure 4.3*a* for leg angle and figure 13.10 for basic body position).

Figure 13.10 Body position in the WAnT.

Step 5: Explain the test protocol to the subject. Emphasize that this is an all-out effort lasting 30 s.

Step 6: Have the subject perform the standardized warm-up and recovery outlined in table 13.11.

Step 7: Initiate the test with the subject pedaling against one-third of the prescribed load at 20 to 50 rev · min^{-1} for 5 to 10 s.

Step 8: Instruct the subject to increase the pedaling rate to maximum. Instruct the force setter to apply the load and the timer to start the timer and yell *"Go"* when the subject reaches a maximal pedaling rate.

Step 9: Upon hearing the *"Go"* command, the counter should begin counting pedal revolutions. (This step is not performed if you are using a computerized system.)

Step 10: The counter should tell the recorder the total number of revolutions completed at the end of each 5 s time interval during the test (i.e., at 5, 10, 15, 20, 25, and 30 s).

Step 11: At the 30 s mark, the timer yells *"Stop,"* and the force setter reduces the load so that the subject can pedal at 25 to 100 W for 2 to 5 min or until recovered. At this time, the counter tells the recorder the number of total revolutions completed for the total 30 s ride. The recorder then determines the number of revolutions completed during each 5 s segment of the test.

Step 12: Convert the prescribed load to newtons by multiplying the load by 9.80665. Calculate the power output from the total number of revolutions, the distance the flywheel travels per rotation (6 m if using a Monark), and the duration of the interval. Calculate the power output for each time interval, and record it in the appropriate location on the individual and group data sheets.

Step 13: Divide the power output calculated for each time interval by body weight to determine a relative time interval. Record these values on the individual data sheet.

Step 14: Determine the absolute and relative peak power outputs. Record these values on the individual and group data sheets.

Step 15: Calculate the absolute and relative work accomplished during the 30 s interval. Record these values on the individual and group data sheets.

Step 16: Calculate the absolute and relative mean power outputs accomplished during the 30 s test. Record these values on the individual and group data sheets.

Step 17: Determine the absolute lowest power output and record this value on the individual data sheet.

Step 18: Calculate the fatigue index using the equation supplied on the individual and group data sheets.

Question Set 13.5

1. What energy system contributes to the ability to perform the WAnT? Bioenergetically, what limits performance during the first 10 s, from 0 to 20 s, and from 0 to 30 s?

2. What is the typical force setting for the WAnT for an untrained person? For an athlete? Would different force values affect the results of the test? If so, how?

3. How does one calculate the fatigue index, and what does it indicate?

4. During which time interval did you achieve your highest power output?

5. How does your absolute peak power output compare with that of the class average and with the norms presented in table 13.12?

6. Using Excel or an equivalent spreadsheet program, graph the power outputs achieved in each of the 5 s time intervals.

7. What was your fatigue index, and how does it compare with the class results?

8. Are there differences in performance results on the WAnT between males and females? If so, are these differences related to the peak power, mean power, or fatigue rate? Explain your answer.

9. How would the performance of someone with a high percentage of Type I fibers compare with that of a person with a high percentage of Type II fibers? What would you expect to see in terms of performance?

Table 13.12 WAnT Percentile Ranks for College-Aged Males and Females

	COLLEGE-AGED MALES (18-28 y; $N = 60$)		COLLEGE-AGED FEMALES (18-28 y; $N = 69$)	
Percentile rank	W	$W \cdot kg^{-1}$	W	$W \cdot kg^{-1}$
90	662	8.2	470	7.3
80	618	8.0	419	7.0
70	600	7.9	410	6.8
60	577	7.6	391	6.6
50	565	7.4	381	6.4
40	548	7.1	367	6.1
30	530	7.0	353	6.0
20	496	6.6	337	5.7
10	471	6.0	306	5.3

Norm data for WAnT test performed on a Monark cycle ergometer with a resistance of 0.075 kp $\cdot$ kg^{-1} body mass. W = Watts; $W \cdot kg^{-1}$ = watts per kilogram body mass.

Reprinted, by permission, from J.R. Hoffman, 2006, *Norms for fitness, performance, and health* (Champaign, IL: Human Kinetics), 54; Adapted, by permission, from SHAPE America, 1989, "Norms for the Wingate Anaerobic Test with comparison to another similar test," *Research Quarterly for Exercise and Sport* Vol. 60(2): 144-151.

LABORATORY ACTIVITY 13.5

INDIVIDUAL DATA SHEET

Name or ID number: _____ Date: _____

Tester: _____ Time: _____

Sex: M / F (circle one)   Age: _____ y

Height: _____ in. _____ m

Weight: _____ lb _____ kg

Temperature: _____ °F _____ °C

Barometric pressure: _____ mmHg

Relative humidity: _____ %

Location of Testing

❏ Outdoor Field

❏ Indoor Field

❏ Indoor Track

❏ Outdoor Track

❏ Gym

❏ Other _____

Footwear

❏ Jogging shoe

❏ Walking shoe

❏ Tennis shoe

❏ Basketball shoe

❏ Running Shoe

❏ Cross Training Shoe

❏ CrossFit Shoe

❏ Other _____

Determination of Prescribed Load

$$\text{Prescribed load} = \underset{\text{body mass (kg)}}{_____} \times \underset{\text{kp per kg body mass*}}{_____} = _____ \text{ kp}$$

*In most instances, the multiplier should be 0.075 kp/kg body mass. Refer to table 13.13 for other options.

$$\text{Force (N)} = \underset{\text{force setting in kp}}{_____} \times 9.80665 = _____ \text{N}$$

Quantification of Pedal Revolutions

Time interval (s)	Revolutions	Time interval (s)	Revolutions
0		0-5	
5		5-10	
10		10-15	
15		15-20	
20		20-25	
25		25-30	

Power

Time interval (s)	Peak power (W)
0-5	Peak power (W) = (_____ × _____ × 6 m) / 5 s = _____ W force (N) revolutions
5-10	Peak power (W) = (_____ × _____ × 6 m) / 5 s = _____ W force (N) revolutions
10-15	Peak power (W) = (_____ × _____ × 6 m) / 5 s = _____ W force (N) revolutions
15-20	Peak power (W) = (_____ × _____ × 6 m) / 5 s = _____ W force (N) revolutions
20-25	Peak power (W) = (_____ × _____ × 6 m) / 5 s = _____ W force (N) revolutions
25-30	Peak power (W) = (_____ × _____ × 6 m) / 5 s = _____ W force (N) revolutions

Time interval (s)	Relative power ($W \cdot kg^{-1}$)
0-5	Relative peak power ($W \cdot kg^{-1}$) = _____ / _____ = _____ $W \cdot kg^{-1}$ peak power (W) body mass (kg)
5-10	Relative peak power ($W \cdot kg^{-1}$) = _____ / _____ = _____ $W \cdot kg^{-1}$ peak power (W) body mass (kg)
10-15	Relative peak power ($W \cdot kg^{-1}$) = _____ / _____ = _____ $W \cdot kg^{-1}$ peak power (W) body mass (kg)
15-20	Relative peak power ($W \cdot kg^{-1}$) = _____ / _____ = _____ $W \cdot kg^{-1}$ peak power (W) body mass (kg)
20-25	Relative peak power ($W \cdot kg^{-1}$) = _____ / _____ = _____ $W \cdot kg^{-1}$ peak power (W) body mass (kg)
25-30	Relative peak power ($W \cdot kg^{-1}$) = _____ / _____ = _____ $W \cdot kg^{-1}$ peak power (W) body mass (kg)

Absolute peak power output = _____ W

Relative peak power output = _____ $W \cdot kg^{-1}$

Absolute lowest power output = _____ W

Work

Time interval (s)	Work
0-30	Total work = ($\underline{\hspace{2cm}}$ × $\underline{\hspace{2cm}}$ × 6 m) = $\underline{\hspace{2cm}}$ J force (N)revolutions
0-30	Relative work = ($\underline{\hspace{2cm}}$ / $\underline{\hspace{2cm}}$) = $\underline{\hspace{2cm}}$ J $\cdot$ kg^{-1} total work (J)body mass (kg)

Mean Power Output

Time interval (s)	Mean power output
0-30	Mean power output = $\underline{\hspace{2cm}}$ / 30 s = $\underline{\hspace{2cm}}$ W total work (J)
0-30	Relative mean power output = $\underline{\hspace{2cm}}$ / $\underline{\hspace{2cm}}$ = $\underline{\hspace{2cm}}$ W $\cdot$ kg^{-1} mean power (W)body mass (kg)

Fatigue Index

Lowest power output = $\underline{\hspace{3cm}}$ W

Highest power output = $\underline{\hspace{3cm}}$ W

Calculating the Fatigue Index

Fatigue index = [($\underline{\hspace{2cm}}$ − $\underline{\hspace{2cm}}$) / $\underline{\hspace{2cm}}$] × 100 = $\underline{\hspace{2cm}}$ %
highest power (W)lowest power (W)highest power (W)

Margaria-Kalamen Stair-Climb Test

Equipment

- Staircase with a 6 m run-up and a minimum of nine stairs that are each 174 to 175 cm high
- Two contact mats interfaced with a computer
- Calibrated beam scale or electronic digital scale
- Tape to mark starting line
- Measuring tape
- Stopwatch
- Calculator
- Individual data sheet
- Group data sheet
- Microsoft Excel or equivalent spreadsheet program

Laboratory Orientation

When performing the Margaria-Kalamen stair test, it is essential to select an appropriate staircase. The stairs must have a 6 m (20 ft) run-up and a minimum of nine stairs, each of which is 174 to 175 cm high (see figure 13.7). The stairway should be well ventilated.

The testing protocol involves the following steps: warm-up, practice trials, and testing (figure 13.11). As with other anaerobic testing protocols, the subject should perform a warm-up that begins with 5 min of general warm-up activities (e.g., jogging, cycling, rope jumping) and moves on to 5 min of dynamic stretching activities (e.g., leg swings, high knees, walking knee tucks, walking lunges, butt kicks, inchworms, power skips). After completing the general warm-up, the subject should perform several practice trials to develop the timing necessary to maximize the accuracy of the test (25). After mastering the technique of running up the stairs three steps at a time, the subject is ready. Wait 2 to 3 min after the completion of the practice trials, and then have the subject perform the test twice with 2 or 3 min between the tests.

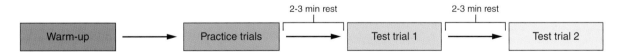

Figure 13.11 Procedure for the Margaria-Kalamen stair test.

Table 13.13 Normative Values for the Margaria-Kalamen Stair Test

AGE (y)	15-20		20-30		30-40		40-50		>50	
Classification	W	kg · m · s⁻¹	W	kg · m · s⁻¹	W	kg · m · s⁻¹	W	kg · m · s⁻¹	W	kg · m · s⁻¹
MEN										
Excellent	>2,197	>224	>2,059	>210	>1,648	>168	>1,226	>125	>961	>98
Good	1,840	188	1,722	176	1,379	141	1,036	106	810	83
Average	1,839	188	1,721	175	1,378	141	1,035	106	809	82
Fair	1,466	149	1,368	139	1,094	112	829	85	642	65
Poor	<1,108	<113	<1,040	<106	<834	<85	<637	<65	<490	<50
WOMEN										
Excellent	>1,789	>182	>1,648	>168	>1,226	>125	>961	>98	>736	>75
Good	1,487	152	1,379	141	1,036	106	810	83	604	62
Average	1,486	152	1,378	141	1,035	106	809	82	603	61
Fair	1,182	121	1,094	112	829	85	642	65	476	49
Poor	<902	<92	<834	<85	<637	<65	<490	<50	<373	<38

Adapted, by permission, from E. Fox, R. Bowers and M. Foss, 1993, *The physiological basis for exercise and sport*, 5th ed. (Dubuque, IA: Wm C. Brown), 676. ©The McGraw-Hill Companies.

Margaria-Kalamen Stair-Climb Test

Step 1: Select a staircase that meets the requirements of a 6 m (20 ft) run-up and a minimum of nine stairs (figure 13.12).

Step 2: Measure the height of each stair and calculate the displacement between the third and ninth stairs.

Figure 13.12 Running in the Margaria-Kalamen test.

Step 3:   Place the switch mats on steps 3 and 9 (see figure 13.7), and put a mark on the floor 6 m from the bottom of the first step. (If a switch mat is unavailable, use a person with a stopwatch to determine this time.)

Step 4:   Gather basic data (e.g., age, height, weight) for the individual data sheet as described in laboratory activity 1.1. Have the subject put on his or her shoes.

Step 5:   Direct the subject through a general 10 min warm-up.

Step 6:   Have the subject perform several practice runs up the stairs to learn how to step on every third step.

Step 7:   Initiate the first test 2 to 3 min after completing the last practice run.

Step 8:   Instruct the subject to run up the stairs as quickly as possible.

Step 9:   The timer starts when the subject steps on the switch mat on step 3 and stops when the subject steps on step 9. (If using a stopwatch, the timer should start the watch when the subject steps on step 3 and stop it when the subject crosses step 9).

Step 10: Record the time on the individual data sheet.

Step 11: Repeat the test 2 to 3 min after completing the first trial. Record the second time value on the individual and group data sheets.

Step 12: Calculate the mean time based upon trial 1 and trial 2. Record this value on the individual data sheet.

Step 13: Calculate power with the following equation:

$$\text{Power}(\text{kg} \cdot \text{m} \cdot \text{s}^{-1}) = \frac{\text{weight (kg)} \times \text{distance (m)}}{\text{time (s)}}$$

Step 14: Convert the power value to watts by multiplying by 9.807. Record this value on the individual and group data sheets.

Step 15: Calculate the relative power by dividing absolute power by body weight. Record this value on the individual and group data sheets.

Question Set 13.6

1. Explain how power is calculated in the Margaria-Kalamen stair test.

2. How does your power output compare with those presented in the norm tables and the averages for the class?

3. Were there any differences between your individual power outputs for trial 1 and 2? If so, why might this be? If not, why is there no difference?

4. Using Excel or an equivalent spreadsheet program, graph the power outputs in W and $W \cdot kg^{-1}$.

5. Are there any differences between the results for the men and the women in the class for absolute power (W) or relative power ($W \cdot kg^{-1}$)? What might be the physiological explanation for this?

6. What are some pros and cons of performing this test?

LABORATORY ACTIVITY 13.6

INDIVIDUAL DATA SHEET

Name or ID number: _____ Date: _____

Tester: _____ Time: _____

Sex: M / F (circle one) Age: _____ y Height: _____ in. _____ m

Weight: _____ lb _____ kg Temperature: _____ °F _____ °C

Barometric pressure: _____ mmHg Relative humidity: _____ %

Measurement of Step Height and Time Between Third and Ninth Steps

Step	=	Height (m)
4	=	
5	=	
6	=	
7	=	
8	=	
9	=	
Total vertical distance	=	

Trial	=	Time (s)
1	=	
2	=	
Mean	=	

Calculation of Power

$$\text{Power} (\text{kg} \cdot \text{m} \cdot \text{s}^{-1}) = \frac{\text{weight (kg)} \times \text{distance (m)}}{\text{time (s)}}$$

Trial 1: Use the time from trial 1 to perform these calculations.

$$\text{Power} (\text{kg} \cdot \text{m} \cdot \text{s}^{-1}) = (\underset{\text{body weight (kg)}}{\underline{\hspace{2cm}}} \times \underset{\text{distance (m)}}{\underline{\hspace{2cm}}}) / \underset{\text{time (s)}}{\underline{\hspace{2cm}}} = \underset{\text{power (kg} \cdot \text{m} \cdot \text{s}^{-1})}{\underline{\hspace{2cm}}}$$

$$\text{Power (W)} = \underset{\text{power (kg} \cdot \text{m} \cdot \text{s}^{-1})}{\underline{\hspace{2cm}}} \times 9.807 = \underset{\text{power (W)}}{\underline{\hspace{2cm}}}$$

$$\text{Power} (\text{W} \cdot \text{kg}^{-1}) = \underset{\text{power (W)}}{\underline{\hspace{2cm}}} / \underset{\text{body weight (kg)}}{\underline{\hspace{2cm}}} = \underset{\text{power (W} \cdot \text{kg}^{-1})}{\underline{\hspace{2cm}}}$$

Trial 2: Use the time from trial 2 to perform these calculations.

$$\text{Power} (\text{kg} \cdot \text{m} \cdot \text{s}^{-1}) = (\underset{\text{body weight (kg)}}{\underline{\hspace{2cm}}} \times \underset{\text{distance (m)}}{\underline{\hspace{2cm}}}) / \underset{\text{time (s)}}{\underline{\hspace{2cm}}} = \underset{\text{power (kg} \cdot \text{m} \cdot \text{s}^{-1})}{\underline{\hspace{2cm}}}$$

$$\text{Power (W)} = \underset{\text{power (kg} \cdot \text{m} \cdot \text{s}^{-1})}{\underline{\hspace{2cm}}} \times 9.807 = \underset{\text{power (W)}}{\underline{\hspace{2cm}}}$$

$$\text{Power} (\text{W} \cdot \text{kg}^{-1}) = \underset{\text{power (W)}}{\underline{\hspace{2cm}}} / \underset{\text{body weight (kg)}}{\underline{\hspace{2cm}}} = \underset{\text{power (W} \cdot \text{kg}^{-1})}{\underline{\hspace{2cm}}}$$

Average of trials 1 and 2: Use the average time from trials 1 and 2 to perform these calculations.

$$\text{Power} (\text{kg} \cdot \text{m} \cdot \text{s}^{-1}) = (\underset{\text{body weight (kg)}}{\underline{\hspace{2cm}}} \times \underset{\text{distance (m)}}{\underline{\hspace{2cm}}}) / \underset{\text{time (s)}}{\underline{\hspace{2cm}}} = \underset{\text{power (kg} \cdot \text{m} \cdot \text{s}^{-1})}{\underline{\hspace{2cm}}}$$

$$\text{Power (W)} = \underset{\text{power (kg} \cdot \text{m} \cdot \text{s}^{-1})}{\underline{\hspace{2cm}}} \times 9.807 = \underset{\text{power (W)}}{\underline{\hspace{2cm}}}$$

$$\text{Power} (\text{W} \cdot \text{kg}^{-1}) = \underset{\text{power (W)}}{\underline{\hspace{2cm}}} / \underset{\text{body weight (kg)}}{\underline{\hspace{2cm}}} = \underset{\text{power (W} \cdot \text{kg}^{-1})}{\underline{\hspace{2cm}}}$$

Pulmonary Function Testing

Objectives

- Introduce the terms associated with respiratory function.
- Execute several pulmonary function tests (PFTs).
- Compare your values with normative values for similar individuals (in terms of age and sex).
- Identify pulmonary disease states and diagnoses.

DEFINITIONS

breathing frequency (bf)—Number of breaths per minute.

chronic obstructive pulmonary disease (COPD)—Disease of the lungs characterized by reduced air flow rates.

exercise-induced asthma—Disease of the lungs characterized by reduced air flow rates during or immediately after exercise.

hyperpnea—Increased rate of ventilation, such as during exercise; does not result in significant changes in blood gases (PaO_2 and $PaCO_2$).

hyperventilation—Increased rate of ventilation that results in a decrease in arterial carbon dioxide ($PaCO_2$).

hypoxia—Reduced atmospheric oxygen.

hypoxemia—Reduced oxygen in the blood (or partial pressure of arterial oxygen, PaO_2).

pulmonary function test (PFT)—Test that measures lung function and encapsulates both lung volumes and flow rates.

restrictive diseases—Diseases of the lungs characterized by reduced total lung capacity or volume.

spirometry—Measurement of expired breath, or lung volume. Considered a type of PFT using a bell spirometer.

tidal volume (V_T)—Amount of air moved per breath; expressed in breaths · min^{-1}.

ventilation ($\dot{V}_E$)—Amount of air expired from the pulmonary system, considered a function of tidal volume and breathing frequency; expressed in L · min^{-1}.

There are many misconceptions about breathing, or **ventilation ($\dot{V}_E$)**. Of course, ventilation is necessary for the exchange of gases between the atmosphere and our metabolism. We use oxygen (O_2) from the atmosphere as the final electron acceptor in the electron transport chain. We must also expire carbon dioxide (CO_2) resulting from macronutrient combustion (refer to lab 5). During exercise, O_2 consumption and CO_2 production increase; therefore, we need to increase ventilation. As you learned in labs 10 and 11, we ventilate much more at higher workloads above the ventilatory threshold. This is the result of greater CO_2 production at higher intensities from the buffering of lactate, or in other words, respiratory compensation for metabolic acidosis. Thus, besides O_2 and CO_2 exchange, the lungs also regulate acid–base balance.

At normal altitudes, ventilation is more sensitive to the production of CO_2 than it is to the consumption of O_2. Very small increases in PCO_2 result in a proportional increase in $\dot{V}_E$, whereas PO_2 can decrease significantly with little effect on $\dot{V}_E$ (figure 14.1, *a* and *b*). At high altitudes (>2,000 m [6,500 ft]), the reduced partial pressure of O_2, or **hypoxia**, is capable of stimulating greater amounts of ventilation. Not only do changes in CO_2 and O_2 stimulate us to breathe differently, but we also have voluntary control over our breathing.

The increased need for ventilation during exercise as the result of increased CO_2 production and O_2 consumption leads exercise physiologists to be interested in its measurement. We move atmospheric gases into our lungs by creating negative pressure in the thoracic cavity. This is accomplished by the skeletal muscles of inspiration (diaphragm and intercostal muscles) and expiration (intercostal and abdominal muscles), which change the volume within the thoracic cavity. The greater the change in volume, the bigger the breath, or **tidal volume (V_T)**. Ventilation is, therefore, a function of **breathing frequency (bf)** multiplied by tidal volume (V_T):

$$\dot{V}_E (L \cdot min^{-1}) = bf (b \cdot min^{-1}) \times V_T (L \cdot b^{-1})$$

Tidal volume is one of the measurements made during **spirometry**. Ventilation can be divided into distinct volumes and capacities, as seen in figure 14.2.

PULMONARY FUNCTION TESTING

A **pulmonary function test (PFT)** measures the amount (volume) and speed (flow rate) of inspiration and expiration. PFTs can be used to identify the health and capacity of the pulmonary system. Most PFTs can measure tidal volume (V_T), as well as the following:

- Forced vital capacity (FVC)—Maximum volume forcibly expired after maximum inspiration (in liters).
- Forced expiratory volume ($FEV_{1.0}$)—Volume of air exhaled in the first second after maximal inhalation; used as a diagnostic tool for limitations in flow rates (expressed in liters but by definition $L \cdot s^{-1}$).
- $FEV_{1.0}$/FVC ratio—Ratio of $FEV_{1.0}$ to FVC, which is a common measure of pulmonary disease (see later in this lab). In healthy people, it is about 70% to 85%; it is reduced in COPD patients but may be normal in people with restrictive diseases.
- Peak expiratory flow (PEF)—Maximum expiratory flow during a forced expiration from the point of maximum inspiration (total lung capacity); expressed in $L \cdot min^{-1}$ or $L \cdot s^{-1}$. PEF is used to provide a measure of

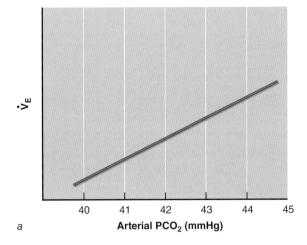

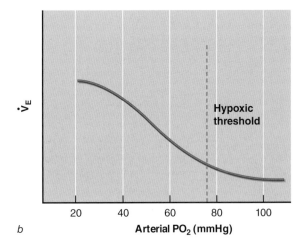

Figure 14.1 The effect of *(a)* arterial PCO_2 and *(b)* PO_2 on ventilation.

airway caliber (diameter) and airflow, yet it is dependent not only on airway caliber but also on lung elastic recoil, patient effort, and patient cooperation. It is generally less specific than $FEV_{1.0}$ as a diagnostic measure.

- Maximum voluntary ventilation (MVV)—Maximum amount of air expired in 1 min ($L \cdot min^{-1}$); measured at rest while breathing maximally for ~12 s.
- Maximum exercise ventilation ($\dot{V}_E max$)—Maximal volume expired during maximal exercise ($L \cdot min^{-1}$).
- Residual volume (RV)—Amount of air remaining in the lungs following a maximal expiration (in liters).
- Total lung capacity (TLC)—Vital capacity and RV combined for the total volume of the lungs (in liters).

Lung volume is largely a function of age, height, and sex. Physical fitness does not significantly affect lung size per se, but it may improve measurement of flow rates, though this idea may be challenged by some evidence concerning swimmers (5, 23).

Volumes for PFT results are usually expressed in terms of BTPS (body temperature and pressure, saturated). You may recall from physics that volume of gas depends on the temperature, barometric pressure, and level of humidity, or water saturation. As a result, expression of gas volume needs a systematic unit of expression. Because the air in our lungs is at body temperature, ambient pressure, and nearly 100% humidity, units of BTPS are common for PFT results. However, as you exhale air and it cools to the temperature of the room and loses humidity or condenses, its volume changes; thus, it may be necessary to convert from ATPS (ambient [or room] temperature, ambient pressure, saturated). Your laboratory instructor can indicate whether this is necessary. The equation for converting ATPS to BTPS is as follows, where V_{ATPS} = volume of gas at ATPS, T_A = temperature in the lab, P_B = barometric pressure in the lab, and P_{H2O} = water vapor pressure at the T_A:

$$V_{BTPS} = V_{ATPS} \times [310 / (273 + T_A)] \times [(P_B - P_{H_2O}) / (P_B - 47)]$$

PULMONARY FUNCTION TESTING AS A TOOL FOR DIAGNOSING PULMONARY DISEASE

PFT results can be used to diagnose respiratory diseases, which are categorized into two types: obstructive and restrictive. An obstructive disease is characterized by an acute or chronic obstruction in the bronchi leading to the alveoli. These obstructions are often referred to as **chronic obstructive pulmonary disease (COPD)**. People with these diseases usually have normal lung volumes but reduced flow rates. COPD can be diagnosed by testing the $FEV_{1.0}$, $FEV_{1.0}/FVC$ ratio, PEF, or other flow rate measures. Examples of COPD include asthma, chronic bronchitis, and emphysema. Asthma differs from the other two in that it can be temporary and reversible. **Exercise-induced asthma** is a temporary inflammatory response during exercise that obstructs ventilatory flow rates. Exercise-induced asthmatics may have normal flow rates at rest but produce a positive test (i.e., a 15% decrease from preex-

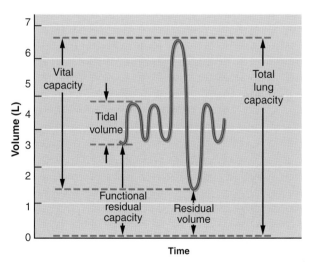

Figure 14.2 Lung volumes from spirometry.

Reprinted from NSCA, 2012, Cardiorespiratory system and gas exchange, by M.H. Malek. In *Essentials of personal training*, 2nd ed. (Champaign, IL: Human Kinetics), 25.

ercise measured $FEV_{1.0}$) following 6 to 8 min of vigorous exercise at a target intensity of 85% to 90% of HRmax. Asthma attacks can be induced by other mechanisms besides exercise, such as cold, allergens, stress, or air pollutants (4).

A **restrictive disease** is characterized by reduced total lung capacities or volumes, such as TLC or vital capacity (VC), but the individual may have normal flow rates. Restrictive diseases can be diagnosed by measuring FVC or a slow VC. Examples of restrictive diseases include pulmonary fibrosis, scar tissue, and tumors.

People with pulmonary disease have increased work of breathing and, if severe enough, limited O_2 delivery to working tissues, including the brain. Under these circumstances, exercise is extremely uncomfortable, and supplemental oxygen may be necessary for activities of daily living. Despite the discomfort of exercise for these individuals, it can alleviate many of their symptoms. For further information on pulmonary rehabilitation, refer to reference sources 1, 2, 3, 11, 12, 19, 20, and 28.

RESPIRATORY LIMITATIONS ON EXERCISE

In healthy individuals without respiratory disease classifications, lungs rarely limit function during rest, activities of daily living, or exercise performance. Many people believe that breathing hard during exercise (**hyperpnea**) means the lungs are limiting performance. In reality, the lungs are extremely overbuilt (7, 8, 15). Their job is to remove CO_2 and maintain O_2 delivery. We know the lungs are succeeding because the pressure of O_2 in the arteries (PaO_2), the pressure of CO_2 in the arteries ($PaCO_2$), and hemoglobin saturation are maintained even during high intensities in the majority of people (25). This is different from **hyperventilation** where heavy breathing results in reduced CO_2 in the arteries ($PaCO_2$), such as during anxiety attacks or excessive sympathetic drive. However, exercise-induced **hypoxemia** (decrease in PaO_2 and hemoglobin saturation) can occur in some elite athletes (25, 27) and in healthy trained women (10, 16, 17), but it is relatively rare. To diagnose exercise-induced hypoxemia, we also need to be able to measure blood gases and hemoglobin saturation.

A potential noninvasive way to estimate ventilatory limitation is to calculate the $\dot{V}_E$max/MVV ratio × 100, which in normal healthy individuals falls somewhere between 50% and 70% (22). In fact, $\dot{V}_E$max can be predicted by taking 72% of MVV. In addition, breathing reserve can be defined as 100% − ($\dot{V}_E$max / MVV × 100) or, more simply, MVV − $\dot{V}_E$, and it can be used to indicate ventilatory limitation (22).

Although the lungs typically succeed in maintaining PaO_2 and $PaCO_2$ during exercise, respiratory musculature may contribute to fatigue in other ways. Respiratory muscles (diaphragm, intercostal muscles, abdominal muscles) are

Pulmonary Disease Classifications

For both obstructive and restrictive pulmonary diseases, PFT results are classified by the following comparisons with predicted results (for restrictive diseases, FVC is comparison; for COPD, percent predicted is for $FEV_{1.0}$) (24):

Good: >100% of predicted

Normal: 100% to 80% of predicted

Mild: 80% to 65% of predicted

Moderate: 65% to 50% of predicted

Moderately severe: 50% to 35% of predicted

Severe: <35% of predicted

skeletal muscles subject to the same requirements of exercising muscles—that is, O_2 and fuel delivery—and to fatigue. It is possible that under certain conditions respiratory muscle failure, or blood flow redirected toward respiratory muscle work, and thus away from exercising muscles, may contribute to fatigue or limit performance (3, 8, 14, 15, 18, 29, 31).

Though possible, it is believed that lung capacity itself does not respond to exercise training; respiratory musculature, however, responds similarly to the training of skeletal muscles, which may allow a trained athlete to reach

greater flow rates ($FEV_{1.0}$ and PEF) and be more resistant to fatigue. Respiratory muscles such as the diaphragm have tremendous oxidative capacities, and although they are not immune to fatigue (or cramps), it is more likely that another mechanism of fatigue would precede the failure of the respiratory musculature (3, 7, 9, 15).

Go to the web study guide to access electronic versions of individual and group data sheets, the question sets from each laboratory activity, and case studies for each laboratory activity.

Lung Volumes and Capacities

Equipment

- Platform scale (or digital or other scale)
- Stadiometer (wall or freestanding) or physician's scale with attached anthropometer
- Calculator
- Individual data sheet
- Group data sheet

Lung Volume Prediction Formulas

Lung volumes are typically predicted using height, weight, and sex. These predictions are usually quite close; in fact, variants are used for clinical diagnoses. The following equations, or ones like them, are built into electronic spirometers to allow percent predicted lung volumes as part of the output data.

Step 1: Begin this lab activity by gathering basic data (e.g., age, height, weight) for the individual data sheet as described in laboratory activity 1.1.

Step 2: Lab members should calculate their own lung volumes given the equations in table 14.1 and place the resulting values in the individual and group data sheets.

Step 3: Norms for lung volumes can be found in table 14.2 (13).

Question Set 14.1

1. What are the main components that predict lung volumes? Flow rates?
2. How do your lung volumes compare with the age-appropriate norms?
3. What is your RV using the equations provided? What is the percent difference between these three estimations? (You will use these values in lab activity 15.3.)
4. In examining the group data sheet, what do you notice about the predicted lung volumes for men and women? For bigger and smaller people?

Table 14.1 Equations for Predicting Lung Volumes

Equation	Reference
WOMEN	
FVC (L) = 0.0491 × H – 0.0216 × A – 3.59	(6)
$FEV_{1.0}$ (L) = 0.0342 × H – 0.0255 × A – 1.578	(6)
RV_A (L) = 0.01812 × H + 0.016 × A – 2.003	(30)
RV_B (L) = 0.023 × H + 0.021 × A – 2.978	(4)
RV_C (L) = 0.28 × FVC	
MEN	
FVC (L) = 0.060 × H – 0.0214 × A – 4.65	(6)
$FEV_{1.0}$ (L) = 0.0414 × H – 0.0244 × A – 2.19	(6)
RV_A (L) = 0.0131 × H + 0.022 × A – 1.232	(30)
RV_B (L) = 0.019 × H + 0.0115 × A – 2.24	(21)
RV_C (L) = 0.24 × FVC	

A = age in y; H = height in cm.

Table 14.2 Normal Values for Lung Volumes

Lung volume	<20 y male	20-40 y male	>40 y male	<20 y female	20-40 y female	>40 y female
FVC (L)	3.22	4.93	4.34	2.82	3.53	2.99
$FEV_{1.0}$ (L)	2.77	4.10	3.37	2.51	3.02	2.36
$FEV_{1.0}$/FVC (%)	86.02	83.16	77.65	89.01	85.55	78.93

LABORATORY ACTIVITY 14.1

INDIVIDUAL DATA SHEET

Name or ID number: _____ Date: _____

Tester: _____ Time: _____

Sex: M / F (circle one) Age: _____ y Height: _____ in. _____ m

Weight: _____ lb _____ kg Temperature: _____ °F _____ °C

Barometric pressure: _____ mmHg Relative humidity: _____ %

Women

$$0.0491 \times \underset{\text{height (cm)}}{\underline{\hspace{2cm}}} - 0.0216 \times \underset{\text{age (y)}}{\underline{\hspace{2cm}}} - 3.59 = \underset{\text{FVC (L)}}{\underline{\hspace{2cm}}}$$

$$0.0342 \times \underset{\text{height (cm)}}{\underline{\hspace{2cm}}} - 0.0255 \times \underset{\text{age (y)}}{\underline{\hspace{2cm}}} - 1.578 = \underset{\text{FEV}_{1.0}}{\underline{\hspace{2cm}}}$$

$$0.01812 \times \underset{\text{height (cm)}}{\underline{\hspace{2cm}}} + 0.016 \times \underset{\text{age (y)}}{\underline{\hspace{2cm}}} - 2.003 = \underset{\text{RV}_A\text{(L)}}{\underline{\hspace{2cm}}}$$

$$0.023 \times \underset{\text{height (cm)}}{\underline{\hspace{2cm}}} + 0.021 \times \underset{\text{age (y)}}{\underline{\hspace{2cm}}} - 2.978 = \underset{\text{RV}_B\text{(L)}}{\underline{\hspace{2cm}}}$$

$$0.28 \times \underset{\text{FVC}}{\underline{\hspace{2cm}}} = \underset{\text{RV}_C\text{(L)}}{\underline{\hspace{2cm}}}$$

Men

$$0.060 \times \underset{\text{height (cm)}}{\underline{\hspace{2cm}}} - 0.0214 \times \underset{\text{age (y)}}{\underline{\hspace{2cm}}} - 4.65 = \underset{\text{FVC (L)}}{\underline{\hspace{2cm}}}$$

$$0.0414 \times \underset{\text{height (cm)}}{\underline{\hspace{2cm}}} - 0.0244 \times \underset{\text{age (y)}}{\underline{\hspace{2cm}}} - 2.19 = \underset{\text{FEV}_{1.0}}{\underline{\hspace{2cm}}}$$

$$0.0131 \times \underset{\text{height (cm)}}{\underline{\hspace{2cm}}} + 0.022 \times \underset{\text{age (y)}}{\underline{\hspace{2cm}}} - 1.232 = \underset{\text{RV}_A\text{ (L)}}{\underline{\hspace{2cm}}}$$

$$0.019 \times \underset{\text{height (cm)}}{\underline{\hspace{2cm}}} + 0.0115 \times \underset{\text{age (y)}}{\underline{\hspace{2cm}}} - 2.24 = \underset{\text{RV}_B\text{ (L)}}{\underline{\hspace{2cm}}}$$

$$0.24 \times \underset{\text{FVC}}{\underline{\hspace{2cm}}} = \underset{\text{RV}_C\text{ (L)}}{\underline{\hspace{2cm}}}$$

Pulmonary Function

Equipment

- Platform scale (or digital or other scale)
- Stadiometer (wall or freestanding) or physician's scale with attached anthropometer
- Spirometer
- Noseclips
- Individual data sheet
- Group data sheet

Pulmonary Function Spirometer Test

Lab spirometers (figure 14.3) that measure pulmonary function may be electronic or use a bell that floats in water. They also differ in the measures they can make; most, however, collect FVC, $FEV_{1.0}$, and PEF. Your laboratory instructor will demonstrate the use of the particular spirometer in your lab. Here are some pretest instructions subjects should follow when testing pulmonary function: avoid exercise, eating, and smoking for 2 h prior to the test; dress appropriately in unrestrictive clothing; and abstain from alcohol for 4 h beforehand.

Step 1: Begin this lab activity by gathering basic data (e.g., age, height, weight) for the individual data sheet as described in laboratory activity 1.1.

Step 2: Use the lab spirometer to measure FVC, $FEV_{1.0}$, and PEF. Spirometers typically provide both the absolute value and % of norm. Because flow rates are being measured, it is important to expire as quickly as possible during these tests.

Step 3: Record the results for the absolute value and % of norm on the individual data sheet (include units). The last four rows of the data sheet provide space for additional pulmonary measures that your particular spirometer records.

Step 4: Repeat each measure for a total of three trials.

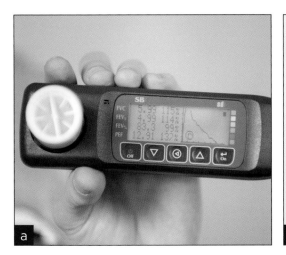 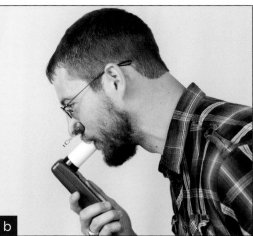

Figure 14.3 *(a)* Electronic spirometer; *(b)* in use with noseclips.

Step 5: Calculate the average for the three attempts for each measure.

Step 6: Record your values on the group data sheet.

Question Set 14.2

1. How does your $FEV_{1.0}$ compare with the norm (see table 14.2)?

2. How might you explain a deviation from the norm?

3. What factors could negatively affect peak expiratory flow?

4. What factors could positively affect forced vital capacity?

5. How is COPD typically diagnosed using PFT? What are the parameters, and what level of change is required for diagnosis?

6. Which of the measures you performed may be subject to changes with greater fitness? Why?

7. Based on the group data sheet, was the flow limited for any of your labmates?

LABORATORY ACTIVITY 14.2

INDIVIDUAL DATA SHEET

Name or ID number: _____ Date: _____

Tester: _____ Time: _____

Sex: M / F (circle one) Age: _____ y Height: _____in. _____m

Weight: _____ lb _____ kg Temperature: _____°F _____°C

Barometric pressure: _____ mmHg Relative humidity: _____ %

Measure	TRIAL 1		TRIAL 2		TRIAL 3		AVERAGE	
	Your value	% of norm	Your value	% of norm	Your value	% of norm	Your value	% of norm
FVC								
$FEV_{1.0}$								
PEF								
$FEV_{1.0}/FVC$								

Exercise-Induced Ventilatory Limitations

Equipment

- Platform scale (or digital or other scale)
- Stadiometer (wall or freestanding) or physician's scale with attached anthropometer
- Spirometer
- Noseclips
- Data from prior $\dot{V}O_2$max tests (lab 10 or 11)
- Individual data sheet
- Group data sheet

MVV Test

One indirect measure of ventilatory limitation involves comparing MVV (maximum voluntary ventilation) with maximum ventilation during an incremental test to exhaustion ($\dot{V}_E$). MVV is the maximum amount of air moved ($L \cdot min^{-1}$) at rest. It should be measured for 10 to 15 s while the subject is seated, and it is most effective when the subject takes modestly deep breaths at a high rate (>30 breaths $\cdot$ min^{-1}). The $\dot{V}_E$/MVV ratio can be used to predict ventilatory limitations.

Step 1: Begin this lab activity by gathering basic data (e.g., age, height, weight) for the individual data sheet as described in laboratory activity 1.1.

Step 2: Select one to three volunteer subjects who have completed a maximal exercise test in a previous lab (lab 10 or 11).

Step 3: Measure MVV on the lab spirometer. It is important for the subjects to be seated because they may become dizzy during and following the test. The subjects should breathe deeply and as rapidly as possible with a noseclip for the entire duration of the test (10-15 s).

Step 4: Record the results in the table on the individual data sheet. Repeat the test after giving the subjects a few minutes to recover.

Step 5: Calculate the average for the two attempts.

Step 6: Record $\dot{V}_E$ from the maximal test reports in lab 10 or 11. Calculate $\dot{V}_E$/MVV ratio, and record on the individual data sheet.

Step 7: If you are not one of the subjects with max data, measure and record your own MVV.

Step 8: Record your results on the group data sheet.

Question Set 14.3

1. Compare the MVVs from the class with the maximum $\dot{V}_E$ attained during the $\dot{V}O_2$max tests. Do you think anyone had ventilatory limitation?

2. Why might you see greater $\dot{V}_E$ than MVV yet remain unsure that ventilation is a limitation?

3. What would you measure to determine whether the lungs truly limit exercise performance?

4. Why might dizziness or syncope accompany the measurement of MVV?

LABORATORY ACTIVITY 14.3

INDIVIDUAL DATA SHEET

Subject A

Name or ID number: _____ Date: _____

Tester: _____ Time: _____

Sex: M / F (circle one) Age: _____ y Height: _____ in. _____ m

Weight: _____ lb _____ kg Temperature: _____ °F _____ °C

Barometric pressure: _____ mmHg Relative humidity: _____ %

Subject B

Name or ID number: _____ Date: _____

Tester: _____ Time: _____

Sex: M / F (circle one) Age: _____ y Height: _____ in. _____ m

Weight: _____ lb _____ kg Temperature: _____ °F _____ °C

Barometric pressure: _____ mmHg Relative humidity: _____ %

Subject C

Name or ID number: _____ Date: _____

Tester: _____ Time: _____

Sex: M / F (circle one) Age: _____ y Height: _____ in. _____ m

Weight: _____ lb _____ kg Temperature: _____ °F _____ °C

Barometric pressure: _____ mmHg Relative humidity: _____ %

Subject	TRIAL 1 MVV	% predicted	TRIAL 2 MVV	% predicted	Average	Max $\dot{V}_E$	$\dot{V}_E$/MVV ratio
A							
B							
C							

Exercise-Induced Asthma

Equipment

- Platform scale (or digital or other scale)
- Stadiometer (wall or freestanding) or physician's scale with attached anthropometer
- Spirometer
- Noseclips
- Treadmill
- HR monitor
- Individual data sheet

> This laboratory activity is supported by a virtual lab experience in the web study guide.
>
> WWW

PFT for Exercise-Induced Asthma

One of the more useful applications of pulmonary function testing is for clinical diagnosis. Exercise-induced asthma is a relatively common ailment resulting from a maladaptive inflammatory response that causes bronchoconstriction. Normally the sympathetic nervous system drive that occurs during exercise results in bronchodilation. However, for several possible reasons, exercise can result in bronchoconstriction in some individuals, which may be exacerbated in cold or dry air. As detailed earlier, people with exercise-induced asthma may have normal flow rates at rest but yield a positive test following 6 to 8 min of vigorous exercise above a certain intensity level.

Step 1: Select one subject with normal lung function and (if possible) one with suspected exercise-induced asthma or other lung dysfunction. (If no one in the class has suspected exercise-induced asthma, you can still perform this test for practice and discuss the results.)

Step 2: Gather basic data (e.g., age, height, weight) for the individual data sheet as described in laboratory activity 1.1.

Step 3: Use the lab spirometer to measure FVC, $FEV_{1.0}$, and PEF at rest. Because flow rates are being measured, it is important for the subjects to expire as quickly as possible during these tests.

Step 4: Record the subjects' absolute values at rest on the individual data sheet (including units and percent predicted).

Step 5: Have the subjects run on the treadmill for 6 to 8 min at 85% to 90% of their age-predicted HRmax.

Step 6: Perform a PFT as quickly as possible after the cessation of exercise.

Step 7: Record the test results on the individual data sheet.

Step 8: Calculate the % change in lung parameters from preexercise to postexercise with the following formula:

$$\% \text{ difference} = (\text{preexercise} - \text{postexercise}) / \text{preexercise}$$

Question Set 14.4

1. What change is required from preexercise to postexercise to produce a positive test for exercise-induced asthma?

2. Did your subject have exercise-induced asthma? Justify your answer.

3. Why might ventilatory flow rates increase following exercise?

LABORATORY ACTIVITY 14.4

INDIVIDUAL DATA SHEET

Subject A

Name or ID number: _____ Date: _____

Tester: _____ Time: _____

Sex: M / F (circle one) Age: _____ y Height: _____ in. _____ m

Weight: _____ lb _____ kg Temperature: _____ °F _____ °C

Barometric pressure: _____ mmHg Relative humidity: _____ %

Subject B

Name or ID number: _____ Date: _____

Tester: _____ Time: _____

Sex: M / F (circle one) Age: _____ y Height: _____ in. _____ m

Weight: _____ lb _____ kg Temperature: _____ °F _____ °C

Barometric pressure: _____ mmHg Relative humidity: _____ %

Measure	PREEXERCISE				POSTEXERCISE				% Δ FROM PRE TO POST	
	Sub A	% predicted	Sub B	% predicted	Sub A	% predicted	Sub B	% predicted	Sub A	Sub B
FVC (L)										
$FEV_{1.0}$ (L)										
PEF (L·s⁻¹)										

Sub A = subject A (with asthma); sub B = subject B (without asthma).

Body Composition Assessments

Objectives

- Know the range of values for percent body fat considered typical for college-aged men and women.
- Know the body mass index (BMI) values that indicate obesity for men and women.
- Explain the principle of underwater weighing (UWW), which is also known as *hydrostatic weighing* or *densitometry*, for determining body density and percent body fat.
- Competently perform BMI, circumference, skinfold, and UWW measurements.
- Understand the importance of estimating or measuring residual volume for UWW validity.
- Calculate body density and percent body fat based on the two-component body composition system.
- Describe common sources of error in body composition measurements.
- Describe the advantages and disadvantages of various techniques for measuring body composition.

DEFINITIONS

body mass index (BMI)—Ratio of an individual's weight (kg) to height squared (m^2).

densitometry—Measurement of body density ($D_{body} = mass_{body} / volume_{body}$).

fat-free mass (FFM)—All body tissue that is not fat, including bone, muscle, organ, and connective tissue; synonymous with lean body mass (LBM). Expressed as a unit of weight (kg).

fat mass (FM)—Body tissue that is made up of fat, including subcutaneous, visceral, cellular membranes, and interstitial fat. Contains both essential (membranes) and nonessential (adipose tissue) components. Expressed as a unit of weight (kg).

relative body fat—Amount of body fat expressed as a percentage of total body weight.

reliability—Reproducibility of a test; high reliability means that if the same measure is tested several times for the same person, the same answer will result.

residual volume (RV)—Amount of air remaining in the lungs after a maximal exhalation.

skinfold caliper—Instrument used to measure the thickness of pinched skinfolds; must provide a standard pinch pressure in order to provide accurate values.

skinfold thickness—Body composition measurement using the thickness of skinfolds, and thus subcutaneous fat, as an estimate of body density; requires skill for correct skinfold placement.

two-component body composition—Commonly used system in which the body is assumed to have two major components: fat and fat-free (or lean) tissue.

underwater weighing (UWW)—Body composition measurement using the displacement of water to estimate body density.

validity—Accuracy of a measure; that is, how close it is to the true value. (In other words, are you *really* measuring what you say you are measuring—in this case, relative body fat?)

vital capacity—Volume of air that can be moved in or out of the lungs in one maximal breath.

Body composition refers to the components that make up the body. Usually, we are interested in the percentage of the body that is adipose (or fat) tissue versus the percentage that is fat free. The primary reason for this interest is that obesity—overfatness—is closely related to a number of health risks prevalent in Western societies (e.g., cardiovascular disease, peripheral vascular disease, hypertension, diabetes). There are also less critical reasons for interest in body composition; for example, athletes wishing to maximize their performance want to optimize their amount of muscle and fat, and many people are interested in body composition due to societal norms that view extreme leanness as aesthetically desirable. In fact, losing weight (and thus fat) is probably the main reason most Americans exercise.

Thus it is important to know not only body weight but also how much of the weight is fat and how much is fat free. Changes resulting from exercise may not be noticeable by a scale weight, but many techniques are available for estimating body composition.

We aim here to discuss some common techniques of evaluating disease risk based on body composition, including circumferences, underwater weighing (UWW), skinfold thickness, body mass index (BMI), waist and hip circumference, and waist-to-hip (W/H) ratio. See table 15.1 for more on methods for estimating body composition. Further explanation of the various body composition models can be found later in this lab.

Table 15.1 Methods for Estimating Body Composition

Method	Model	Description
Anthropometry	2	Measures girths to estimate body fat, such as waist and hip (W/H) ratio.
Body mass index (BMI)	0	Determines ratio of weight to height ($kg \cdot m^{-2}$). Does not separate fat from lean mass but is used to estimate obesity.
Bioelectrical impedance analysis (BIA)	2	Measures resistance to electric current to estimate body water content, lean mass, and body fat.
Skinfold thickness	2	Measures subcutaneous fat to estimate body fat and lean mass.
Plethysmography	2	Whole-body plethysmography measures air displacement and calculates body density (comparable to water displacement used in UWW).
Underwater weighing (UWW; also known as *hydrodensitometry*)	2	UWW technique based on Archimedes' principle estimates body fat and lean mass from body density.
Total body water	2	Measures total body water by isotope dilution techniques to estimate body fat and lean mass.
Ultrasound	2	High-frequency ultrasound waves pass through tissues to image subcutaneous fat and estimate body fat and lean mass.
Dual-energy X-ray absorptiometry (DEXA, DXA)	3	X-ray technique used at two energy levels to estimate body fat, lean mass, and bone mass.
Near-infrared interactance	2	Infrared light passes through tissues to predict body fat versus lean mass.
Magnetic resonance imaging (MRI)	3	Magnetic-field and radio-frequency waves image body tissues (similar to computed tomography scan); useful for imaging deep abdominal fat.
Computed tomography (CT)	3	X-ray images body tissues; used to determine subcutaneous and deep fat to estimate body fat, lean mass, and bone mass.
Neutron activation analysis	3	Neutrons pass through tissues, permitting analysis of nitrogen and mineral content in the body; used to estimate body fat and lean mass.
Dual-photon absorptiometry (DPA)	3	Beam of photons passes through tissues, differentiating soft tissue from bone tissue; used to estimate body fat, lean mass, and bone mass.
Total body potassium	2	Measures total body potassium, the main intracellular ion, to estimate body fat and lean mass.

The order of methods in the table roughly implies ease of use, expense, and thus commonality.

Adapted, by permission, from M. Williams, 2009, *Nutrition for health, fitness, and sport*, 9th ed. (New York: McGraw-Hill Companies), 407. ©The McGraw-Hill Companies; Heyward and Gibson 2014.

These techniques differ by sophistication of equipment, validity, reliability, general assumptions, and ease of use. Some (BMI, circumferences, W/H ratio) do not estimate body composition at all but do correlate strongly with obesity and other lifestyle diseases.

Knowledge of **relative body fat** (%BF) can be useful in pursuing goals of weight gain or loss. It is possible to calculate a desired body weight from a target relative body fat using a simple equation:

$$\text{Desired body weight} = \frac{\text{BW} - (\text{BW} \times \frac{\%\text{BF}}{100})}{1 - (\frac{\text{desired }\%\text{BF}}{100})}$$

This equation can be rewritten as follows (LBM indicates lean body mass):

$$\text{Desired body weight (BW)} = \frac{\text{LBM}}{\text{desired }\%\text{LBM}}$$

These weight prediction goals assume that the person is capable of reducing body fat without changes in LBM or **fat-free mass (FFM)**. Whether or not this is possible may depend on the person's chosen methods for weight loss or gain. In fact, this is the motivation for knowing body composition for many people attempting to lose weight. Bodybuilders, wrestlers, mixed martial arts (MMA) fighters, and others are often attempting to lose body fat while minimizing the loss of FFM.

Of course, some fat in the body is necessary, such as for cellular membranes, vitamins, hormones, and Schwann cells (surrounding the nerves). Storage fat in adipose tissue is available for use as fuel; nonetheless, it is often called *nonessential* or storage fat. It is this fat that people are interested in reducing for reasons of health, athletic performance, or vanity. Essential fat is thought to make up about 3% of body weight in men and 12% in women (6). The remaining fat is considered nonessential or storage fat despite its usefulness as a dense fuel source. Different classifications exist for relative body fat (combined essential and nonessential) dependent on sex and age. They differ by percentile ranks, at-risk cutoffs, and target recommendations.

Table 15.2 and table 15.3 list some examples of these classifications.

BODY COMPOSITION MODELS

Several models exist for describing differences in body composition (see figure 15.1). The **two-component body composition** model (such as UWW and skinfold thickness) splits the body into two parts—fat mass **(FM)** and FFM (fat-free mass). FM is self-evident, whereas FFM is made up of protein, water, and bone. Other multicomponent models separate one or more of the components within the FFM, such as bone density (by DXA or MRI) or total body water (by stable radioisotopes).

The multicomponent models are considered more accurate, or valid, because they remove some assumptions made in the two-component models. FM and FFM in the two-component models are assumed to have consistent densities within a certain population or age group. For FM in white adults, that density is $0.90 \text{ g} \cdot \text{ml}^{-1}$, and for FFM the density is $1.1 \text{ g} \cdot \text{ml}^{-1}$. Siri used these values to derive the equation used to estimate percent body fat in adult white males (11):

$$\% \text{ body fat} = (4.95 / \text{density}) - 4.50$$

Because of slight variances in lean mass (FFM) densities, different equations exist based on age, sex, and race (see table 15.4).

The specificity of these equations increases the accuracy of the body composition measurement. However, the term *accuracy* is often misused. We assume that if a body composition measure is accurate, it is close to the real value, whereas *precision* refers to the repeatability of the measurement. In body composition, the terms **validity** and **reliability** are often used to replace the terms *accuracy* and *precision*, respectively. A body composition technique is considered valid if it measures what it says it measures—relative body fat. To be reliable, the body composition measurement must generate reproducible results. For example, weighing someone on a scale is a simple, reliable measure, but in reference to relative body fat, it is not particularly valid. UWW, on the other hand,

Table 15.2 Body Composition (Normative References)

%	20-29 y	30-39 y	40-49 y	50-59 y	60-69 y	70-79 y
MALE						
90	7.9	11.9	14.9	16.7	17.6	17.8
80	10.5	14.5	17.4	19.1	19.7	20.4
70	12.7	16.5	19.1	20.7	21.3	21.6
60	14.8	18.2	20.6	22.1	22.6	23.1
50	16.6	19.7	21.9	23.2	23.7	24.1
40	18.6	21.3	23.4	24.6	25.2	24.8
30	20.6	23.0	24.8	26.0	25.4	26.0
20	23.1	24.9	26.6	27.8	28.4	27.6
10	26.3	27.8	29.2	30.3	30.9	30.4
FEMALE						
90	14.8	15.6	17.2	19.4	19.8	20.3
80	16.5	17.4	19.8	22.5	23.2	24.0
70	18.0	19.1	21.9	25.1	25.9	26.2
60	19.4	20.8	23.8	27.0	27.9	28.6
50	21.0	22.6	25.6	28.8	29.8	30.4
40	22.7	24.6	27.6	30.4	31.3	31.8
30	24.5	26.7	29.6	32.5	33.3	33.9
20	27.1	29.1	31.9	34.5	35.4	36.0
10	31.4	33.0	35.4	36.7	37.3	38.2

Adapted, by permission, from Cooper Institute, *Physical fitness assessments and norms for adults and law enforcement* (Dallas, TX: The Cooper Institute), 52, 53. For more information: www.cooperinstitute.org.

Table 15.3 Percent Body Fat Standards for Adults, Children, and Physically Active Adults

RECOMMENDED %BF LEVELS FOR ADULTS AND CHILDREN					
	NR	Low	Middle	High	Obesity
MALE					
6-17 y	<5	5-10	11-25	26-31	>31
18-34 y	<8	8	13	22	>22
35-55 y	<10	10	18	25	>25
>55 y	<10	10	16	23	>23
FEMALE					
6-17 y	<12	12-15	16-30	31-36	>36
18-34 y	<20	20	28	35	>35
35-55 y	<25	25	32	38	>38
>55 y	<25	25	30	35	>35

RECOMMENDED %BF LEVELS FOR PHYSICALLY ACTIVE ADULTS			
	Low	Middle	Upper
MALE			
18-34 y	5	10	15
35-55 y	7	11	18
>55 y	9	12	18
FEMALE			
18-34 y	16	23	28
35-55 y	20	27	33
>55 y	20	27	33

NR = not recommended; %BF = percent body fat.

Reprinted, by permission, from V. Heyward and A.L. Gibson, 2014, *Advanced fitness assessment and exercise prescription*, 7th ed. (Champaign, IL: Human Kinetics), 220; Data from T.G. Lohman, L. Houtkooper, and S. Going, 1997, "Body fat measurement goes high-tech: Not all are created equal," *ACSM's Health & Fitness Journal* 7: 30–35.

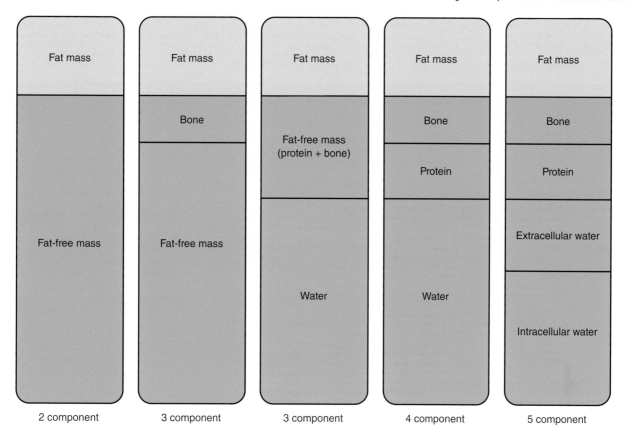

Figure 15.1 Body composition models.

© Charles Dumke

estimates relative body fat and is one of the most valid measurements, but in comparison to body weight measurement, it is not as reliable. Consider using these terms in relation to body composition techniques to reduce confusion.

BMI FOR CATEGORIZING BODY COMPOSITION

Body mass index (BMI) is a simple and reliable measure commonly used in clinical situations and epidemiological research to categorize individuals with regard to obesity (1). BMI is calculated with the following equation:

$$BMI = Wt \ (kg) \ / \ Ht^2 \ (m^2)$$

The theory behind this method is that weight-to-height ratios across the general population have a positive relationship with percent body fat. The NIH used BMI to create definitions of overweight and obesity (8) (table 15.5). Obesity is associated with a threefold greater risk of diabetes and hypertension and a twofold greater

risk of hypercholesterolemia (high blood cholesterol). Individuals classified as obese warrant an intervention such as calorie restriction (8). The popularity of BMI is based in part on its ease of measurement and the potential database of BMI values from doctor's visits. However, BMI does not separate how much weight is fat or LBM, so a muscular person could be falsely classified as obese. BMI is therefore a very reliable measure but has questionable validity as a measure of body composition.

CIRCUMFERENCE MEASUREMENTS AND HEALTH RISK

Another simple measurement associated with increased health risk is the W/H (waist-to-hip) ratio, which is the circumference of the waist divided by the circumference of the hips. As W/H ratio increases, the risk increases twofold for heart attack, stroke, hypertension, diabetes mellitus, gallbladder disease, and death (1).

Table 15.4 Population-Specific Formulas for Converting Body Density to Percent Body Fat Based on Two-Component Model

Population	Age (y)	Sex	%BF[a]	FFB$_d$ (g · ml^{-1})*
RACE OR ETHNICITY				
African American	9-17	Female	(5.24 / Db) – 4.82	1.088
	19-45	Male	(4.86 / Db) – 4.39	1.106
	24-79	Female	(4.85 / Db) – 4.39	1.106
American Indian	18-62	Male	(4.97 / Db) – 4.52	1.099
	18-60	Female	(4.81 / Db) – 4.34	1.108
Japanese native	18-48	Male Female	(4.97 / Db) – 4.52 (4.76 / Db) – 4.28	1.099 1.111
	61-78	Male Female	(4.87 / Db) – 4.41 (4.95 / Db) – 4.50	1.105 1.100
Singaporean (Chinese, Indian, Malay)	Adult	Male Female	(4.94 / Db) – 4.48 (4.84 / Db) – 4.37	1.102 1.107
White	8-12	Male Female	(5.27 / Db) – 4.85 (5.27 / Db) – 4.85	1.086 1.086
	13-17	Male Female	(5.27 / Db) – 4.85 (5.27 / Db) – 4.85	1.092 1.090
	18-59	Male Female	(4.95 / Db) – 4.50 (4.96 / Db) – 4.51	1.100 1.101
	60-90	Male Female	(4.97 / Db) – 4.52 (5.02 / Db) – 4.57	1.099 1.098
Hispanic	20-40	Male Female	NA (4.87 / Db) – 4.41	NA 1.105
ATHLETES				
Resistance-trained	24 ± 4	Male	(5.21 / Db) – 4.78	1.089
	35 ± 6	Female	(4.97 / Db) – 4.52	1.099
Endurance-trained	21 ± 2	Male	(5.03 / Db) – 4.59	1.097
	21 ± 4	Female	(4.95 / Db) – 4.50	1.100
All sports	18-22	Male Female	(5.12 / Db) – 4.68 (4.97 / Db) – 4.52	1.093 1.099
CLINICAL POPULATIONS				
Anorexia nervosa	15-44	Female	(4.96 / Db) – 4.51	1.101
Obesity	17-62	Female	(4.95 / Db) – 4.50	1.100
Spinal cord injury (paraplegic or quadriplegic)	18-73	Male Female	(4.67 / Db) – 4.18 (4.70 / Db) – 4.22	1.116 1.114

FFB$_d$ = fat-free body density; Db = body density; %BF = percent body fat; NA = no data available for this group.

[a] Multiply value by 100 to calculate %BF.

*FFB$_d$ is based on average values reported in selected research articles.

Reprinted, by permission, from V. Heyward and D. Wagner, 2004, *Applied body composition assessment*, 2nd ed. (Champaign, IL: Human Kinetics), 9.

Table 15.5 Classification of Weight by BMI, Waist Circumference, and Associated Disease Risks

| | | RISK LEVEL BY WAIST CIRCUMFERENCE* | | | |
| | | MEN | | WOMEN | |
BMI	Classification	≤102 cm	>102 cm	≤88 cm	>88 cm
<18.5	Underweight				
18.5-24.9	Normal				
25.0-29.9	Overweight	Increased	High	Increased	High
30.0-34.9	Class 1 obesity	High	Very high	High	Very high
35.0-39.9	Class 2 obesity	Very high	Very high	Very high	Very high
≥40.0	Class 3 obesity	Extremely high	Extremely high	Extremely high	Extremely high

*Disease risk for type 2 diabetes, hypertension, and coronary heart disease.

Adapted from the NHLBI 1998.

Waist circumference is the smallest circumference below the ribcage and above the umbilicus while the individual is standing with the abdominal muscles relaxed. Hips are measured at the greatest circumference of the buttocks in a level horizontal plane. Measure the waist while facing the subject, and measure the hips from the side. Record the results to the nearest 0.5 cm. Repeated measures should yield results within 0.5 cm. If possible, use a spring-loaded tape to normalize tape tension. Ratios above 0.90 for men and 0.80 for women relate to sharp increases in disease risk for hypertension, high cholesterol, cardiovascular disease, and diabetes (1) (see table 15.8).

SKINFOLD THICKNESS AS A MEASURE OF BODY FAT

Estimation of body fat based on **skinfold thickness** has become quite common, largely because of the relative ease of measurement with minimal equipment. Some **skinfold calipers** are available for a low price, but many are grossly inaccurate. Instruments such as the ones that you will likely use in this lab can cost several hundred dollars. As you will learn in this lab, the skill of the person making the measurements is important for obtaining meaningful measurements.

The rationale for using skinfold thickness is that an age-dependent proportion of body fat is deposited subcutaneously. Measuring the amount of adipose tissue that can be pinched provides some indication of the amount of overall body fat (figure 15.2).

The process of turning this information (i.e., skinfold thickness at several sites) into a number representing body density is based on research results. This type of research compares the sum of skinfolds from a number of anatomical sites with a gold-standard measure of body density, such as hydrostatic (or underwater) weighing (4). It has been shown that curvilinear or quadratic relationships better predict body density from the sum of skinfolds compared with linear regression (4), which is why the equations require squaring the sum of several skinfolds.

The relationship between subcutaneous fat and total body fat varies with race, age, and sex. Consequently, most equations that have been developed to predict body fatness are population specific and should be used for subjects of a specific race, age, or sex (table 15.6). In general, estimation of percent body fat from skinfold measures has an error of about 3.5% (9).

The different equations require different skinfold sites to be measured. It is important to

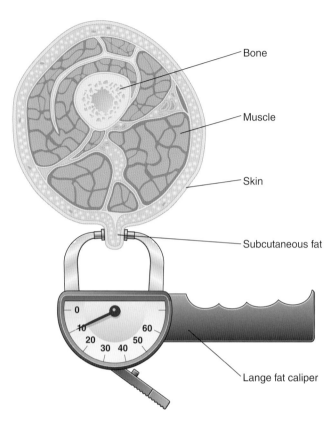

Figure 15.2 Subcutaneous skinfold measurement.

understand the technique of measuring different skinfold sites and which equations require which sites. Although it may seem intuitive that body fat prediction from the sum of skinfolds is improved by using more sites, this is not necessarily the case. Validity is not improved by using more than three sites (9). This lab demonstrates the skinfold sites used in several of the more popular equations.

Unless otherwise appropriate, we choose the commonly used equations developed by Jackson and Pollock for men (4) and Jackson, Pollock, and Ward for women (5), as shown in "Generalized Body Density Formulas" sidebar. These equations are considered generalized equations because the development populations included subjects across a wide age span. Consequently, these equations can be used across various age groups and have been validated for both athletic and nonathletic populations. Separate equations are used for men and women to estimate body density. Once body density has been determined, body composition is estimated from the Siri equation mentioned earlier (11):

% body fat = (4.95 / density) − 4.50

This holds true unless another population-specific equation is more appropriate (see population-specific equations in table 15.4).

Underwater Weighing

Underwater weighing (UWW), also called *hydrostatic weighing* or *hydrodensitometry*, is often considered the most valid method for estimating relative body fat (figure 15.3). The technique is called **densitometry** because fat and lean tissues have different densities. UWW depends on the fact that water has a density of approximately $1 \text{ g} \cdot \text{ml}^{-1}$ (see table 15.7 for the exact density of water at various temperatures) and that fat floats because its density is $<1 \text{ g} \cdot \text{ml}^{-1}$ or approximately $0.9 \text{ g} \cdot \text{ml}^{-1}$. On the other hand, lean tissue (that is, all other tissues excluding fat) sinks because its density is greater than that of water (about $1.1 \text{ g} \cdot \text{ml}^{-1}$). The UWW technique, then, is based on the two-component body composition model in which the body is simply divided into fat and fat-free (or lean) tissue. Because every person is neither total FM nor total FFM, all people have a body density between 0.9 and 1.1 $\text{g} \cdot \text{ml}^{-1}$. This should impress upon practitioners the need to carry out decimal places (at least to the thousands) to preserve validity.

Figure 15.3 UWW method.

Table 15.6 Skinfold Prediction Equations

Skinfold sites	Population subgroups	Equation
∑7SKF (chest + abdomen + thigh + triceps + subscapular + suprailiac + midaxillary)	Black or Hispanic women, 18-55 y	D_b (g · cc^{-1})[a] = 1.0970 − (0.00046971 × ∑7SKF) + (0.00000056 × ∑7SKF^2) − (0.00012828 × age)
	Black men or male athletes, 18-61 y	D_b (g · cc^{-1})[a] = 1.1120 − (0.00043499 × ∑7SKF) + (0.00000055 × ∑7SKF^2) − (0.00028826 × age)
∑4SKF (triceps + anterior suprailiac + abdomen + thigh)	Female athletes, 18-29 y	D_b (g · cc^{-1})[a] = 1.096095 − (0.0006952 × ∑4SKF) + (0.0000011 × ∑4SKF^2) − (0.0000714 × age)
∑3SKF (triceps + suprailiac + thigh)	White or anorexic women, 18-55 y	D_b (g · cc^{-1})[a] = 1.0994921 − (0.0009929 × ∑3SKF) + (0.0000023 × ∑3SKF^2) − (0.0001392 × age)
∑3SKF (chest + abdomen + thigh)	White men, 18-61 y	D_b (g · cc^{-1})[a] = 1.109380 − (0.0008267 × ∑3SKF) + (0.0000016 × ∑3SKF^2) − (0.0002574 × age)
∑3SKF (abdomen + thigh + triceps)	Black or white collegiate male and female athletes, 18-34 y	%BF = 8.997 + (0.2468 × ∑3SKF) − (6.343 × gender[b]) − (1.998 × race[c])
∑2SKF (triceps + calf)	Black or white boys, 6-17 y Black or white girls, 6-17 y	%BF = (0.735 × ∑2SKF) + 1.2 %BF = (0.610 × ∑2SKF) + 5.1

∑SKF = sum of skinfolds (mm).

[a] Use population-specific conversion formulas to calculate %BF (percent body fat) from Db (body density).

[b] Male athletes = 1; female athletes = 0.

[c] Black athletes = 1; white athletes = 0.

Reprinted, by permission, from V. Heyward and A.L. Gibson, 2014, *Advanced fitness assessment and exercise prescription*, 7th ed. (Champaign, IL: Human Kinetics), 237.

Generalized Body Density Formulas

Body Density Equations for Men (4)

$$D_b = 1.10938 − (0.0008267 × ∑3SKF) + (0.0000016 × ∑3SKF^2) − (0.0002574 × age)$$

(Sum of three skinfolds [mm] = chest + abdomen + thigh)

$$D_b = 1.1120 − (0.00043499 × ∑7SKF) + (0.00000055 × ∑7SKF^2) − (0.00028826 × age)$$

(Sum of seven skinfolds [mm] = chest + midaxillary + subscapular + triceps + abdomen + suprailium + thigh)

Body Density Equations for Women (5)

$$D_b = 1.0994921 − (0.0009929 × ∑3SKF) + (0.0000023 × ∑3SKF^2) − (0.0001392 × age)$$

(Sum of three skinfolds [mm] = triceps + suprailium + thigh)

$$D_b = 1.0970 − (0.00046971 × ∑7SKF) + (0.00000056 × ∑7SKF^2) − (0.00012828 × age)$$

(Sum of seven skinfolds [mm] = chest + midaxillary + subscapular + triceps + abdomen + suprailium + thigh)

Table 15.7 Water Density at Various Temperatures

Water (°C)	Temp (°F)	D_w (g · ml⁻¹)	Water (°C)	Temp (°F)	D_w (g · ml⁻¹)
0	32	0.9990	30	86	0.9957
4	39	1.0000	31	88	0.9954
22	72	0.9978	32	89.5	0.9950
23	73	0.9975	33	91	0.9947
24	75	0.9973	34	93	0.9944
25	77	0.9971	35	95	0.9941
26	79	0.9968	36	97	0.9937
27	81	0.9965	37	99	0.9934
28	82	0.9963	38	100	0.9930
29	84	0.9960	39	102	0.9926

When estimating percent body fat by UWW, it is necessary to determine body density, which, by definition, equals mass divided by volume: $D_B = M / V$. Body volume is calculated using Archimedes' principle, which states that when an object is placed in water, it is buoyed by a counterforce equal to the water it displaces. The volume of water displaced would equal the loss of weight while the object is totally submerged. By subtracting the weight measured in water (M_W) from that measured in air (M_A), we obtain the weight of water displaced ($M_A - M_W$). This weight value is converted to a volume by dividing it by the density of water (D_W).

$$\text{Density} = \frac{M}{V} = \frac{M_A}{(M_A - M_W)/D_W}$$

The determined volume must be corrected for the amount of gas in the body, both in the lungs and in the gastrointestinal tract. Without a method of measurement, the gas in the gastrointestinal (V_{GI}) system is estimated to be 100 ml for all subjects. The subject will remove as much air from the respiratory system as possible with a maximal expiration. However, even after a maximal expiration, a certain amount of air remains in the lungs; this volume is called the **residual volume (RV)**. If RV has not been measured, we can estimate it from **vital capacity**. In fact, in lab 14 you estimated RV with three different

equations (RV_A, RV_B, and RV_C). Thus, the body density equation can be rewritten and expressed as follows:

$$\text{Density} = \frac{M}{V} = \frac{M_A}{[(M_A - M_W)/D_W] - RV - V_{GI}}$$

It is important to use consistent density units, such as grams per ml (g · ml⁻¹) or kilograms per liter (kg · L⁻¹). Once the body density has been determined, body composition is estimated from the Siri equation mentioned earlier (11).

% body fat = (4.95 / density) − 4.50

This holds true unless another population-specific equation is more appropriate (see population-specific equations in table 15.4).

UWW involves several potential sources of error. Recall that this technique provides an *estimate* of percent body fat based on body density. The equations depend on the assumed densities for fat and nonfat tissue, which were originally developed from relatively few samples. Any error in these assumptions can affect the estimates. We know that the densities of different tissues in the nonfat component may vary between individuals. For example, bone density varies depending on age, disease, and ethnicity. Older adults and children have lower bone densities than do young adults, and bone density in blacks is greater than in whites. Another source of error lies in the correction for RV and gastrointestinal gas volume. The former can cause error when RV is simply estimated rather than measured by nitrogen washout or helium dilution rebreathing (7). Finally, as you will see during this lab, skill is required both of the subject and of the technician in obtaining accurate readings for underwater weight. Many subjects are uncomfortable submerging themselves after a maximal exhalation. However, when the measurements are performed by skilled and experienced personnel under well-controlled conditions, the error has been estimated to be ±3% (7). Furthermore, when the measurements are made before and after an intervention (e.g., diet or exercise program with the goal of weight loss), the technique can provide useful information about changes in percent body fat or fat-free mass.

These labs focus on some of the more common and accessible body composition techniques. However, you should also be aware of others, including BIA, DXA, and plethysmography (see table 15.1).

Bioelectrical Impedance Analysis

Bioelectrical impedance analysis (BIA) is based on the principle that lean tissue conducts electrical current better than fat tissue. A low-level electrical current is passed through the body, and impedance is determined by the BIA analyzer. BIA also can estimate total body water that is proportional to the amount of lean tissue. Many assumptions and guidelines are needed with this body composition analysis, such as the following: no eating or drinking for 4 h prior, no exercise for 12 h prior, no alcohol consumption for 48 h prior, no diuretic medications. All of these can affect total body water. In addition, results can be affected by sex, menstrual cycle, age, fitness, and race. Population-specific equations are available, but the user of a BIA unit is often at the mercy of the code programmed into the software. Typically, electrodes are placed on the wrists and ankles. In many commercially available BIA units, clients stand on electrodes (much like standing on a bathroom scale) or hold onto electrodes with their hands. Electrical current moves in the route of least resistance, so units that use only two points of contact make assumptions on resistance through the rest of the body.

Dual Energy X-Ray Absorptiometry

Dual energy X-ray absorptiometry (DXA or DEXA) is a three-component model, providing estimates of bone, fat, and lean tissue densities. It requires an expensive piece of equipment (more than $75,000) that is typically used in research and clinical settings in the evaluation of bone mineral density. It produces a full-body X-ray that is evaluated by the manufacturer's software for the various tissue densities. For this reason, it is difficult to evaluate the validity of DXA; the algorithms are not usually available to the user. Even so, many researchers consider DXA to be the gold standard in body composition (2, 10, 12). It offers additional benefits in that it can provide regional body composition assessment, such as relative body fat of the right or left arm or leg.

Air-Displacement Plethysmography

Like UWW, air-displacement plethysmography (ADP), sometimes referred to by the manufacturer's name, BOD POD, estimates body density and is thus a two-component model. Instead of using the displacement of water, however, it uses the displacement of air in a sealed compartment. It requires an expensive piece of equipment (more than $50,000) but is considered quite valid (3). Its benefits include the ability to measure elderly adults and others who may be uncomfortable being submerged in water.

> Go to the web study guide to access electronic versions of individual and group data sheets, the question sets from each laboratory activity, and case studies for each laboratory activity.
>
> www

BMI and Circumference Data

Equipment

- Stadiometer or measuring tape
- Scale (accurate to a tenth of a kilogram)
- Anthropometric tape (preferably with a spring)
- Individual data sheet
- Group data sheet

BMI Calculation

Step 1:   Each subject should be prepared in the appropriate attire (light, nonbulky clothing to avoid mismeasurement of weight and girth).

Step 2:   Students should record their own data in their individual data sheets and work as evaluators for their labmates. Gather basic data (e.g., age, height, weight) for the individual data sheet as described in laboratory activity 1.1.

Step 3:   Remove your shoes, jewelry, wallet, keys, and any excess clothing prior to weighing yourself on the scale. Record your weight in kg to the preciseness allowed by your equipment.

Step 4:   Measure the subject's height using a stadiometer or a measuring tape attached to a wall. Have the subject stand with his or her back to the stadiometer or wall with the heels on the floor. Use a lever arm or similar device placed on the top of the head at the highest point. Be sure that the arm or similar device is parallel to the ground before recording the height in centimeters to the nearest millimeter.

Step 5:   To demonstrate the reliability of this measure, repeat for a total of three trials. You should measure each member of your group once and then repeat in order to demonstrate true reliability.

Step 6:   Record your data on the individual data sheet and collect data for the group data sheet.

Circumference Measurements

Waist and hip circumferences are best measured with a spring-loaded tape measure such as the Gulick measuring tape. This allows correct (4 oz or 113 g) and reproducible tension to be placed on the tape.

Step 1:   To measure the circumference of the waist, wrap the tape measure around the subject from the anterior side. Place the tape at the narrowest circumference between the umbilicus and the lowest rib. Resolve any twists in the tape. Cross the tape and switch hands to avoid having your arms cross each other. Confirm that the tape is parallel to the ground all the way around the subject. (Doing this measure in front of a mirror can be helpful.) Have the subject relax the abdominal muscles. If you have a spring-loaded tape, pull the tape to the calibrated tension. Record the measurement in centimeters to the nearest millimeter at the end of a normal expiration.

Step 2: To measure the circumference of the hips, wrap the tape measure around the subject from the lateral side. Place the tape at the largest circumference of the buttocks. Resolve any twists in the tape. Cross the tape and switch hands to avoid having your arms cross each other. (Doing this measure in front of a mirror can be helpful.) Confirm that the tape is parallel to the ground all the way around the subject. If you have a spring-loaded tape, pull the tape to the calibrated tension. Record the measurement in centimeters to the nearest millimeter.

Step 3: Repeat steps 1 and 2 two more times to complete the individual datasheet.

Step 4: Fill out the body composition group data sheet, calculating the averages, BMI, and W/H ratio using the individual data sheet.

Question Set 15.1

1. What are potential sources of error for BMI and W/H ratio?
2. For which subject population would the measurements of BMI and W/H ratio be particularly useful?
3. What sex-specific differences do you notice for W/H ratio on the group data sheet?
4. What are the risk stratifications for BMI and W/H ratio (see table 15.8)?

Table 15.8 W/H Ratio Norms for Men and Women

| Age (y) | RISK | | | |
	Low	Moderate	High	Very high
MEN				
20-29	<0.83	0.83-0.88	0.89-0.94	>0.94
30-39	<0.84	0.84-0.91	0.92-0.96	>0.96
40-49	<0.88	0.88-0.95	0.96-1.00	>1.00
50-59	<0.90	0.90-0.96	0.97-1.02	>1.02
60-69	<0.91	0.91-0.98	0.99-1.03	>1.03
WOMEN				
20-29	<0.71	0.71-0.77	0.78-0.82	>0.82
30-39	<0.72	0.72-0.78	0.79-0.84	>0.84
40-49	<0.73	0.73-0.79	0.80-0.87	>0.87
50-59	<0.74	0.74-0.81	0.82-0.88	>0.88
60-69	<0.76	0.76-0.83	0.84-0.90	>0.90

Adapted from Bray and Gray 1988.

LABORATORY ACTIVITY 15.1

INDIVIDUAL DATA SHEET

Name or ID number: _____ Date: _____

Tester: _____ Time: _____

Sex: M / F (circle one) Age: _____ y Height: _____ in. _____ m

Weight: _____ lb _____ kg Temperature: _____ °F _____ °C

Barometric pressure: _____ mmHg Relative humidity: _____ %

Trial	Weight (kg)	Height (m)	Waist (cm)	Hip (cm)	Evaluator (initials)
1					
2					
3					
Average					

Calculation of BMI: Use average weight and height measurements.

BMI = _____ $kg \cdot m^{-2}$

Calculation of W/H: Use average waist and hip measurements.

W/H = _____ cm / _____ cm = _____

Techniques for Measuring Skinfold Thickness

Equipment

- Skinfold calipers (handle the calipers carefully—they are delicate and expensive)
- Stadiometer or measuring tape
- Scale (accurate to a tenth of a kilogram)
- Individual data sheet
- Group data sheet

Standard Skinfold Test Procedures

Accurate skinfold measurement requires practice. Because of the variability present even in measurements taken by experienced technicians, three separate measures are made at each site, all on the same side of the body (e.g., the right side). Once a technician is experienced, it is common to repeat the measurements of sites only twice (provided they are within 10% of each other).

This is an opportunity to practice professionalism. Subjects will need to move clothing to allow access to skinfold sites. Always have subjects move their clothing themselves—never reach under clothes. Making jokes is rarely useful in relieving tension; instead, foster a confident, professional clinical manner to put subjects at ease. Caution the subject not to try to assist you or lean away, because doing either can change anatomical orientations. It is common to measure and mark the skinfold site with a marker. However, this can affect the learning experience when several evaluators will be measuring the same subject; for this reason, we suggest measuring but not marking the subject in this lab.

Step 1:   Subjects should wear loose clothing that facilitates skinfold measurement at the appropriate sites.

Step 2:   Each student should serve as a subject for two evaluators and serve as an evaluator for two subjects (if possible, one male and one female). Gather basic data (e.g., age, height, weight) for the individual data sheet as described in laboratory activity 1.1.

Step 3:   Avoid making measurements through clothing. If this is not possible (e.g., as with a sports bra), subtract the thickness of the clothing from your skinfold measurement.

Step 4:   Perform repeated measures at multiple sites by making one measure at each site (following the order on the individual data sheet) and repeating this sequence in a circuit. See the individual data sheet and the following list of anatomical-site procedures for specialized instructions for each measurement. (The black Xs on the subjects in figure 15.4 are for an effective anatomical visual. Students should not repeat this practice because it removes the purpose of repeated measures.)

Step 5: Pinch the appropriate skinfold using the thumb and index finger 1 cm (0.4 in.) proximal from where the jaws of the calipers will be placed. Inexperienced technicians often make the mistake of having their fingers too close together when taking a skinfold. Start wide and bring your fingers together to create a fold. It is impossible to make too big a fold, but it is possible to underestimate the fold by pinching too closely.

Step 6: Place the jaws of the calipers half the distance between the base of the skinfold that is not held with the fingers and the crest of the fold (see figure 15.2). The calipers should be perpendicular to the fold. Do not twist the calipers so they are easier to read.

Step 7: Record the reading from the scale on the calipers. Most calipers are accurate to the nearest 0.5 mm. Do not release the skinfold with your fingers until the calipers have been removed. Do not allow the calipers to snap shut!

Step 8: Proceed through the rest of the anatomical skinfold sites (descriptions follow these steps) and record all results for two evaluators at each of the nine sites.

Step 9: Skinfold trial outliers (>30% different from other trials) can be eliminated. Calculate averages for the remaining three trials for a given evaluator on the individual data sheet.

Step 10: Students should fill out their individual data sheet with their own skinfold measurements.

Step 11: Complete the sex-specific three- and seven-site equations for body density given earlier. Estimate body fat using the Siri equation. Use these data to answer the lab activity questions, and report them in the group data sheet.

Anatomical-Site Skinfold Test Procedures

These site-specific procedures complement the standard skinfold procedures already given. They provide the necessary details for working with the sites used by the Jackson and Pollock (4) and Jackson, Pollock, and Ward (5) equations.

Triceps

- Have the subject stand facing away from you with the right arm hanging straight down.
- Standing behind the subject, pinch a site 1 cm (0.4 in.) above the midpoint between the shoulder (acromion process of the scapula) and the tip of the elbow (inferior portion of the olecranon process of the ulna) on the posterior aspect of the triceps.
- Place the jaws of the calipers 1 cm below and perpendicular to the vertical fold. See figure 15.4a.

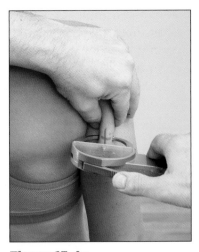

Figure 15.4a

Suprailium

- Grasp the skinfold just above the iliac crest at the level of the anterior axillary line along the natural cleavage of the skinfold (running diagonally down the crest toward the umbilicus).
- Place the jaws of the calipers 1 cm (0.4 in.) distal and perpendicular to the diagonal fold. See figure 15.4*b*.

Figure 15.4*b*

Abdomen

- Grasp a vertical skinfold 2 cm (0.8 in.) to the left of and 1 cm (0.4 in.) above the umbilicus.
- Place the jaws of the calipers 1 cm below and perpendicular to the vertical fold (to the right of the umbilicus). See figure 15.4*c*.

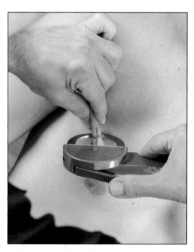

Figure 15.4*c*

Chest

- For men, the chest skinfold is the midpoint between the anterior axillary line (the front of the armpit) and the nipple. Grasp the skinfold 1 cm (0.4 in.) above the midpoint in a diagonal line along the angle of the pectoralis major muscle. Place the jaws of the calipers 1 cm below and perpendicular to the diagonal fold.
- For women, take a diagonal fold as high as possible on the anterior axillary fold.
- Place the jaws of the calipers 1 cm below and perpendicular to the diagonal fold. See figure 15.4*d*.

Figure 15.4*d*

Thigh

- Have the subject stand with all weight on the left leg and a slight bend at the knee in the right leg.
- Have the subject lift up his or her shorts to gain access to the anterior aspect of the thigh. Technicians often take folds too distally because of longer shorts.
- Take a vertical fold on the anterior aspect of the thigh 1 cm (0.4 in.) above the halfway point between the inguinal crease and the proximal border of the patella.
- Place the jaws of the calipers 1 cm below and perpendicular to the vertical fold. See figure 15.4e.

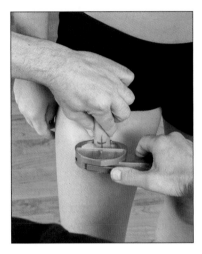

Figure 15.4e

Midaxillary

- Take a vertical fold along the midaxillary line at the level of the xiphoid process of the sternum. The subject's arm is routinely placed on the tester's shoulder to best gain access to the midaxillary line.
- Place the jaws of the calipers 1 cm (0.4 in.) below and perpendicular to the vertical fold. See figure 15.4f.

Figure 15.4f

Subscapular

- Take a diagonal fold 1 to 2 cm (0.4-0.8 in.) from the inferior angle of the scapula. The subject can fold the arm behind the back to expose the scapula; once the spot is located, the arm is returned to the normal position.
- Place the jaws of the calipers 1 cm below and perpendicular to the diagonal fold. See figure 15.4g.

Figure 15.4g

Biceps

- Take a vertical fold on the anterior aspect of the arm over the belly of the biceps muscle 1 cm (0.4 in.) above the level used to mark the triceps. The subject should rotate the palm anteriorly.
- Place the jaws of the calipers 1 cm below and perpendicular to the vertical fold. See figure 15.4h.

Figure 15.4h

Calf

- Take a vertical fold on the medial aspect at the level of the maximal calf circumference with the knee and the hip flexed to 90° (rest the foot on a stool or chair).
- Place the jaws of the calipers 1 cm (0.4 in.) below and perpendicular to the vertical fold. See figure 15.4i.

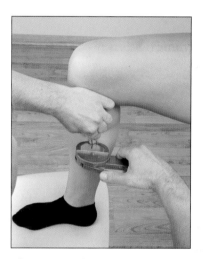

Figure 15.4i

Question Set 15.2

1. Calculate body fat using the numbers from the two evaluators. How did the skinfold measurements compare between the evaluators? How did the reproducibility turn out within each evaluator's measurements? What does this suggest about the reliability of skinfold measures? What does it suggest about the development of skinfold measurement skill?

2. What are potential sources of error for skinfold measurement?

3. Place all of the body composition measurements you performed on a continuum from least to most reliable. Do the same for validity. Briefly defend your answers.

4. What explanations might there be for differences in the %BF values between UWW and skinfolds?

5. Briefly describe how you (as an exercise specialist, coach, researcher, athletic trainer, or physical therapist) might be able to use the information obtained from estimating body composition with your student, athlete, subject, or patient.

6. How do your own values for each of the measurements compare with the norms? What is the mean of all the measurements for your percent body fat? Which one do you believe is the most valid? Why?

7. Calculate your desired body weight. How do you plan to attain (or maintain) this ideal weight? What assumptions are you making?

LABORATORY ACTIVITY 15.2

INDIVIDUAL DATA SHEET

Name or ID number: _____ Date: _____

Tester: _____ Time: _____

Sex: M / F (circle one) Age: _____ y Height: _____in. _____m

Weight: _____ lb _____ kg Temperature: _____°F _____°C

Barometric pressure: _____ mmHg Relative humidity: _____ %

Measurement	Trial 1	Trial 2	Trial 3	Average	Evaluator
Chest					
Midaxillary					
Subscapular					
Triceps					
Biceps					
Abdomen					
Suprailium					
Thigh					
Calf					

Skinfold measurements are in mm.

Calculation of body density and body fat from Σ3SKF and Σ7SKF equations:

D_b (Σ3SKF) = _____ $g \cdot ml^{-1}$

D_b (Σ7SKF) = _____ $g \cdot ml^{-1}$

%BF (Σ3SKF) = _____ %

%BF (Σ7SKF) = _____ %

Estimating Relative Body Fat Using Hydrodensitometry

Equipment

- UWW tank (custom tub in your exercise science lab or a swimming pool)
- Scale within the tank (suspended balance or load cell) that is able to support a chair or bench on which the subject sits to be suspended in the water
- Scale (accurate to a tenth of a kilogram)
- Stadiometer or measuring tape
- Individual data sheet

> This laboratory activity is supported by a virtual lab experience in the web study guide.
>
> www

UWW

Step 1: Gather basic data (e.g., age, height, weight) for the individual data sheet as described in laboratory activity 1.1. Pretest instructions include a 3 h fast and abstinence from exercise for 24 h.

Step 2: Measure the subject's weight in the air. The subject should wear a swimsuit, ideally one that minimizes air trapped in the suit, and remove all jewelry. Weigh the subject while dry.

Step 3: Have the subject shower to remove oils and lotions that can contaminate the UWW tank.

Step 4: The water temperature should be between 33 °C and 36 °C (for comfort).

Step 5: Calibrate the scale before the subject gets in the tank. This is done simply by hanging known weights on the scale above the water level.

Step 6: The subject should get into the tank and submerge in order to get completely wet, including the hair. Air bubbles should be removed from the bathing suit, hair, and skin, as well as from the swing and any other part of the equipment.

Step 7: Have the subject submerge up to the neck without touching the scale. Note the weight on the scale at this point. This is the tare weight. The subject displaces water, which will change the weight of the scale. If the subject is wearing a weight during the UWW, the weight should be secured to the swing when the tare weight is determined. Some systems allow you to zero out at this tare weight. This process should be performed for every subject who is weighed because different-sized bodies can alter the tare weight.

Step 8: The subject should sit in the chair or swing and practice the UWW procedure. It may be necessary to have the subject wear a weight (2-3 kg), such as a dive belt, in order to submerge fully. The subject should expel most of the air in the lungs before submerging, then submerge and expel any additional air. It is essential that neither the subject nor the swing touch the side of the tank. The subject should try to be as still as possible. Slight movements cause large oscillations in the scale, making it difficult to read. Have the subject attempt to count to 10 slowly and stay submerged during this time if possible. Record the underwater weight on the individual data sheet.

Step 9: Repeat the procedure with the subject three to five times until at least three stable readings are obtained. A classmate should make the readings of the scale. The scale may oscillate significantly; it should stabilize as the subject becomes suspended. The scale reader should determine the extremes of the needle oscillations when the fluctuations are stabilized and record the midpoint between these extremes. The goal is to read the scale to the nearest 20 g, but it may be possible only to read to the nearest 50 g. This level of precision may vary between systems. If your lab has load cells, even greater precision is possible.

Step 10: Fill out the body composition lab table in the group data sheet for laboratory 15.2, including your calculations for the subjects' percent body fat. (Use all three predicted RVs. Be sure to retain significant digits.)

Question Set 15.3

1. What are potential sources of error for UWW?
2. What was the difference between the three RVs used for the calculations? What effect did this difference have on the calculated percent body fat? What percentage difference does this make in body fat?
3. For each of your UWW subjects, calculate the total amount (kg) of fat weight and LBM based on your UWW results. Show your calculations.
4. Why might separate equations be developed for resistance-trained individuals in the conversion from Db to %BF?

LABORATORY ACTIVITY 15.3

INDIVIDUAL DATA SHEET

Name or ID number: _____ Date: _____

Tester: _____ Time: _____

Sex: M / F (circle one) Age: _____ y Height: _____ in. _____ m

Weight: _____ lb _____ kg Temperature: _____ °F _____ °C

Barometric pressure: _____ mmHg Relative humidity: _____ %

Mass in air (M_A) _____ kg

RV_A = _____ L

RV_B = _____ L

RV_C = _____ L

Water temp (T_w) _____ °C

D_w (table 15.7): _____

Tare weight: _____ g

Trial	1	2	3	4	5
Mass in water (M_w; g)					

Circle the three highest M_w values and use the average of those three: _____ g

$$\underset{\text{mean } M_W \text{ (g)}}{\underline{\hspace{2cm}}} - \underset{\text{tare weight (g)}}{\underline{\hspace{2cm}}} = \underset{\text{net } M_W \text{ (g)}}{\underline{\hspace{2cm}}}$$

Calculations Using RV_A

$$\left[\left(\underset{M_A \text{(g)}}{\underline{\hspace{1.5cm}}} - \underset{M_W \text{(g)}}{\underline{\hspace{1.5cm}}} \right) \Big/ \frac{\overline{M_A \text{(g)}}}{\underset{D_W \text{(g} \cdot \text{ml}^{-1})}{\underline{\hspace{1.5cm}}}} \right] - \underset{RV_A \text{(ml)}}{\underline{\hspace{1.5cm}}} - 100 \text{ ml}^\star = \underset{D_b \text{(g} \cdot \text{ml}^{-1})}{\underline{\hspace{1.5cm}}}$$

$$(4.95 / \underset{D_b \text{(g} \cdot \text{ml}^{-1})}{\underline{\hspace{1.5cm}}}) - 4.50 = \underset{\% \text{fat}}{\underline{\hspace{1.5cm}}}$$

Calculations Using RV_B

$$\left[\left(\underset{M_A \text{(g)}}{\underline{\hspace{1.5cm}}} - \underset{M_W \text{(g)}}{\underline{\hspace{1.5cm}}} \right) \Big/ \frac{\overline{M_A \text{(g)}}}{\underset{D_W \text{(g} \cdot \text{ml}^{-1})}{\underline{\hspace{1.5cm}}}} \right] - \underset{RV_B \text{(ml)}}{\underline{\hspace{1.5cm}}} - 100 \text{ ml}^\star = \underset{D_b \text{(g} \cdot \text{ml}^{-1})}{\underline{\hspace{1.5cm}}}$$

$$(4.95 / \underset{D_b \text{(g} \cdot \text{ml}^{-1})}{\underline{\hspace{1.5cm}}}) - 4.50 = \underset{\% \text{fat}}{\underline{\hspace{1.5cm}}}$$

Calculations Using RV$_C$

$$\frac{\left[\left(\dfrac{\rule{2cm}{0.4pt}}{M_A\,(g)} - \dfrac{\rule{2cm}{0.4pt}}{M_W\,(g)}\right) \Big/ \dfrac{\overline{\dfrac{\rule{2cm}{0.4pt}}{M_A\,(g)}}}{\dfrac{\rule{2cm}{0.4pt}}{D_W\,(g \cdot ml^{-1})}}\right] - \dfrac{\rule{2cm}{0.4pt}}{RV_C\,(ml)} - 100\ ml^{\star}}{} = \dfrac{\rule{3cm}{0.4pt}}{D_b\,(g \cdot ml^{-1})}$$

$$\left(4.95 \Big/ \dfrac{\rule{3cm}{0.4pt}}{D_b\,(g \cdot ml^{-1})}\right) - 4.50 = \dfrac{\rule{3cm}{0.4pt}}{\%\,fat}$$

*This 100 ml is the estimated volume of gas in the gastrointestinal tract.

Electrocardiograph Measurements

Objectives

- Introduce the terminology associated with electrocardiography.
- Understand the electrical activity of the heart and the three-dimensional view of the heart through ECG leads.
- Learn the procedure for electrode placement for a 12-lead ECG and observe a resting ECG.
- Learn to determine heart rate (HR) and heart axis from an ECG recording.
- Demonstrate a 12-lead ECG during submaximal exercise.
- Learn basic interpretation of ECG strips.

DEFINITIONS

12-lead ECG—Set of leads originating from 10 electrodes placed on a subject to record electrical activity of the heart.

ectopic pacemaker—Heart cell that is not in the sinoatrial (SA) node but acts as the pacemaker for the electrical activity of the heart.

electrocardiograph (ECG or EKG)—Graphical recording of the electrical activity of the heart.

heart axis—Angular position of the heart within the chest cavity.

mean vector—Sum of the electrical vectors of the heart, which determines the heart's placement in the body, or axis.

P wave—ECG recording representing atrial depolarization.

pacemaker—Heart cell or group of cells originating the electrical activity for the entire heart (normally, the Sinoatrial [SA] node).

QRS complex—ECG recording representing ventricular depolarization.

T wave—ECG recording representing ventricular repolarization.

All cells in the body can conduct a current of electricity, and two of the most electrically active tissues are the brain and heart. Currents conducted in these tissues can be recorded through electrodes placed on the skin and connected to an **electrocardiograph (ECG or EKG)**. The resulting three-dimensional graphical recording provides detailed information about the heart's rate and rhythm, current performance, history of injury, and potential risk of injury. Many clinicians spend a lifetime pouring over ECGs to characterize a subject's heart performance. Clinical exercise physiologists often use the stress of exercise to identify people who may be at risk for a heart injury. Certifications in this area may allow a student to seek employment within the growing field of clinical exercise physiology and cardiac rehabilitation (1).

ELECTRICAL ACTIVITY OF THE HEART

The electrical activity in a normal heart (figure 16.1) originates in the sinoatrial (SA) node, which serves as the **pacemaker** of the heart. However, any heart cell, or myocyte, can initiate pacemaking, and when a cell not in the SA node acts as the pacemaker, it is called an **ectopic pacemaker**. From the SA node, electrical activity spreads through the atria, taking advantage of the electrical gap junctions or intercalated discs between myocytes. These specialized junctions allow the spread of depolarization without the need for a chemical synapse. The atria and ventricles of a heart are divided by thin connective tissue; therefore, they act as separate electrical units. The wave of atrial depolarization is funneled through the atrioventricular (AV) node. Upon reaching the AV node, there is a slight pause before a rapid depolarization is sent down the AV bundle, bundle branches, and Purkinje fibers (3).

These conductive fibers in the ventricles are specialized to rapidly conduct the electrical signal. This is particularly important to the ventricles, which are responsible for ejecting blood to the entire body, including (against the force of gravity) to the brain. This rapid transmission of the depolarization through the ventricles allows a coordinated contraction to improve blood ejection. The wave of depolarization from cell to cell using intercalated discs travels at about 0.3 m · s^{-1}, whereas the bundle branches and Purkinje fibers can conduct the electrical current at >3 m · s^{-1}. An ECG records this electrical activity and provides information about the quantity and quality of the heart's rhythm.

An ECG readout has three main parts: the **P wave**, the **QRS complex**, and the **T wave**. These waves represent the electrical activity of the various heart chambers and thus the contraction of these areas (see figure 16.2).

Electrical impulses travel from the SA node in the right atrium throughout both atria to generate the first wave of the ECG—the P wave,

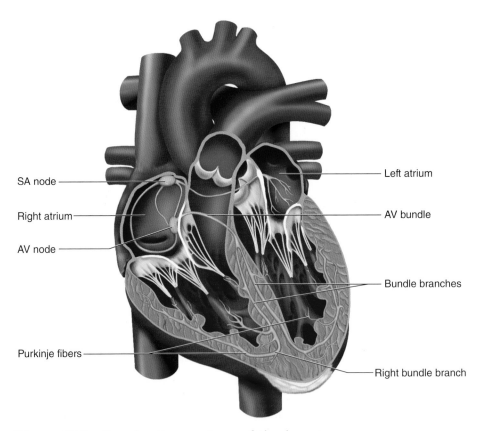

SA node

Right atrium

AV node

Purkinje fibers

Left atrium

AV bundle

Bundle branches

Right bundle branch

Figure 16.1 Conduction system of the heart.

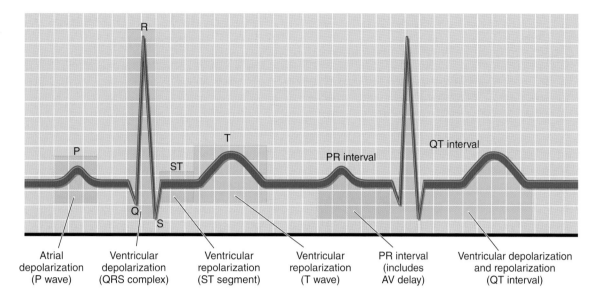

| Atrial depolarization (P wave) | Ventricular depolarization (QRS complex) | Ventricular repolarization (ST segment) | Ventricular repolarization (T wave) | PR interval (includes AV delay) | Ventricular depolarization and repolarization (QT interval) |

Figure 16.2 ECG waves.

Reprinted, by permission, from W.L. Kenney, J.H. Wilmore, and D.L. Costill, 2015, *Physiology of sport and exercise*, 6th ed. (Champaign, IL: Human Kinetics), 159.

which thus represents atrial depolarization. The impulse then travels to the AV node, where there is a slight pause as the ventricles fill, and then travels through the AV bundle to the Purkinje fibers located in the walls of the ventricles. As the wave of depolarization spreads from the AV node throughout the ventricles, a wave of deflection is shown on the ECG. Known as the *QRS complex*, it represents ventricular depolarization. The final wave of the ECG tracing, the T wave, represents repolarization of the ventricles. The atrial repolarization is obscured on the tracing by the QRS complex; therefore, it cannot be seen.

In the heart, the electrical events of depolarization precede the mechanical events of contraction. However, much information about the heart's contraction can be determined from the electrical activity. The axes for an ECG recording are in millivolts (mV) on the y-axis and time on the x-axis, which allows a clinician to determine time delays and quantity of electrical activity. For example, large P waves or QRS complexes indicate greater amounts of muscle in the atria or ventricles, respectively. This type of hypertrophy may indicate valvular defects or even adaptations to exercise. The time

for atrial depolarization (P wave) takes longer than ventricular depolarization (QRS complex). This fact relates to the previously mentioned specialized fibers within the ventricles. In addition, the direction of the electrical signal can be determined. ECG leads have direction, so if a wave is in the direction of an ECG lead, then the wave will be upright, or positive. Conversely, if the wave is in the opposite direction of the lead, it will be downward, or negative. As you might then expect, if a wave is equally upward and downward in its inflection, it is considered to be at 90° to the ECG lead.

PLACEMENT OF ECG LEADS

Although an ECG can be done with just one electrode and is commonly done with four, a **12-lead ECG** is completed with 10 carefully placed electrodes (see figure 16.3). Together, the electrodes make up leads, which can be thought of as electrical vectors that have direction from a negative to positive pole. This allows the creation of a three-dimensional picture of the heart and the direction of its electrical activity. There are 6 limb leads from the 4 torso or limb electrodes and

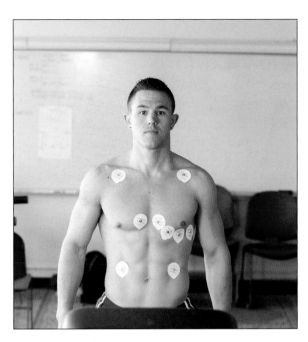

Figure 16.3 Placement of 12-lead electrodes.

6 chest leads from the 6 chest electrodes. Leads are the voltage difference between one or more electrodes. The bipolar limb leads (I, II, and III) are called such because they use the combination of positive and negative poles between limb electrodes. For example, lead I is the voltage between the left arm (LA) and right arm (RA) electrodes. The other 3 limb leads are called the *augmented limb leads*—aVR, aVL, and aVF. These leads are unipolar leads, even though they also have negative and positive poles; the negative pole is a composite of several other electrodes. They are considered augmented because the negative pole is a combination of electrodes, which augments the signal strength. For example, aVR uses the RA as the positive electrode, and the negative pole is a combination of the LA and left leg (LL) electrodes. Note that the right leg (RL) acts as an electrical ground. See figure 16.4.

The chest or precordial leads (V1-V6) are considered unipolar as well, and they view the heart in a horizontal plane due to their close proximity. Their placement across the chest allows additional electrical vectors to aid in the determination of electrical direction, heart rate (HR), or axis.

INTERPRETING THE ECG RECORDING

Much can be determined from an ECG recording, or strip, including HR, heart axis, clinical diagnoses, and performance of heart contractions. We focus here on HR and axis as relatively easy determinations. Normal sinus rhythm occurs when the SA node is acting as the pacemaker of the heart. Sinus bradycardia involves a normal heart rhythm but slow HR at rest (>60 beats · min⁻¹). Sinus tachycardia involves a normal heart rhythm but elevated HR at rest (>100 beats · min⁻¹).

Calculating HR from an ECG strip involves relatively easy math. Normally, ECG paper speed is 25 mm · s⁻¹ or 1,500 mm · min⁻¹. Therefore, in order to calculate HR, one simply needs to measure the millimeters between R waves and use this simple equation:

$$HR \text{ (beats} \cdot \text{min}^{-1}) = 1{,}500 \text{ / R-to-R distance (mm)}$$

This measurement can be made with a metric ruler or with the knowledge that each small box on the ECG strip is 1 mm (0.04 s) and each bigger box is 5 mm (0.2 s). It is common to measure the distance from R-to-R across several cardiac cycles to get an average of the R-to-R distances because HR variability (HRV) is significant in some people. For example, imagine we measure 50 mm between the four R waves in a normal sinus rhythm ECG. This would be an average of 16.67 mm · beat⁻¹ because it covers three R-to-R distances. The HR could be calculated as follows: 1,500 mm · min⁻¹ / 16.67 mm · beat⁻¹ = 90 beats · min⁻¹. In figure 16.2, the 18 mm between the R waves would yield an HR of 83.33 beats · min⁻¹. Clearly, this method depends on the assumed paper speed time of 25 mm · s⁻¹, so it is essential to confirm that this holds true for your ECG machine.

Other methods exist for calculating HR from the ECG strip, such as dividing 300 by the number of larger boxes (5 mm). This approach may save time and give the interpreter a rough estimate of HR. Some ECG strips provide tick marks in 1, 3, or 6 s intervals to allow a rough

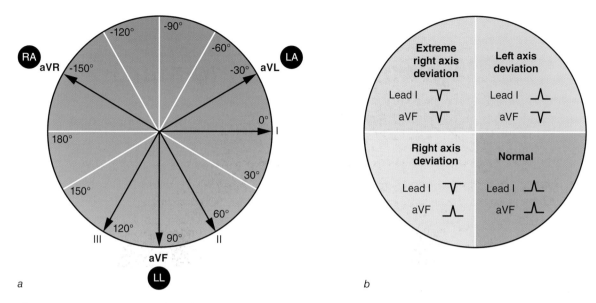

Figure 16.4 *(a)* ECG lead vectors; *(b)* determining heart axis.

b) © Charles Dumke

estimate of HR by multiplying the number of R intervals within the tick marks by 60, 20, or 10, respectively. However, these estimates require the interpreter to estimate the number of partial beats between the time intervals, which introduces bias and potential error. Measuring the distance between R waves remains the most accurate, albeit time-consuming, method of attaining HR from an ECG strip. In reality, ECG machines often automatically calculate HR; however, we suggest that the lab instructor mask this output so students learn how to perform this calculation.

The **heart axis** is the angular position of the heart within the chest cavity. Heart axis can be affected by body position, cardiac hypertrophy, pregnancy, and some disease states. Determining heart axis from the direction of the electrical vectors is also useful in finding the specific location of damage in a diseased heart. The mean direction of the total electrical activity of the heart gives it its axis, or **mean vector**. Because of the mass of the ventricles, the direction of the QRS complex in the limb leads (I, II, III, aVL, aVR, and aVF) is used to determine the mean vector (see figure 16.4). Because of the wave of depolarization from the SA node throughout the heart, a normal healthy heart under sinus

rhythm has a 60° axis, which causes a positive (upward) QRS deflection in leads I, II, and aVF. (Refer to figure 16.4*b* for degrees of the heart axis in relation to the electrical vectors.) This axis causes an equally positive and negative deflection (isoelectric) in the leads perpendicular to this axis, in this case the aVL lead, and an almost completely negative deflection in the aVR lead, which is in a near-opposite direction. Normal heart axes range from 0° to 90°, whereas a right heart axis ranges from 90° to 180°, and a left heart axis ranges from −90° to 0°. Thus, most hearts have an axis between leads I, II, and aVF, and a positive QRS deflection in all or most of these leads suggests a normal heart axis. A more specific mean vector can be found by examining isoelectric and negative QRS leads. For example, say an ECG result with positive QRS deflections in the I, II, and aVF leads suggests a normal heart axis. For further refinement, you identify an isoelectric lead III and a negative aVR. Together, this would suggest a mean vector of 30°, which is perpendicular to the most isoelectric signal. Deviations from normal axis can indicate left or right ventricular hypertrophy, right or left bundle branch blocks, other electrical abnormalities, or a myocardial infarction (heart attack).

ECG AS A TOOL FOR DIAGNOSING CARDIAC ABNORMALITIES

This laboratory introduces concepts of the ECG. It is not meant to provide complete training regarding the diagnostic uses of ECG; however, several simple abnormalities are discussed. For a more complete discussion of ECG interpretation, see this lab's reference list (2, 4). The following sections briefly describe a few ECG abnormalities that differ from a normal sinus rhythm.

ST Segment Abnormalities

The period of time between S and the beginning of the T wave can indicate abnormalities in blood flow to the heart. ST segment elevation indicates an already existing heart attack or some level of necrotic heart tissue. On the other hand, reduced coronary blood flow, or ischemia, puts a person at risk for a future heart attack. Heart ischemia appears as ST segment depression on an ECG. The severity of the discrepancy of the ST segment from the isoelectric line is proportional to the severity of the disease. ST segment depression may be unapparent at rest yet appear during the stress of exercise. See figure 16.5.

Preventricular Contractions

Preventricular contractions (PVCs) occur when a myocyte in the ventricles initiates the heart contraction, thereby acting as an ectopic pacemaker.

PVCs are identified by a wide QRS (figure 16.6a). They are relatively common at rest, but if the frequency increases significantly during exercise they may result in poor ventricular performance. A sequence of several PVCs indicates greater severity of this abnormality. The term *bigeminy* refers to an abnormal heartbeat, such as a PVC, occurring every other cardiac cycle (figure 16.6b). *Trigeminy* refers to abnormal beats occurring every third cardiac cycle.

Ventricular Tachycardia

Ventricular tachycardia, or V-tach, is a run of consecutive PVCs. If V-tach is sustained, it requires immediate medical attention, typically including CPR (cardiopulmonary resuscitation) or use of an AED (automated external defibrillator) to shock the heart back into sinus rhythm.

Ventricular Fibrillation

Ventricular fibrillation, or V-fib, is a continual spasm of the ventricles disallowing appropriate blood ejection. This abnormality requires immediate medical attention, typically including cardiopulmonary resuscitation (CPR) or use of an automated external defibrillator (AED) to shock the heart back into sinus rhythm. See figure 16.7.

Asystole

Asystole involves a flat line ECG requiring immediate medical attention, typically including

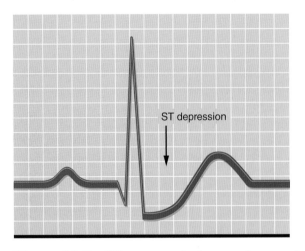
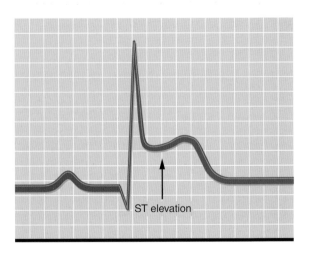

Figure 16.5 ST segment depression and elevation.

use of an AED to shock the heart back into sinus rhythm.

Premature Atrial Contractions

Premature atrial contractions (PACs) are similar to PVCs in that they are ectopic contractions of the atria. However, because they are in the atria, they are far less worrisome than PVCs. PACs

normally go untreated; however, they may lead to atrial flutter or fibrillation. See figure 16.8.

Atrial Flutter and Fibrillation

Atrial flutter is an uncoordinated, very rapid (>250 beats · min⁻¹) contraction of the atria. Though not necessarily dangerous in itself, it jeopardizes complete filling of the ventricles

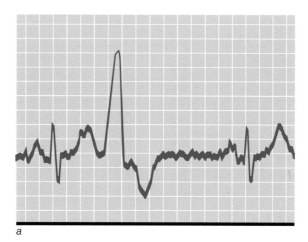

a

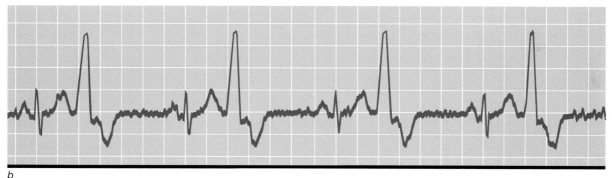

b

Figure 16.6   *(a)* PVC; *(b)* bigeminy.

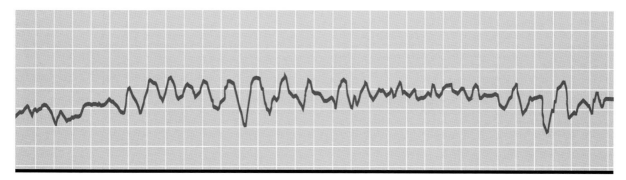

Figure 16.7   Ventricular fibrillation.

and is normally found in older adults with other heart disease or chronic obstructive pulmonary disease (COPD). It may be temporary but may also lead to atrial fibrillation. Flutter is different from fibrillation in that the atria beat regularly in flutter but irregularly in fibrillation. Atrial flutter and fibrillation may go unnoticed by the patient; however, they may also lead to more serious ventricular problems. Atrial fibrillation is more likely to cause an irregular contraction of the ventricles and may also predispose a patient for stroke. See figure 16.9.

HEART RATE RESPONSE TO EXERCISE

As you have likely observed in previous labs, HR increases during graded exercise in direct proportion to the intensity of the exercise. This regulation of HR during exercise is controlled by the autonomic nervous system (3). HRs below 100 beats · min⁻¹ are due to parasympathetic innervations of the SA node by the vagus nerve. Increasing the HR from rest to exercise involves both parasympathetic withdrawal and sympathetic stimulation of the SA node. See figure 16.10.

With each increase in intensity during a graded exercise test (GXT), HR increases for the first 1 or 2 min and then levels off at steady state as it meets the new metabolic demands. The higher the intensity, the longer it takes the HR to reach steady state. At maximum intensity (exhaustion), the HR may plateau; this is then considered the maximum HR.

Aside from changes in HR, ECGs may look different during exercise of increasing intensity. Often, a person with occasional PVCs at rest may have a normal sinus rhythm during exercise when under sympathetic control. Conversely, and more dangerously, the frequency of PVCs or PACs may increase with exercise. In addition, changes may occur in the ST segment; of particular interest is ST segment depression, which indicates myocardial ischemia. The intensity at which ST segment depression appears is

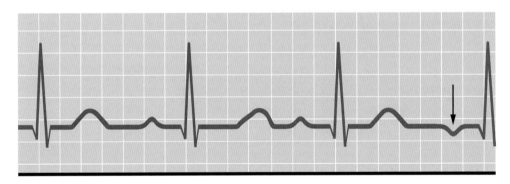

Figure 16.8   PACs.

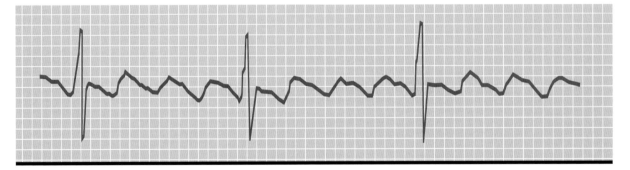

Figure 16.9   Atrial flutter and fibrillation.

referred to as an *ischemic threshold*, and it constitutes reason to immediately stop a test (1). In your introduction to ECGs, you are unlikely to observe such changes, which are less common among young, apparently healthy individuals

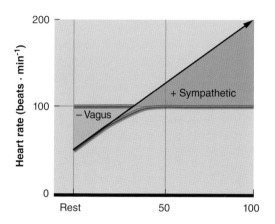

Figure 16.10 Autonomic nervous system control of HR.

Reprinted, by permission, from W.L. Kenney, J.H. Wilmore, and D.L. Costill, 2015, *Physiology of sport and exercise*, 6th ed. (Champaign, IL: Human Kinetics), 159; Adapted from L.B. Rowell, 1993, *Human cardiovascular control* (Oxford, UK: Oxford University Press).

such as your labmates. Changes you may observe include increased level of noise from the recruitment of respiratory muscles and jostling of the electrodes, which make ECG interpretation difficult at higher intensities. This noise can be reduced by keeping the electrode connections as still as possible and having the subject refrain from overly swinging the arms.

Athletes are often misdiagnosed with pathological hearts due to the effect of exercise training on their ECGs (2). Common changes in an athlete's heart can include sinus bradycardia, changes in ST segment and T waves (often appearing as ST segment elevation due to the increased size of the T wave), ventricular hypertrophy, bundle branch or AV blocks, and various arrhythmias. These are thought of not as clinically significant but as by-products of the heart's adaptations to exercise training.

Go to the web study guide to access electronic versions of the individual data sheets, the question sets from each laboratory activity, and case studies for each laboratory activity.

WWW

Resting ECG

Equipment

- Platform scale (or digital or other scale)
- Stadiometer (wall or freestanding) or physician's scale with attached anthropometer
- ECG machine
- Disposable ECG electrodes
- Alcohol prep pads, gauze, scrubbing pad, disposable razor
- Individual data sheet

6-Lead Resting ECG Test

Step 1: Begin this lab activity by gathering basic data (e.g., age, height, weight) for the individual data sheet as described in laboratory activity 1.1.

Step 2: Prep all students to obtain a resting 6-lead ECG of their heart.

Step 3: ECG electrodes measure electricity of the heart; therefore, any impedance to conductivity impairs the electrical recording. Remove all hair, body oil, dead skin, and lotion from the electrode sites prior to placing each disposable electrode.

Step 4: Prep your labmate for the four limb electrodes (LA, RA, LL, RL). Locate the correct site for electrode placement (see the following list) and prep the area by shaving (if necessary), scrubbing with an abrasive pad, and cleaning with alcohol. Women should be instructed ahead of time to wear a swimsuit top or sports bra. Use discretion with female subjects and maintain a professional approach at all times. The limb electrodes are often placed on the ankles and wrists in clinical arenas; however, because we are exercise physiologists, we will prep the limb electrodes on the torso as if the subject were going to exercise. Follow these anatomical locations (figure 16.11):

- *LA (left arm):* Just below the middle of the subject's left clavicle above the pectoralis muscle, medial to the anterior deltoid.
- *RA (right arm):* Just below the middle of the subject's right clavicle above the pectoralis muscle, medial to the anterior deltoid.
- *LL (left leg):* Between the subject's left external oblique and rectus abdominis muscles, level with the umbilicus.
- *RL (right leg):* Between the subject's right external oblique and rectus abdominis muscles, level with the umbilicus.

Step 5: Connect the labeled ECG wires to the correct ECG labeled electrodes.

Step 6: Obtain ECGs on the subjects while they sit quietly in a chair.

Step 7: Be sure to note the ECG paper speed.

Step 8: Complete the individual data sheet.

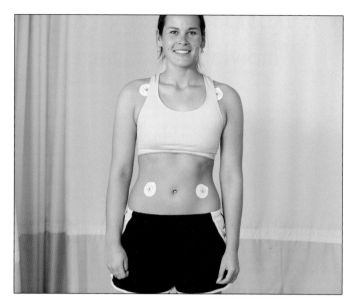

Figure 16.11 Placement of 4 limb electrodes.

Question Set 16.1

1. Using the data from your personal ECG, calculate your HR with the three methods described (1,500 / R-to-R distance; # of R-to-R in tick marks; 300 / # big boxes). Show your work on the individual data sheet and attach your ECG.

2. Did these three methods result in the same HR? Why or why not?

3. Describe the cascade of electrical activity through the heart with a normal sinus rhythm. What controls HR?

4. Label one cardiac cycle identifying the P wave, QRS complex, T wave, PR interval, and ST segment. Explain what each represents in terms of heart action.

5. What is the purpose of using a 6-lead ECG as opposed to only 1 lead?

6. Determine the axis of your heart. Does your heart axis fall within the normal range (see figure 16.4 and accompanying text)?

LABORATORY ACTIVITY 16.1

INDIVIDUAL DATA SHEET

Name or ID number: _____ Date: _____

Tester: _____ Time: _____

Sex: M / F (circle one) Age: _____ y Height: _____ in. _____ m

Weight: _____ lb _____ kg Temperature: _____ °F _____ °C

Barometric pressure: _____ mmHg Relative humidity: _____ %

HR

A. Distance between 4 R waves _____ mm / 3 = _____ mm · beat^{-1}

1,500 mm · min^{-1} / _____ mm · beat^{-1} = _____ beats · min^{-1}

B. 300 big boxes · min^{-1} / _____ big boxes · beat^{-1} = _____ beats · min^{-1}

C. ECG tick-mark distance = _____ s

\# of R waves · tick^{-1} _____ × _____ ticks · min^{-1} = _____ beats · min^{-1}

Heart Axis

A. Lead with the most positive defection _____ = _____ °

B. Lead with the nearest isoelectric defection _____ = _____ °

C. Lead with the most negative defection _____ = _____ °

D. Estimated heart axis = _____ °

Paste ECG here:

Effects of Body Position on the Heart Axis

Equipment

- Platform scale (or digital or other scale)
- Stadiometer (wall or freestanding) or physician's scale with attached anthropometer
- ECG machine
- Disposable ECG electrodes
- Alcohol prep pads, gauze, scrubbing pad, disposable razor
- Individual data sheet

6-Lead Body Position ECG Test

Step 1: Collect subject characteristics such as age, height, and weight for one student in your group and then prep this student for a 6-lead ECG of the heart. See laboratory activity 1.1 for information on gathering basic subject data.

Step 2: ECG electrodes measure electricity of the heart; therefore, any impedance to conductivity impairs the electrical recording. Remove hair, body oil, dead skin, and lotion from the electrode sites prior to placing each disposable electrode.

Step 3: Prep your subject for the four limb electrodes (LA, RA, LL, RL). Locate the correct site for electrode placement (see the following list) and prep the area by shaving (if necessary), scrubbing with an abrasive pad, and cleaning with alcohol. Women should be instructed ahead of time to wear a swimsuit top or sports bra. Use discretion with female subjects and maintain a professional approach at all times. The limb electrodes are often placed on the ankles and wrists; however, because we are exercise physiologists, we will prep the limb electrodes as if the subject were going to exercise. Follow these anatomical locations (see correct limb electrode placement in figure 16.11):

  - *LA (left arm):* Just below the middle of the subject's left clavicle above the pectoralis muscle, medial to the anterior deltoid.
  - *RA (right arm):* Just below the middle of the subject's right clavicle above the pectoralis muscle, medial to the anterior deltoid.
  - *LL (left leg):* Between the subject's left external oblique and rectus abdominis muscles, level with the umbilicus.
  - *RL (right leg):* Between the subject's right external oblique and rectus abdominis muscles, level with the umbilicus.

Step 4: Connect the labeled ECG wires to the correct ECG labeled electrodes.

Step 5: Have the subject lie in a supine position. Allow 2 to 3 min for the subject to relax and come to steady state, and then obtain an ECG recording.

Step 6: Have the subject sit in a chair in a slumped position. Allow about 2 min in this position, and then obtain an ECG recording.

Step 7: Have the subject stand erect. Allow about 2 min in this position, and then obtain an ECG recording.

Step 8: Note the ECG paper speed.

Step 9: Complete the individual data sheet.

Question Set 16.2

1. Using the ECG recordings from your subject in the three body positions, calculate HR while supine, sitting, and standing. Show the techniques used on the individual data sheet and attach the ECG.

2. How does body position affect HR? What is occurring physiologically to cause these changes?

3. Using the ECG recordings from your subject in the three body positions, determine heart axis while supine, sitting, and standing (see figure 16.4 and accompanying text). How does body position affect heart axis? What causes these changes?

LABORATORY ACTIVITY 16.2

INDIVIDUAL DATA SHEET

Name or ID number: _____ Date: _____

Tester: _____ Time: _____

Sex: M / F (circle one) Age: _____ y Height: _____ in. _____ m

Weight: _____ lb _____ kg Temperature: _____ °F _____ °C

Barometric pressure: _____ mmHg Relative humidity: _____ %

HR

A. Supine: Distance between 4 R waves _____ mm /3 = _____ mm · beat^{-1}

1,500 mm · min^{-1} / _____ mm · beat^{-1} = _____ beats · min^{-1}

B. Sitting: Distance between 4 R waves _____ mm/3 = _____ mm · beat^{-1}

1,500 mm · min^{-1} / _____ mm · beat^{-1} = _____ beats · min^{-1}

C. Standing: Distance between 4 R waves _____ mm /3 = _____ mm · beat^{-1}

1,500 mm · min^{-1} / _____ mm · beat^{-1} = _____ beats · min^{-1}

Heart Axis

A. Supine: Lead with the most positive defection _____ = _____ °

Lead with the nearest isoelectric defection _____ = _____ °

Lead with the most negative defection _____ = _____ °

Estimated heart axis = _____ °

B. Sitting: Lead with the most positive defection _____ = _____ °

Lead with the nearest isoelectric defection _____ = _____ °

Lead with the most negative defection _____ = _____ °

Estimated heart axis = _____ °

C. Standing: Lead with the most positive defection _____ = _____ °

Lead with the nearest isoelectric defection _____ = _____ °

Lead with the most negative defection _____ = _____ °

Estimated heart axis = _____ °

Paste ECG here:

Submaximal Exercise Effects With the 12-Lead ECG

Equipment

- Platform scale (or digital or other scale)
- Stadiometer (wall or freestanding) or physician's scale with attached anthropometer
- ECG machine
- Treadmill
- Disposable ECG electrodes
- Alcohol prep pads, gauze, scrubbing pad, disposable razor
- Individual data sheet

12-Lead ECG

Step 1: Collect characteristics such as age, height, and weight for the one student from your group on whom you will perform a 12-lead ECG of the heart. See laboratory activity 1.1 for information on gathering basic subject data.

Step 2: ECG electrodes measure electricity of the heart; therefore, any impedance to conductivity impairs the electrical recording. Remove all hair, body oil, dead skin, and lotion from the electrode sites prior to placing each disposable electrode.

Step 3: Prep your subject for 10 electrodes (LA, RA, LL, RL, V1-V6). Locate the correct site for electrode placement (see the following list) and prep the area by shaving (if necessary), scrubbing with an abrasive pad, and cleaning with alcohol. Use discretion with female subjects and maintain a professional approach at all times. Women should be instructed ahead of time to wear a swimsuit top or sports bra. Women may want to find the electrode sites themselves. A curtained area in the lab can provide a convenient space for women to prepare for a 12-lead ECG, perhaps with the aid of a female labmate. The limb electrodes are often placed on the ankles and wrists; however, because we are exercise physiologists, we will prep the four limb electrodes for exercise. Follow these anatomical locations (figure 16.12):

- *LA (left arm):* Just below the middle of the subject's left clavicle above the pectoralis muscle, medial to the anterior deltoid.
- *RA (right arm):* Just below the middle of the subject's right clavicle above the pectoralis muscle, medial to the anterior deltoid.
- *LL (left leg):* Between the subject's left external oblique and rectus abdominis muscles, level with the umbilicus.
- *RL (right leg):* Between the subject's right external oblique and rectus abdominis muscles, level with the umbilicus.

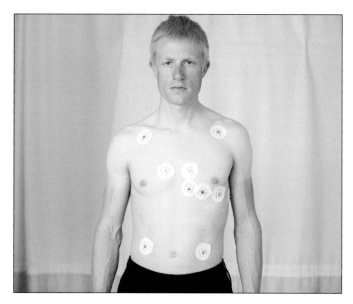

Figure 16.12 Placement of 12-lead electrodes.

For the six chest leads, place the electrodes in the order given in the following list. You must palpate for the intercostal spaces between the ribs (you cannot rely on sight). Begin by palpating just below the clavicle left of the sternum. The first space is the first intercostal space. Then count downward to the fourth intercostal space for V2. *Left* and *right* refer to the subject's left and right.

V2: On the left sternal border on the fourth intercostal space

V1: On the right sternal border on the fourth intercostal space

V4: On the midclavicular line in the fifth intercostal space

V3: At the midpoint of a straight line between V2 and V4

V5: On the anterior axillary line horizontal to V4

V6: On the midaxillary line horizontal to V4 and V5

Note that V1, V2, and V4 are in the spaces between the ribs, not on the rib bones themselves. This difference is important because bone does not conduct electricity well.

Step 4: Connect the labeled ECG wires to the correct ECG labeled electrodes.

Step 5: Collect the ECG wires and secure them to be safe and out of the subject's way during exercise on the treadmill. This is most frequently done with a soft Velcro strap.

Step 6: Have the subject attain steady state while in the standing position for 2 to 3 min. It is advisable to get a resting ECG in the position in which the subject will be exercising—that is, standing for a treadmill or sitting for a bike. This is useful due to the effects of body position on the ECG, as demonstrated in lab activity 16.2. If a treadmill is not available, it is possible to use a bicycle ergometer; however, clinical testing is more frequently done on treadmills. Obtain a 12-lead resting ECG recording for your subject after reaching steady state.

Step 7: Have the subject straddle the treadmill. Bring the speed and grade of the treadmill to 1.7 mph (2.7 km · h⁻¹) and 10% grade. This is the first stage of the Bruce protocol that is often used in exercise stress testing (see lab 10). Allow 3 min for the subject to come to steady state. Obtain a 12-lead exercise ECG recording for the subject as well as an RPE. (See figure 16.13.)

Step 8: Increase speed and grade to 2.5 mph (4.0 km · h⁻¹) and 12% grade. Allow 3 min for the subject to come to steady state. Obtain a 12-lead exercise ECG recording as well as an RPE.

Step 9: Increase speed and grade to 3.4 mph (5.5 km · h⁻¹) and 14% grade. Allow 3 min for the subject to come to steady state. Obtain a 12-lead exercise ECG recording as well as an RPE.

Step 10: Stop the treadmill and allow the subject to recover while sitting in a chair. Monitor the 12-lead ECG. Following 3 min of recovery, obtain a 12-lead ECG recording.

Step 11: Fill out the individual data sheet with data from the test.

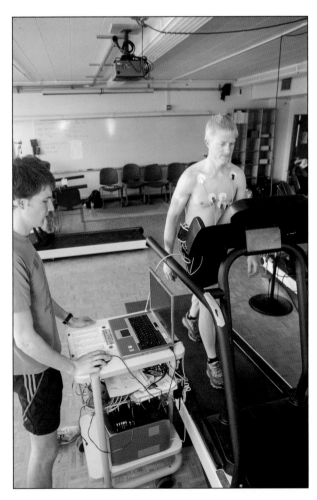

Figure 16.13 Exercising ECG.

Question Set 16.3

1. Using the ECG recordings, calculate the subject's HR while at rest, during each stage of exercise, and during recovery in order to complete the individual data sheet. Show the technique you used and attach the ECGs.

2. How does exercise of increasing intensity change HR? What controls this change?

3. Using the ECG recordings, determine the subject's heart axis while at rest, during each stage of exercise, and during recovery in order to complete the individual data sheet (see figure 16.4 and accompanying text).

4. How does exercise of increasing intensity change the heart axis?

5. What changes are anticipated in an ECG recording in the transition from rest to exercise? What abnormal or pathological changes may occur in the transition from rest to exercise?

LABORATORY ACTIVITY 16.3

INDIVIDUAL DATA SHEET

Name or ID number: _____ Date: _____

Tester: _____ Time: _____

Sex: M / F (circle one) Age: _____ y Height: _____in. _____m

Weight: _____ lb _____ kg Temperature: _____°F _____°C

Barometric pressure: _____ mmHg Relative humidity: _____%

Time (min)	Speed (mph)	Grade (%)	Power (kg · m · min⁻¹)	Avg R-to-R (mm)	HR (beats · min⁻¹)	Heart axis (°)	RPE
0	0	0	Rest				_____
3	1.7	10					
6	2.5	12					
9	3.4	14					
12	0	0	Recovery				_____

Paste ECG here:

APPENDIX A
Units of Measure Conversions

Standard Length Conversions

1 in. = 0.0254 m = 2.54 cm = 25.4 mm
1 ft = 0.3048 m = 30.48 cm = 304.8 mm
1 yd = 0.914 m = 91.44 cm = 914.4 mm
1 mi = 1.609 km = 1,609.34 m = 160,934.4 cm
1 mm = 0.1 cm = 0.001 m = 0.000001 km
1 cm = 10 mm = 0.01 m = 0.00001 km
1 km = 1,000,000 mm = 100,000 cm = 1,000 m
1 mm = 0.0394 in. = 0.00328 ft
1 cm = 0.3937 in. = 0.03281 ft
1 km = 39,370.1 in. = 3,280.8 ft = 0.62137 mi

Note: cm = centimeter, ft = foot, in. = inch, km = kilometer, m = meter, mi = mile, mm = millimeter.

Standard Mass Conversions

1 kg = 1,000 g = 100,000 cg = 1,000,000 mg
1 g = 100 cg = 1,000 mg
1 kg = 2.2046 lb = 35.274 oz
1 g = 0.0022046 lb = 0.035274 oz
1 lb = 16 oz = 0.4536 kg = 453.6 g
1 oz = 0.0283495 kg = 28.3495 g

Note: cg = centigram, g = gram, kg = kilogram, lb = pound, mg = milligram, oz = ounce.

Common Volume Conversions

1 L = 10 dl = 100 cl = 1,000 ml
1 dl = 0.1 L = 10 cl = 100 ml
1 cl = 0.01 L = 0.1 dl = 10 ml
1 ml = 0.001 L = 0.01 dl = 0.1 cl
1 cm^3 = 1 ml = 0.01 dl = 0.001 L
1 L = 1.0567 qt
1 L = 1 kg

Note: cl = centiliter, cm^3 = cubic centimeter, dl = deciliter, kg = kilogram, L = liter, ml = milliliter, qt = quart.

Length (Distance)

1 m = 39.370 in. = 3.281 ft = 1.094 yd
1 km = 0.621 mi
1 cm = 0.394 in.
1 in. = 2.540 cm
1 ft = 12 in. = 30.480 cm
1 yd = 3 ft = 0.914 m
1 mi = 5,280 ft = 1,760 yd = 1,609.344 m

Velocity

1 mi·h⁻¹ (mph) = 26.822 m·min⁻¹ = 1.467 ft·s⁻¹ = 0.447 m·s⁻¹
1 m·s⁻¹ = 2.237 mi·h⁻¹ (mph) = 3.281 ft·s⁻¹

Force

1 N = 0.225 lb of force = 0.102 kg of force
1 lb force = 4.448 N
1 kg force = 9.81 N

Torque

1 N·m = 0.738 ft-lb

Mass and Weight

1 kg = 2.205 lb
1 kp = 1 kg
1 g = 0.035 oz
1 lb = 16 oz = 0.454 kg
1 oz = 28.350 g
(1 L of water weighs 1 kg)

Energy

1 kcal = 4.186 kJ = 426.935 kg·m = 1.163 W·h
1 BTU = 0.252 kcal = 1.055 kJ = 107.586 kg·m
1 J = 1 N·m = 0.102 kg-m = 0.239 cal
1 kg·m = 1 kp·m = 9.807 J = 2.342 cal
1 L of oxygen consumed = 5.05 kcal = 21.143 kJ (at an RER of 1.00)

Power

1 W = 1 J·s⁻¹ = 6.118 kg·m·min⁻¹ = 0.00134 hp
1 kg·m·min⁻¹ = 1 kp·m·min⁻¹ = 0.163 W
1 W = 6.118 kg·m·min⁻¹

Pressure

1 atm = 760 mmHg = 101.325 kPa = 14.696 psi
1 mmHg = 1 torr = 0.0193 psi = 133.322 Pa = 0.00132 atm
1 kPa = 0.01 mbar

Temperature

°C = 0.555 × [(°F) − 32]
°F = 1.8 × [(°C) + 32]

Volume

1 L = 1.057 qt
1 qt = 0.946 L = 2 pt = 32 oz
1 (U.S.) gal = 4 qt = 128 oz = 3.785 L
1 c = 8 oz = 0.237 L
1 oz = 2 tbsp = 6 tsp = 29.574 ml
1 tbsp = 3 tsp = 14.787 ml
1 tsp = 4.929 ml

Quantity of a Substance

1 mol of a gas = 22.4 L (standard conditions) = 6.022×10^{23} molecules (Avogadro's number)

1 L of gas (standard conditions) = 44.6 mmol

mol = mass (g) / molecular mass

molarity of a solution = mol of substance / L of solvent

APPENDIX B

Estimation of the O$_2$ Cost of Walking, Running, and Leg Ergometry

Use the following equations to estimate the O$_2$ requirement for walking and running.

TREADMILL WALKING

The oxygen cost of horizontal treadmill walking can be estimated for treadmill speeds of 50 to 100 m $\cdot$ min^{-1} (~1.5-4 mi $\cdot$ h^{-1} or 2.4-6.4 km $\cdot$ h^{-1}) using a formula based on the linear relationship between treadmill speed and oxygen consumption ($\dot{V}O_2$ ml $\cdot$ kg^{-1} $\cdot$ min^{-1}). The relationship possesses a slope of 0.1 and a y-intercept of 3.5 ml $\cdot$ kg^{-1} $\cdot$min^{-1} (resting $\dot{V}O_2$).

$$\dot{V}O_2 \text{ (horizontal component)} =$$
$$\frac{0.1 \text{ ml} \cdot \text{kg}^{-1} \cdot \text{min}^{-1}}{\text{m} \cdot \text{min}^{-1}} \times$$
$$\text{speed} (\text{m} \cdot \text{min}^{-1}) + 3.5 \text{ ml} \cdot \text{kg}^{-1} \cdot \text{min}^{-1}$$

The oxygen cost of graded treadmill walking has a vertical component as follows:

$$\dot{V}O_2 \text{ (vertical component)} =$$
$$\frac{1.8 \text{ ml} \cdot \text{kg}^{-1} \cdot \text{min}^{-1}}{\text{m} \cdot \text{min}^{-1}} \times$$
$$\text{speed} (\text{m} \cdot \text{min}^{-1}) \times \text{fractional grade}$$

The $\dot{V}O_2$ required for graded treadmill walking is the sum of the horizontal and vertical O$_2$ costs. Thus, the O$_2$ cost of walking at 90 m $\cdot$ min^{-1} up a 7.5% grade would be as follows:

Horizontal $\dot{V}O_2$ = (0.1 ml $\cdot$ kg^{-1} $\cdot$ min^{-1} × 90 m $\cdot$ min^{-1}) + 3.5 ml $\cdot$ kg^{-1} $\cdot$ min^{-1} = 12.5 ml $\cdot$ kg^{-1} $\cdot$ min^{-1}

Vertical $\dot{V}O_2$ = 1.8 ml $\cdot$ kg^{-1} $\cdot$ min^{-1} × (90 m $\cdot$ min^{-1} × 0.075) = 12.2 ml $\cdot$ kg^{-1} $\cdot$ min^{-1}

Total $\dot{V}O_2$ = 12.5 ml $\cdot$ kg^{-1} $\cdot$min^{-1} + 12.2 ml $\cdot$ kg^{-1} $\cdot$ min^{-1} = 24.7 ml $\cdot$ kg^{-1} $\cdot$ min^{-1}

TREADMILL RUNNING

The oxygen cost for horizontal and graded treadmill running for speeds greater than 134 m $\cdot$ min^{-1} (5 mi $\cdot$ h^{-1} or 8 km $\cdot$ h^{-1}) can be calculated in a manner similar to that used for treadmill walking. The $\dot{V}O_2$ (ml $\cdot$ kg^{-1} $\cdot$ min^{-1}) of horizontal treadmill running is calculated using the following formula:

$$\dot{V}O_2 \text{ (horizontal component)} =$$
$$\frac{0.2 \text{ ml} \cdot \text{kg}^{-1} \cdot \text{min}^{-1}}{\text{m} \cdot \text{min}^{-1}} \times$$
$$\text{speed} (\text{m} \cdot \text{min}^{-1}) + 3.5 \text{ ml} \cdot \text{kg}^{-1} \cdot \text{min}^{-1}$$

The $\dot{V}O_2$ for graded treadmill running has a vertical component as follows:

$$\dot{V}O_2 \text{ (vertical component)} =$$
$$\frac{0.9 \text{ ml} \cdot \text{kg}^{-1} \cdot \text{min}^{-1}}{\text{m} \cdot \text{min}^{-1}} \times$$
$$\text{speed} (\text{m} \cdot \text{min}^{-1}) \times \text{fractional grade}$$

Total $\dot{V}O_2$ = horizontal $\dot{V}O_2$ + vertical $\dot{V}O_2$

LEG ERGOMETRY

The oxygen cost for leg ergometry, or stationary cycling, can be calculated in a similar manner:

$\dot{V}O_2$ (ml $\cdot$ kg^{-1} $\cdot$ min^{-1}) = 10.8 × workload (W) × 1 / body weight (kg) + 7 ml $\cdot$ kg^{-1} $\cdot$ min^{-1}

or

$\dot{V}O_2$ (ml $\cdot$ kg^{-1} $\cdot$ min^{-1}) = 1.8 × workload (kg $\cdot$ m $\cdot$ min^{-1}) × 1 / body weight (kg) + 7 ml $\cdot$ kg^{-1} $\cdot$ min^{-1}

APPENDIX C

Haldane Transformation

Current-day metabolic carts often take away from students the knowledge of what is being collected and calculated. In reality, they only measure O_2 and CO_2 in the expired breath (F_EO_2 and F_ECO_2) and the volume of air expired ($\dot{V}_E$), while the rest of the metabolic cart output is generated by calculations based on the Haldane transformation. The following calculations assume expired gases from an exercising individual were collected in Douglas bags and analyzed with appropriate equipment for F_EO_2, F_ECO_2, and $\dot{V}_E$.

The Haldane transformation is based on the fact that N_2 is neither produced nor utilized by the human body. Thus, the volume of N_2 inhaled must be equivalent to the volume of N_2 exhaled. This fact is expressed by the following formula (see definitions at the end of this appendix):

$$\dot{V}_I \times F_IN_2 = \dot{V}_E \times F_EN_2$$

We can rearrange this formula to solve for V_I as follows:

$$\dot{V}_I = (\dot{V}_E \times F_EN_2) / F_IN_2$$

We know that $F_IN_2 = 0.7904$ and $F_EN_2 = 1.0 - F_EO_2 - F_ECO_2$. Both F_EO_2 and F_ECO_2 can be measured by gas analyzers, while $\dot{V}_E$ can be measured by a Tissot or other gas volume analyzer. Thus, $\dot{V}_I$ can be calculated from these measured values and constants.

CONSTANTS

$$F_ICO_2 = 0.0003$$
$$F_IO_2 = 0.2093$$
$$F_IN_2 = 0.7904$$

From Douglas bag collection and gas analysis, you will know the other three unknown parameters (fraction of expired CO_2 and O_2 and expired gas volume or $\dot{V}_E$). With these, along with ambi-ent temperature (T_A) and barometric pressure (P_A), you can determine energy expenditure.

CONVERSION OF V_{ATPS} TO V_{STPD}

In all metabolic calculations, gas volumes are expressed at standard temperature, pressure, dry (STPD). Thus, it is necessary to reduce a gas volume to STPD before plugging $\dot{V}_E$ into the $\dot{V}_I$ equation. To do so, use the following formula:

$$\dot{V}_{E\,STPD} = \dot{V}_{E\,ATPS} \times [273 / (273 + T_A)] \, [(P_A - P_{H2O}) / 760]$$

CALCULATION OF OXYGEN CONSUMPTION AND CARBON DIOXIDE PRODUCTION

Calculation of oxygen consumption ($\dot{V}O_2$) is relatively simple. The amount of oxygen exhaled is subtracted from the amount of oxygen inhaled:

$$\dot{V}O_2 = (\dot{V}_I \times F_IO_2) - (\dot{V}_E \times F_EO_2)$$

Carbon dioxide production ($\dot{V}CO_2$) is the difference between the CO_2 expired and the CO_2 inspired. It can be calculated as follows:

$$\dot{V}CO_2 = (\dot{V}_E \times F_ECO_2) - (\dot{V}_I \times F_ICO_2)$$

From these results, RER can be easily calculated as $\dot{V}CO_2 / \dot{V}O_2$.

This may seem like a confusing array of equations, and for this reason we suggest the following order of operations:

1. From the expired gas collected, express $\dot{V}_E$ in $L \cdot min^{-1}$ (from Tissot or similar device to measure total gas expired and knowledge of collection time).

2. Using the equations presented earlier in this appendix, convert to STPD.

3. Using the relevant equations above, calculate $\dot{V}_I$.

4. Using the relevant equations above and the results from F_EO_2 and F_ECO_2, calculate $\dot{V}O_2$.

5. Using the relevant equations above and the results from F_EO_2 and F_ECO_2, calculate $\dot{V}CO_2$.

6. Using the relevant equations above, calculate RER.

DEFINITIONS

$\dot{V}CO_2$—Amount of carbon dioxide produced ($L \cdot min^{-1}$).

$\dot{V}O_2$—Amount of oxygen consumed ($L \cdot min^{-1}$).

$\dot{V}_I$—Volume of gas inspired ($L \cdot min^{-1}$).

$\dot{V}_E$—Volume of gas expired ($L \cdot min^{-1}$).

F_IN_2—Fraction of inspired nitrogen.

F_EN_2—Fraction of expired nitrogen.

F_ICO_2—Fraction of inspired carbon dioxide.

F_ECO_2—Fraction of expired carbon dioxide.

F_IO_2—Fraction of inspired oxygen.

F_EO_2—Fraction of expired oxygen.

T_A—Ambient temperature (°C).

P_A—Atmospheric pressure (mmHg).

P_{H2O}—Partial pressure of water vapor (mmHg).

APPENDIX D
Metabolic Cart Information

Several companies provide machines capable of measuring oxygen consumption, carbon dioxide production, and ventilation. They include the following:

Parvo Medics

COSMED

New Leaf

AEI Technologies

VacuMed

These carts are based on a relatively simple concept. They measure the amount of O_2 and CO_2 in the room air and in expired breath, as well as the volume of expired breath. They do so by means of O_2 and CO_2 analyzers and a flow meter. Knowing these results, along with the atmospheric temperature and pressure, enables you to do simple calculations to determine $\dot{V}O_2$ and $\dot{V}CO_2$ (see appendix C).

It is essential to properly calibrate these carts. To do so, your lab needs reference gases of known concentration (usually around 16.0% O_2 and 4.0% CO_2) and a turbine of known volume (usually 3 L). It may also be necessary to purchase a portable weather station for measuring ambient barometric pressure, humidity, and temperature. You can then calibrate the cart (figure D.1). to the room air concentrations of O_2 and CO_2 (20.93% O_2 and 0.03% CO_2) and the reference gases. You can also calibrate its flow meter with the known volume turbine. Your instructor will lead you through the metabolic cart calibration in your lab. You can also see a short video on metabolic cart calibration in the web study guide.

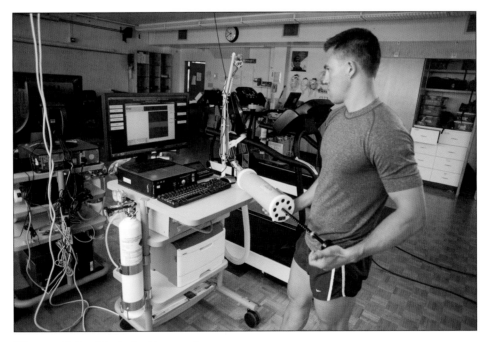

Figure D.1 Metabolic cart.

APPENDIX E
Calibration of Equipment

The field of exercise physiology is based on the measurement of metabolic responses to exercise at given absolute or relative workloads. Equipment such as bicycle ergometers and treadmills are essential in eliciting measureable workloads. Because of this, it is important to be able to determine if a workload is accurately reflected by what is dialed in for a piece of equipment. For example, one must be able to determine whether a treadmill set to 6 mph (approximately 10 km · h^{-1}) and 8% grade is actually moving at that speed and grade. This appendix aids students in calibrating the treadmills and bicycle ergometers in their labs.

TREADMILL CALIBRATION

Countless brands of treadmills exist on the market today, and it is not unusual for treadmills to be significantly wrong in speed or grade. Because of this, it is important to calibrate the speed and grade as described here. To do this, you will need some tape, a measuring device, a stopwatch, and a calculator.

Speed

To check the speed of a treadmill, place a piece of tape on the belt perpendicular to its movement. Measure the belt length by measuring from the front edge of this piece of tape to where the belt loops under the machine. Then, place another piece of tape where this measurement stopped (you may want to label tape pieces as *1, 2, 3 . . .* or *A, B, C . . .*). Move the belt along by standing on the treadmill and pushing with your feet, but do not start the treadmill at this point. Once the belt length is measured as accurately as possible (we suggest meters), remove all but one piece of tape. Start the treadmill at a relatively slow speed (0.5-2 mph [0.8-3.2 km · h^{-1}]). With a stopwatch, measure the time (to the nearest hundredth of a second) it takes the piece of tape to make 20 revolutions. To calculate actual speed of the treadmill, take the belt length (m) multiplied by

number of revolutions divided by the time. Or, you can use this formula:

Treadmill speed (distance / time) = (belt length in m × 20) / time for 20 revolutions

This process should be repeated a minimum of two times at this speed to reduce human error. Also, repeat this procedure at four different speeds to see if any difference in actual speed is a function of the speed the treadmill is running. When timing revolutions at higher speeds, it is important to use more than 20 revolutions to increase the validity and reliability of the measurement. It may also be advised to repeat this procedure with someone walking on the treadmill; the additional weight of a person may slow the motor depending on the horsepower of the treadmill.

If significant differences are found between treadmill speed and actual speed, check the treadmill instruction manual for calibration specifications. Your lab may also have a service contract for the treadmill if it is under warranty. Even if the treadmill cannot be fixed, corrections can be made while performing exercise tests. For example, if a reading of 4 mph corresponds to an actual speed of 3.8 mph, then calculations for workload can be adjusted accordingly.

Grade

It is also important to calibrate the grade of your lab treadmill. Most lab treadmills express grade as a percentage; that is, they express grade as rise over run times 100. This is similar to degree slope; for example, a 10% grade is equal to a 5.7° slope. Because treadmills and prediction equations are expressed in *% grade,* these same units are used here.

Because of different treadmills and different floors, a treadmill reading of 0% grade may not be correct. To check this, use a level on the treadmill. With the treadmill off and set to 0% grade, the tread should be level to the ground. Surprisingly, this is often not the case because floors settle and move with time. If you don't have a level, another way to check this is to measure from the ground to the top of the tread in the back of the treadmill and at the front. The measurements should be identical. If the treadmill is not at zero, it may be possible to shim the treadmill up to zero.

To check the percent grade, you must first know the *run*. Measure the run from the front edge to the back edge of the treadmill at a convenient clear point, such as the roller. Set the treadmill at 0 mph and 1% to 3% grade. Now measure the rise by measuring the difference between the front and back of the treadmill. However, by raising the front of the treadmill, the run has shortened. To account for this, you should remeasure the run using a carpenter's square aligned with the front of the treadmill, ensuring a right angle. Otherwise, you can use the Pythagorean theorem to calculate the run:

$$\text{Rise}^2 + \text{Run}^2 = \text{Hypotenuse}^2$$

Rewritten to solve for run, it is

$$\text{run} = \sqrt{\text{hypotenuse}^2 - \text{rise}^2}$$

where the hypotenuse is the original length of the treadmill at 0% grade. Once rise and run are determined, calculate actual percent grade:

$$\% \text{ grade} = (\text{rise} / \text{run}) \times 100$$

Repeat this procedure to ensure reliability of the measurement. Repeat this procedure at several grades to ensure correct calibration.

If the percent grade of your treadmill is off but the zero is correct, your treadmill instruction manual may have calibration specifications. You may also have a service contract if your treadmill is under warranty. At the very least, on-site corrections can be made while performing exercise tests. For example, if 10% grade is really 9.8%, calculations for workload can be adjusted accordingly.

BICYCLE ERGOMETER CALIBRATION

In most bicycle ergometers, flywheel resistance is provided by a strap or rope. Thus, under the premise of a fixed pedal cadence, the tighter the strap, the greater the resistance to the flywheel, and thus the greater the exercise intensity (see lab 7 for directions on how to calculate kg · m · min^{-1} on a bicycle). Some new ergometers are electronically braked, which makes calibration difficult or impossible. However, the following is a description of how to calibrate a bicycle ergometer similar to a Monark, which is often used in exercise physiology labs. It involves three steps: zeroing, calibrating resistance, and checking the strap condition.

Zeroing

The first thing to check is that when there is zero resistance to a flywheel, the scale reads *0*. Many of the scale boards that allow the pendulum to indicate resistance in kilograms (kg or kp) actually move themselves. There may be a thumbscrew that can be loosened, which will allow you to move the 0 over the pendulum; then retighten the screw. This is often done with the strap or rope completely removed from the flywheel to ensure zero resistance. Be careful not to allow the strap to fall into the enclosed flywheel, because it can be difficult to retrieve. See figure E.1.

Resistance Calibration

The next step of calibration also requires the strap to be disconnected. From the top strap, hang a weight between 0.25 and 7.0 kg. With this known weight on the strap, the resistance indicator on the bike should display the appropriate amount of weight. Ideally, several weights are hung to ensure calibration between the ranges used in an exercise test. Be sure the weight does not rub or hook on anything while hanging from the strap. If the resistance indicator on the bicycle is not displaying the correct weight, several things can be done. Once several weights across a range have been tested, a linear regression can be completed. This allows you to calculate the real resistance when a particular weight is indicated on the bike. Many models of bicycle ergometers allow you to adjust the pendulum as well. Adjusting the pendulum might involve

Figure E.1 Scale board, pendulum, and flywheel of a bicycle ergometer.

using a hex wrench or other equipment to move weight toward or away from the pendulum in order to achieve the correct weight calibration.

Strap Condition

Because resistance on the flywheel is a function of strap friction, the nylon straps can become worn, develop burrs, or even melt over time, especially if used for Wingate tests. Examine your straps regularly to ensure good condition and replace them when needed. Strap life can be affected by a dirty flywheel. Clean the flywheel with a dry cloth during calibration, paying special attention to burrs or other surface irregularities that may affect contact with the strap.

Monark provides an online version of the instruction manual for the popular 828E model. It is available at www.manualslib.com/manual/837200/Monark-828e.html.

APPENDIX F

Certifications in Exercise Science

Not all certifications are created equal. Many personal trainer exams can be taken online and completed without qualifications. Recognized certifications also vary between countries. To list all the governing bodies that provide fitness certifications is beyond the scope of this manual. However, some of the more reputable certifications in the United States are offered by two main governing bodies—the American College of Sports Medicine (ACSM; www.acsm.org) and the National Strength and Conditioning Association (NSCA; http://nsca-lift.org). In Canada, certifications of the Canadian Society for Exercise Physiology (CSEP; www.csep.ca) are also recognized. In some countries accreditation is required, including Great Britain (British Association of Sport and Exercise Sciences [BASES]; www.bases.org.uk) and Australia (Exercise and Sports Science Australia [ESSA]; www.essa.org.au).

ACSM CERTIFICATIONS

ACSM has three tracks of certification: health and fitness certifications focused on apparently healthy adults, clinical certifications focused on clients with disease, and specialty certifications for working with certain populations. For more information, see ACSM's certification web page: http://certification.acsm.org/.

ACSM Health Fitness Certifications

• **Certified Group Exercise Instructor (GEI):** These are fitness professionals who teach, lead, and motivate individuals through intentionally designed exercise classes. ACSM GEIs are high achievers inside and outside the group studio. Not only do they excel at planning effective, exercise science-based group sessions for various fitness levels, but they also possess a wealth of motivational and leadership techniques that help their classes achieve fitness goals.

• **Certified Personal Trainer (CPT):** These are fitness professionals who develop and implement personalized exercise programs for people with a diverse set of health and fitness backgrounds, from professional athletes to individuals only recently cleared to exercise.

• **Certified Exercise Physiologist (EP-C):** These are fitness professionals with a minimum of a bachelor's degree in exercise science who are qualified to pursue a career in university, corporate, commercial, hospital, and community settings. Beyond training, ACSM EP-Cs not only conduct complete physical assessments, but they also interpret the results to prescribe personalized exercise programs.

Clinical Certifications

• **Certified Clinical Exercise Physiologist (CEP):** These are health care professionals with a minimum of a bachelor's degree in exercise science and practical experience who provide exercise-related consulting for research, public health, and other clinical and nonclinical services and programs. In addition to prescribing exercise and lifestyle management, ACSM CEPs are qualified to manage cardiac and pulmonary risks, training, and rehabilitation.

• **Registered Clinical Exercise Physiologist (RCEP):** These are health care professionals (with a graduate degree) who use scientific rationale to design, implement, and supervise exercise programming for those with chronic diseases, conditions, or physical limitations. In addition to performing clinical assessments, ACSM RCEPs may oversee programs and departments related to exercise testing, prescription, and training, such as exercise rehabilitation, weight management, and more.

Specialty Certifications

Certified Cancer Exercise Trainer (CET)

Certified Inclusive Fitness Trainer (CIFT)

Physical Activity in Public Health Specialist (PAPHS)

Exercise is Medicine Credential

CSEP CERTIFICATIONS

CSEP offers the following certifications.

- **CSEP Certified Personal Trainers (CSEP-CPT):** The CSEP-CPT works with healthy populations and develops and implements tailored physical activity, fitness, and lifestyle plans.

- **CSEP Certified Exercise Physiologist (CSEP-CEP):** The CSEP-CEP has the ability to perform assessments, develop and prescribe exercise, supervise exercise, and counsel apparently healthy individuals or populations with medical conditions or disabilities associated with cardiopulmonary, metabolic, musculoskeletal, neuromuscular, and aging conditions.

BASES ACCREDITATIONS AND CERTIFICATIONS

BASES offers the following accreditations.

- **BASES Accredited Sport and Exercise Scientist:** This accreditation is provided to members who have achieved a minimum knowledge, skills, and understanding necessary to practice sport and exercise science. Accreditation is achieved via work in applied sport and exercise science, support, and research.

- **BASES High Performance Sport Accreditation (HPSA):** This accreditation is designed for people who provide sport science services for high-performance sport programs. It is recognized by the British Olympic Association.

- **BASES Certified Exercise Practitioner (CEP):** This certification serves as a professional quality assurance for those wishing to use sport and exercise science degrees to establish credibility.

ESSA ACCREDITATIONS

ESSA offers several accreditations:

Accredited Exercise Scientist (AES)

Accredited Exercise Physiologist (AEP)

Accredited Sports Scientist (ASp)

Accredited High Performance Manager (AHPM)

NSCA CERTIFICATIONS

The NSCA focuses its certifications on working with apparently healthy populations. For more information, see the NSCA's certification website at www.nsca.com/certification.

- **Certified Strength and Conditioning Specialist (CSCS):** These are professionals who apply scientific knowledge to train athletes for the primary goal of improving athletic performance. They conduct sport-specific testing sessions, design and implement safe and effective strength training and conditioning programs, and provide guidance regarding nutrition and injury prevention.

- **NSCA-Certified Personal Trainer (NSCA-CPT):** These are health and fitness professionals who use an individualized approach to assess, motivate, educate, and train clients regarding their personal health and fitness needs. They design safe and effective exercise programs, provide guidance to help clients achieve their personal health and fitness goals, and respond appropriately in emergency situations.

- **Certified Special Population Specialist (CSPS):** These are fitness professionals who use an individualized approach to assess, motivate, educate, and train clients of all ages in special populations regarding their health and fitness needs, both preventively and in collaboration with health care professionals. Special populations include those with chronic and temporary health conditions.

- **Tactical Strength and Conditioning Facilitator (TSAC-F):** These professionals apply scientific knowledge to physically train military, fire and rescue, law enforcement, protective services, and other emergency personnel to improve performance, promote wellness, and decrease injury risk. They conduct needs analyses and physical testing sessions, design and implement safe and effective strength training and conditioning programs, and provide general information regarding nutrition.

REFERENCES

Laboratory 1

1. Acevedo EO and Starks MA. *Exercise Testing and Prescription Lab Manual*. 2nd ed. Champaign, IL: Human Kinetics, 2011.

2. Adams V, Jiang H, Yu J, Mobius-Winkler S, Fiehn E, Linke A, Weigl C, Schuler G, and Hambrecht R. Apoptosis in Skeletal Myocytes of Patients With Chronic Heart Failure Is Associated With Exercise Intolerance. *J Am Coll Cardiol* 33: 959-965, 1999.

3. American College of Sports Medicine. ACSM Position Stand: The Recommended Quantity and Quality of Exercise for Developing and Maintaining Cardiorespiratory and Muscular Fitness, and Flexibility in Healthy Adults. *Med Sci Sports Exerc* 30: 975-991, 1998.

4. American College of Sports Medicine. *ACSM's Guidelines for Exercise Testing and Prescription*. 10th ed. Philadelphia: Wolters Kluwer, 2018.

5. American College of Sports Medicine. *ACSM's Health-Related Physical Fitness Assessment Manual*. 3rd ed. Baltimore: Lippincott Williams & Wilkins, 2010.

6. Beam WC and Adams GM. *Exercise Physiology Laboratory Manual*. 5th ed. Boston: McGraw-Hill, 2007.

7. Berg KE and Latin RW. *Essentials of Research Methods in Health, Physical Education, Exercise Science, and Recreation*. Baltimore: Lippincott Williams & Wilkins, 2004.

8. Bompa TO and Haff GG. *Periodization: Theory and Methodology of Training*. 5th ed. Champaign, IL: Human Kinetics, 2009.

9. Brooks GA, Fahey TD, White TP, and Baldwin KM. *Exercise Physiology: Human Bioenergetics and Its Applications*. 4th ed. Mountain View, CA: Mayfield, 2004.

10. Brown FM, Neft EE, and LaJambe CM. Collegiate Rowing Crew Performance Varies by Morningness-Eveningness. *J Strength Cond Res* 22: 1894-1900, 2008.

11. Buchheit M. Le 30-15 Intermittent Fitness Test: 10 Year Review. *Myorobie J* 1: 1-9, 2010.

12. Dolezal BA, Thompson CJ, Schroeder CA, Haub MD, Haff GG, Comeau MJ, and Potteiger JA. Laboratory Testing to Improve Athletic Performance. *Strength and Cond J* 19: 20-24, 1997.

13. Flanagan EP. The Effect Size Statistic-Applications for the Strength and Conditioning Coach. *Strength Cond J* 35: 37-40, 2013.

14. Gonzalez-Alonso J, Teller C, Andersen SL, Jensen FB, Hyldig T, and Nielsen B. Influence of Body Temperature on the Development of Fatigue During Prolonged Exercise in the Heat. *J Appl Physiol* 86: 1032-1039, 1999.

15. Haff GG, Ruben RP, Lider J, Twine C, and Cormie P. A comparison of methods for determining the rate of force development during isometric midthigh clean pulls. *J Strength Cond Res* 29: 386-395, 2015.

16. Heyward VH and Gibson AL. *Advanced Fitness Assessment and Exercise Prescription*. 7th ed. Champaign, IL: Human Kinetics, 2014. p. 552.

17. Hoffman JR. *Norms for Fitness, Performance, and Health*. Champaign, IL: Human Kinetics, 2006.

18. Hoffman JR. *Physiological Aspects of Sport Training and Performance*. 2nd ed. Champaign, IL: Human Kinetics, 2014.

19. Hopkins WG. Measures of Reliability in Sports Medicine and Science. *Sports Med* 30: 1-15, 2000.

20. Hopkins WG, Marshall SW, Batterham AM, and Hanin J. Progressive Statistics for Studies in Sports Medicine and Exercise Science. *Med Sci Sports Exerc* 41: 3-13, 2009.

21. Housh TJ, Cramer JT, Weir JP, Beck TW, and Johnson GO. *Physical Fitness Laboratories on a Budget*. Scottsdale, AZ: Holcomb-Hathaway, 2009.

22. Hruda KV, Hicks AL, and McCartney N. Training for Muscle Power in Older Adults: Effects on Functional Abilities. *Can J Appl Physiol* 28: 178-189, 2003.

23. Kenney WL, Wilmore JH, and Costill DL. *Physiology of Sport and Exercise*. 6th ed. Champaign, IL: Human Kinetics, 2015.

24. Komi PV. *Strength and Power in Sport*. Oxford, UK: Blackwell Scientific, 1991.

25. Komi PV. *Strength and Power in Sport*. 2nd ed. Malden, MA: Blackwell Scientific, 2003.

26. Kreighbaum E and Barthels KM. *Biomechanics: A Qualitative Approach for Studying Human Movement*. New York: MacMillan, 1990.

27. Maud PJ and Foster C, eds. *Physiological Assessment of Human Fitness*. 2nd ed. Champaign, IL: Human Kinetics, 2006.

28. McArdle WD, Katch FI, and Katch VL. *Exercise Physiology: Energy, Nutrition, and Human Performance*. Baltimore: Lippincott Williams & Wilkins, 2007.

29. McBride, J.M. Biomechanics of Resistance Exercise. In: Haff GG and Triplett NT, eds., *Essentials of Strength Training and Conditioning*. 4th ed. Champaign, IL: Human Kinetics, 2016, pp. 19-42.

30. McGuigan M. Administration, Scoring, and Interpretation of Selected Tests. In: Haff GG and Triplett NT, eds., *Essentials of Strength Training and Conditioning*. 4th ed. Champaign, IL: Human Kinetics, 2016, pp. 259-316.

31. National Institute of Diabetes and Digestive and Kidney Diseases. *Understanding Adult Obesity.* Bethesda, MD: National Institutes of Health, U.S. Department of Health and Human Services, 2008.

32. Nieman DC. *Exercise Testing and Prescription: A Health-Related Approach.* New York: McGraw-Hill, 2003.

33. Pandolf KB, Cafarelli E, Noble BJ, and Metz KF. Hyperthermia: Effect on Exercise Prescription. *Arch Phys Med Rehabil* 56: 524-526, 1975.

34. Pyne D. Data Collection and Analysis. In: Tanner RK and Gore CJ, eds., *Physiological Tests for Elite Athletes.* Champaign, IL: Human Kinetics, 2013, pp. 35-42.

35. Souissi N, Bessot N, Chamari K, Gauthier A, Sesboue B, and Davenne D. Effect of Time of Day on Aerobic Contribution to the 30-s Wingate Test Performance. *Chronobiol Int* 24: 739-748, 2007.

36. Souissi N, Gauthier A, Sesboue B, Larue J, and Davenne D. Circadian Rhythms in Two Types of Anaerobic Cycle Leg Exercise: Force-Velocity and 30-s Wingate Tests. *Int J Sports Med* 25: 14-19, 2004.

37. Stone MH, Stone ME, and Sands WA. *Principles and Practice of Resistance Training.* Champaign, IL: Human Kinetics, 2007.

38. Tanner RK and Gore CJ, eds. *Physiological Tests for Elite Athletes.* 2nd ed. Champaign, IL: Human Kinetics, 2013.

39. Thomas JR, Nelson JK, and Silverman SJ. *Research Methods in Physical Activity.* 5th ed. Champaign, IL: Human Kinetics, 2005.

40. Thompson A and Taylor BN. NIST Special Publication 811, 2008 edition: Guide for the Use of International System of Units (SI). Gaithersburg, MD: United States Department of Commerce, National Institute of Standards and Technology, 2008.

41. Winter EM, Abt G, Brookes FB, Challis JH, Fowler NE, Knudson DV, Knuttgen HG, Kraemer WJ, Lane AM, Mechelen W, Morton RH, Newton RU, Williams C, and Yeadon MR. Misuse of "Power" and Other Mechanical Terms in Sport and Exercise Science Research. *J Strength Cond Res* 30: 292-300, 2016.

42. Young DS. Implementation of SI Units for Clinical Laboratory Data. Style Specifications and Conversion Tables. *Ann Intern Med* 106 (1): 114-129, 1987. [Published errata appear in *Ann Intern Med* 107 (2): 265 (1987, Aug); *Ann Intern Med* 110 (4): 328 (1989, Feb 15); *Ann Intern Med* 114 (2):172 (1991, Jan 15), and *Ann Intern Med* 148 (9): 715 (2008, May 6)].

Laboratory 2

1. American College of Sports Medicine. *ACSM'S Guidelines for Exercise Testing and Prescription.* 8th ed. Baltimore: Lippincott Williams & Wilkins, 2010.

2. American College of Sports Medicine. *ACSM's Guidelines for Exercise Testing and Prescription.* 10th ed. Philadelphia: Wolters Kluwer, 2018.

3. American College of Sports Medicine. *ACSM's Health-Related Physical Fitness Assessment Manual.* 3rd ed. Baltimore: Lippincott Williams & Wilkins, 2009.

4. Combs J and Williams A. Cardiovascular Health. In: Combes J and Skinner T., eds., *ESSA's Student Manual for Health, Exercise, and Sport Assessment.* Chatswood, Australia: Elsevier Australia, 2014, pp. 29-58.

5. Gibbons RJ et al. 2002. *ACC/AHA 2002 Guideline Update for Exercise Testing. A Report of the American College of Cardiology/American Heart Association Task Force on Practice Guidelines.* https://doi.org/10.1161/01.CIR.0000034670.06526.15

6. Heyward VH and Gibson AL. *Advanced Fitness Assessment and Exercise Prescription.* 7th ed. Champaign, IL: Human Kinetics, 2014.

7. Nieman DC. *Exercise Testing and Prescription: A Health-Related Approach.* 5th ed. New York: McGraw-Hill, 2003.

8. Magal M and Riebe D. New Preparticipation Health Screening Recommendations: What Exercise Professionals Need to Know. *ACSMs Health Fit J* 20: 22-27, 2016.

9. Pate RR, Pratt M, Blair SN, et al. Physical Activity and Public Health. A Recommendation from the Centers for Disease Control and Prevention and the American College of Sports Medicine. *JAMA* 273: 402-407, 1995.

10. Riebe D, Franklin BA, Thompson PD, Garber CE, Whitfield GP, Magal M, and Pescatello LS. Updating ACSM's Recommendations for Exercise Preparticipation Health Screening. *Med Sci Sports Exerc* 47: 2473-2479, 2015.

11. Whaley, MH, Brubacker PH, and Otto RM, eds. *ACSM's Guidelines for Exercise Testing and Prescription.* 6th ed. Baltimore: Lippincott Williams & Wilkins, 2000.

12. Wilson PW, D'Agostino RB, Levy D, Belanger AM, Silbershatz H, and Kannel WB. Prediction of Coronary Heart Disease Using Risk Factor Categories. *Circulation* 97: 1837-1847, 1998.

Laboratory 3

1. Acevedo EO and Starks MA. *Exercise Testing and Prescription Lab Manual.* 2nd ed. Champaign, IL: Human Kinetics, 2011.

2. Adams GM. *Exercise Physiology Laboratory Manual.* 3rd ed. Boston: McGraw-Hill, 1998.

3. American Alliance for Health, Physical Education, Recreation and Dance (AAHPERD). *Health Related Physical Fitness Test Manual.* Reston, VA: AAHPERD, 1980.

4. American College of Sports Medicine. ACSM Position Stand. The Recommended Quantity and Quality of Exercise for Developing and Maintaining Cardiorespiratory and Muscular Fitness, and Flexibility in Healthy Adults. *Med Sci Sports Exerc* 30: 975-991, 1998.

5. American College of Sports Medicine. *ACSM's Guidelines for Exercise Testing and Prescription.* 8th ed. Philadelphia: Lippincott, Williams & Wilkins, 2010.

6. American College of Sports Medicine. *ACSM's Health-Related Physical Fitness Assessment Manual.* 3rd ed. Baltimore: Lippincott Williams & Wilkins, 2010.

7. Baltaci G, Un N, Tunay V, et al. Comparison of Three Different Sit and Reach Tests for Measurement of Hamstring Flexibility in Female University Students. *Br J Sports Med* 37: 59-61, 2003.

8. Boone DC, Azen SP, Lin CM, et al. Reliability of Goniometric Measurements. *Phys Ther* 58: 1355-1390, 1978.

9. Cailliet R. *Low Back Pain Syndrome.* 5th ed. Philadelphia: Davis, 1995.

10. Chapman EA, de Vries HA, and Swezey R. Joint Stiffness: Effects of Exercise on Young and Old Men. *J Gerontol* 27: 218-221, 1972.

11. Cooper Institute for Aerobics Research (CIAR). *The Prudential FITNESSGRAM Test Administration Manual.* Dallas: CIAR, 1992.

12. de Vries HA, Housh TJ, and Weir LL. *Physiology of Exercise for Physical Education, Athletics and Exercise Science.* 5th ed. Dubuque, IA: Brown, 1995.

13. Goeken LN and Hof AL. Instrumental Straight-Leg Raising: Results in Healthy Subjects. *Arch Phys Med Rehabil* 74: 194-203, 1993.

14. Golding LA, ed. *YMCA Fitness Testing and Assessment Manual.* 4th ed. Champaign, IL: Human Kinetics, 2000.

15. Haff GG, Cramer JT, Beck DT, et al. Roundtable Discussion: Flexibility Training. *Strength and Cond J* 28: 64-85, 2006.

16. Heyward VH and Gibson AL. *Advanced Fitness Assessment and Exercise Prescription.* 7th ed. Champaign, IL: Human Kinetics, 2014.

17. Hoeger WW and Hopkins DR. A Comparison of the Sit and Reach and the Modified Sit and Reach in the Measurement of Flexibility in Women. *Res Q Exerc Sport* 63: 191-195, 1992.

18. Hoffman JR. *Norms for Fitness, Performance, and Health.* Champaign, IL: Human Kinetics, 2006.

19. Housh TJ, Cramer JT, Weir JP, Beck TW, and Johnson GO. *Physical Fitness Laboratories on a Budget.* Scottsdale, AZ: Holcomb-Hathaway, 2009.

20. Hui SC, Yuen PY, Morrow JR, Jr., et al. Comparison of the Criterion-Related Validity of Sit-and-Reach Tests With and Without Limb Length Adjustment in Asian Adults. *Res Q Exerc Sport* 70: 401-406, 1999.

21. Hui SS and Yuen PY. Validity of the Modified Back-Saver Sit-and-Reach Test: A Comparison With Other Protocols. *Med Sci Sports Exerc* 32: 1655-1659, 2000.

22. Institute FaL. *Fitness and Lifestyle in Canada.* Ottawa: Fitness and Amateur Sport, 1983.

23. Jackson AM and Langford NJ. The Criterion-Related Validity of the Sit and Reach Test: Replication and Extension of Previous Findings. *Res Q Exerc Sport* 60: 384-387, 1989.

24. Jackson AM and Baker AA. The Relationship of the Sit and Reach Test to Criterion Measures of Hamstring and Back Flexibility in Young Females. *Res Q Exerc Sport* 57: 183-186, 1986.

25. Jeffreys I. *Total Soccer Fitness.* Monterey: Coaches Choice, 2007.

26. Jeffreys I. Warm-Up and Flexibility Training. In: Haff GG and Triplett NT, eds., *Essentials of Strength Training and Conditioning.* 4th ed. National Strength and Conditioning Association. Champaign, IL: Human Kinetics, 2016, pp. 317-350.

27. Kell RT, Bell G, and Quinney A. Musculoskeletal Fitness, Health Outcomes and Quality of Life. *Sports Med* 31: 863-873, 2001.

28. Kofotolis N and Kellis E. Effects of Two 4-Week Proprioceptive Neuromuscular Facilitation Programs on Muscle Endurance, Flexibility, and Functional Performance in Women With Chronic Low Back Pain. *Phys Ther* 86: 1001-1012, 2006.

29. Lark S, Brancato T, and Skinner T. Flexibility. In: Coombes J and Skinner T, eds., *ESSA's Student Manual for Health, Exercise and Sport Assessment.* Chatswood, Australia: Mosby, 2014, pp. 174-199.

30. Lea RD and Gerhardt JJ. Range-of-Motion Measurements. *J Bone Joint Surg Am* 77: 784-798, 1995.

31. Liemohn W, Sharpe GL, and Wasserman JF. Criterion Related Validity of the Sit-and-Reach Test. *J Strength Cond Res* 8: 91-94, 1994.

32. Macrae IF and Wright V. Measurement of Back Movement. *Ann Rheum Dis* 28: 584-589, 1969.

33. Maud PJ and Kerr KM. Static Techniques for the Evaluation of Joint Range of Motion and Muscle Length. In: Maud PJ and Foster C, eds., *Physiological Assessment of Human Fitness.* 2nd ed. Champaign, IL: Human Kinetics, 2006, pp. 227-252.

34. McAuley E, Hudash G, Shields K, et al. Injuries in Women's Gymnastics. The State of the Art. *Am J Sports Med* 16 (Suppl 1): S124-S131, 1988.

35. McNeal JR and Sands WA. Stretching for Performance Enhancement. *Curr Sports Med Rep* 5: 141-146, 2006.

36. Nieman DC. *Exercise Testing and Prescription: A Health-Related Approach.* 5th ed. New York: McGraw-Hill, 2003.

37. Sands WA, McNeal JR, Stone MH, et al. Effect of Vibration on Forward Split Flexibility and Pain Perception in Young Male Gymnasts. *Int J Sports Physiol Perf* 3: 469-481, 2008.

38. Shephard RJ, Berridge M, and Montelpare W. On the Generality of the "Sit and Reach" Test: An Analysis of Flexibility Data for an Aging Population. *Res Q Exerc Sport* 61: 326-330, 1990.

39. Spring T, Franklin B, and deJong A. Muscular Fitness and Assessment. In: *ACSM's Resource Manual for Guidelines for Exercise Testing and Prescription.* Baltimore: Lippincott, Williams & Wilkins, 2010, pp. 332-348.

40. Warner JJ, Micheli LJ, Arslanian LE, et al. Patterns of Flexibility, Laxity, and Strength in Normal Shoulders and Shoulders With Instability and Impingement. *Am J Sports Med* 18: 366-375, 1990.

41. Wilmore JH, Parr RB, Girandola RN, et al. Physiological Alterations Consequent to Circuit Weight Training. *Med Sci Sports* 10: 79-84, 1978.

42. Winters MV, Blake CG, Trost JS, et al. Passive Versus Active Stretching of Hip Flexor Muscles in Subjects With Limited Hip Extension: A Randomized Clinical Trial. *Phys Ther* 84: 800-807, 2004.

Laboratory 4

1. Whelton PK, Carey RM, Aronow WS, Casey DE Jr, Collins KJ, Dennison Himmelfarb C, DePalma SM, Gidding S, Jamerson KA, Jones DW, MacLaughlin EJ, Muntner P, Ovbiagele B, Smith SC Jr, Spencer CC, Stafford RS, Taler SJ, Thomas RJ, Williams KA Sr, Williamson JD, Wright JT Jr. 2017 ACC/AHA/AAPA/ABC/ACPM/AGS/APhA/ASH/ASPC/NMA/PCNA guideline for the prevention, detection, evaluation, and management of high blood pressure in adults: a report of the American College of Cardiology/American Heart Association Task Force on Clinical Practice Guidelines. *J Am Coll Cardiol.* 2017; epub before print. DOI: 10.1016/j.jacc.2017.11.006

2. American College of Sports Medicine. *ACSM's Guidelines for Exercise Testing and Prescription.* 10th ed. Philadelphia: Wolters Kluwer, 2018.

3. Brooks GA, Fahey TD, and Baldwin KM. *Exercise Physiology: Human Bioenergetics and Its Applications.* 4th ed. New York: McGraw-Hill, 2005.

4. Kahn JF, Jouanin JC, Bussiere JL, Tinet E, Avrillier S, Ollivier JP, and Monod H. The Isometric Force That Induces Maximal Surface Muscle Deoxygenation. *Eur J Appl Physiol Occup Physiol* 78(2): 183-187, 1998.

5. Kenney WL, Wilmore JH, and Costill DL. *Physiology of Sport and Exercise.* 6th ed. Champaign, IL: Human Kinetics, 2015.

6. Perloff D, Grim C, Flack J, Frohlich ED, Hill M, McDonald M, and Morgenstern BZ. Human Blood Pressure Determination by Sphygmomanometry. *Circulation* 88: 2460-2470, 1993.

Laboratory 5

1. Brooks GA, Butte NF, Rand WM, Flatt JP, and Caballero B. Chronicle of the Institute of Medicine Physical Activity Recommendation: How a Physical Activity Recommendation Came to Be Among Dietary Recommendations. *Am J Clin Nutr* 79: 921S-930S, 2004.

2. Carpenter WH, Poehlman ET, O'Connell M, and Goran MI. Influence of Body Composition and Resting Metabolic Rate on Variation in Total Energy Expenditure: A Meta-Analysis. *Am J Clin Nutr* 61: 4-10, 1995.

3. Compher C, Frankenfield D, Keim N, and Roth-Yousey L. Best Practice Methods to Apply to Measurement of Resting Metabolic Rate in Adults: A Systematic Review. *J Am Diet Assoc* 106: 881-903, 2006.

4. Flack KD, Siders WA, Johnson L, and Roemmich JN. Cross Validation of Resting Metabolic Rate Prediction Equations. *J Acad Nutr Diet* 116(9): 1413-1422, 2016.

5. Harris JA and Benedict FG. *A Biometric Study of Basal Metabolism in Man.* Washington, DC: Carnegie Institution of Washington, 1919.

6. Hasson RE, Howe CA, Jones BL, and Freedson PS. Accuracy of Four Resting Metabolic Prediction Equations: Effects of Sex, Body Mass Index, Age, and Race/Ethnicity. *J Sci Med Sport* 14: 344-351, 2011.

7. Haugen HA, Melanson EL, Tran ZV, Kearney JT, and Hill JO. Variability of Measured Resting Metabolic Rate. *Am J Clin Nutr* 78: 1141-1145, 2003.

8. Heshka S, Yang MU, Wang J, Burt P, and Pi-Sunyer FX. Weight Loss and Change in Resting Metabolic Rate. *Am J Clin Nutr* 52: 981-986, 1990.

9. Jorgensen JO, Vahl N, Dall R, and Christiansen JS. Resting Metabolic Rate in Healthy Adults: Relation to Growth Hormone Status and Leptin Levels. *Metabolism* 47: 1134-1139, 1998.

10. Kenney WL, Wilmore JH, and Costill DL. *Physiology of Sport and Exercise.* 6th ed. Champaign, IL: Human Kinetics, 2015.

11. Lemmer JT, Ivey FM, Ryan AS, Martel GF, Hurlbut DE, Metter JE, Fozard JL, Fleg JL, and Hurley BF. Effect of Strength Training on Resting Metabolic Rate and Physical Activity: Age and Gender Comparisons. *Med Sci Sports Exerc* 33: 532-541, 2001.

12. Mifflin MD, St Jeor ST, Hill LA, Scott BJ, Daugherty SA, and Koh YO. A New Predictive Equation for Resting Energy Expenditure in Healthy Individuals. *Am J Clin Nutr* 51: 241-247, 1990.

13. Owen OE, Kavle E, Owen RS, Polansky M, Caprio S, Mozzoli MA, Kendrick ZV, Bushman MC, and Boden G. A Reappraisal of Caloric Requirements in Healthy Women. *Am J Clin Nutr* 44: 1-19, 1986.

14. Schmidt WD, Hyner GC, Lyle RM, Corrigan D, Bottoms G, and Melby CL. The Effects of Aerobic and Anaerobic Exercise Conditioning on Resting Metabolic Rate and the Thermic Effect of a Meal. *Int J Sport Nutr* 4: 335-346, 1994.

15. Sparti A, DeLany JP, de la Bretonne JA, Sander GE, and Bray GA. Relationship Between Resting Metabolic Rate and the Composition of the Fat-Free Mass. *Metabolism* 46: 1225-1230, 1997.

16. Speakman JR and Selman C. Physical Activity and Resting Metabolic Rate. *Proc Nutr Soc* 62: 621-634, 2003.

17. World Health Organization (WHO). *Energy and Protein Requirements: Report of a Joint FAO/WHO/UNU Expert Consultation.* WHO Technical Report Series No. 724. Geneva, 1985.

Laboratory 6

1. Dolezal BA, Potteiger JA, Jacobsen DJ, and Benedict SH. Muscle Damage and Resting Metabolic Rate After Acute Resistance Exercise With an Eccentric Overload. *Med Sci Sports Exer* 32: 1202-1207, 2000.

2. Gaesser GA and Brooks GA. Metabolic Bases of Excess Post-Exercise Oxygen Consumption: A Review. *Med Sci Sports Exer* 16: 29-43, 1984.

3. Grassi B. Regulation of Oxygen Consumption at Exercise Onset: Is It Really Controversial? *Exer Sport Sci Rev* 29: 134-138, 2001.

4. Kenney WL, Wilmore JH, and Costill DL. *Physiology of Sport and Exercise.* 6th ed. Champaign, IL: Human Kinetics, 2015.

5. Korzeniewski B and Zoladz JA. Biochemical Background of the $\dot{V}O_2$ On-Kinetics in Skeletal Muscles. *J Physiol Sci* 56: 1-12, 2006.

6. Maizes JS, Murtuza M, and Kvetan V. Oxygen Transport and Utilization. *Respir Care Clin N Am* 6: 473-500, 2000.

7. Poole DC, Barstow TJ, Gaesser GA, Willis WT, and Whipp BJ. VO2 Slow Component: Physiological and Functional Significance. *Med Sci Sports Exer* 26: 1354-1358, 1994.

8. Whipp BJ. The Slow Component of O_2 Uptake Kinetics During Heavy Exercise. *Med Sci Sports Exer* 26: 1319-1326, 1994.

9. Xu F and Rhodes EC. Oxygen Uptake Kinetics During Exercise. *Sports Med* 27: 313-327, 1999.

10. Zoladz JA and Korzeniewski B. Physiological Background of the Change Point in VO_2 and the Slow Component of Oxygen Uptake Kinetics. *J Physiol Pharmacol* 52: 167-184, 2001.

Laboratory 7

1. American College of Sports Medicine. *Guidelines for Exercise Testing and Prescription.* 4th ed. Philadelphia: Lea & Febiger, 1991.

2. American College of Sports Medicine. *ACSM's Guidelines for Exercise Testing and Prescription.* 10th ed. Philadelphia: Wolters Kluwer, 2018.

3. Ebbeling CB, Ward A, Puleo EM, Widrick J, and Rippe JM. Development of a Single-Stage Submaximal Treadmill Walking Test. *Med Sci Sports Exer* 23: 966-973, 1991.

4. Kaminsky LA, Arena R, Myers J. Reference Standards for Cardiorespiratory Fitness Measured With Cardiopulmonary Exercise Testing: Data From the Fitness Registry and the Importance of Exercise National Database (FRIEND). *Mayo Clin Proc* 90 (11): 1515-1523, 2015.

5. Kaminsky LA, Imboden MT, Arena R, and Myers J. Reference Standards for Cardiorespiratory Fitness Measured With Cardiopulmonary Exercise Testing Using Cycle Ergometry: Data From the Fitness Registry and the Importance of Exercise National Database (FRIEND) Registry. *Mayo Clin Proc* 92 (2): 228-233, 2017.

6. McArdle WD, Katch FI, Pechar GS, Jacobson L, and Ruck S. Reliability and Interrelationships Between Maximal Oxygen Intake, Physical Work Capacity and Step-Test Scores in College Women. *Med Sci Sports* 4: 182-186, 1972.

7. McArdle WD, Pechar GS, Katch FI, and Magel JR. Percentile Norms for a Valid Step Test in College Women. *Res Q* 44: 498-500, 1973.

8. Noble BJ, Borg GA, Jacobs I, Ceci R, and Kaiser P. A Category-Ratio Perceived Exertion Scale: Relationship to Blood and Muscle Lactates and Heart Rate. *Med Sci Sports Exer* 15: 523-528, 1983.

9. Robertson RJ and Noble BJ. Perception of Physical Exertion: Methods, Mediators, and Applications. *Exer Sport Sci Rev* 25: 407-452, 1997.

10. Sharkey BJ and Gaskill S. *Fitness and Health.* 6th ed. Champaign, IL: Human Kinetics, 2007.

11. Tanaka H, Monahan KD, and Seals DR. Age-Predicted Maximal Heart Rate Revisited. *J Am Coll Cardiol* 37: 153-156, 2001.

Laboratory 8

1. American College of Sports Medicine. *ACSM's Guidelines for Exercise Testing and Prescription.* 8th ed. Philadelphia: Lippincott Williams & Wilkins, 2010.

2. American College of Sports Medicine. *ACSM's Guidelines for Exercise Testing and Prescription.* 10th ed. Philadelphia: Wolters Kluwer, 2018.

3. Balke B. *A Simple Field Test for the Assessment of Physical Fitness.* Civil Aeromedical Research Institute Report, 63-18. Oklahoma City: Federal Aviation Agency, 1963.

4. Beam WC and Adams GM. *Exercise Physiology Laboratory Manual.* 6th ed. Boston: McGraw-Hill, 2011.

5. Billat V and Lopes P. Indirect Methods for Estimation of Aerobic Power. In: Maud PJ and Foster C, eds., *Physiological Assessment of Human Fitness.* 2nd ed. Champaign, IL: Human Kinetics, 2006, pp. 19-38.

6. Cooper KH. A Means of Assessing Maximal Oxygen Intake: Correlation Between Field and Treadmill Testing. *JAMA* 203: 201-204, 1968.

7. Dolgener FA, Hensley LD, Marsh JJ, and Fjelstul JK. Validation of the Rockport Fitness Walking Test in College Males and Females. *Res Q Exerc Sport* 65: 152-158, 1994.

8. Fenstermaker K, Plowman SA, and Looney M. Validation of the Rockport Walking Test in Females 65 Years and Older. *Res Q Exer Sport* 63: 322-327, 1992.

9. George JD, Fellingham GW, and Fisher AG. A Modified Version of the Rockport Fitness Walking Test for College Men and Women. *Res Q Exer Sport* 69: 205-209, 1998.

10. George JD, Vehrs PR, Allsen PE, Fellingham GW, and Fisher AG. $\dot{V}O_2$max Estimation From a Submaximal 1-Mile Track Jog for Fit College-Age Individuals. *Med Sci Sports Exerc* 25: 401-406, 1993.

11. Heyward VH and Gibson AL. *Advanced Fitness Assessment and Exercise Prescription.* 7th ed. Champaign, IL: Human Kinetics, 2014.

12. Hoffman JR. *Norms for Fitness, Performance, and Health.* Champaign, IL: Human Kinetics, 2006.

13. Housh TJ, Cramer JT, Weir JP, Beck TW, and Johnson GO. *Physical Fitness Laboratories on a Budget.* Scottsdale, AZ: Holcomb-Hathaway, 2009.

14. Kline GM, Porcari JP, Hintermeister R, Freedson PS, Ward A, McCarron RF, Ross J, and Rippe JM. Estimation of $\dot{V}O_2$max From a One-Mile Track Walk, Gender, Age, and Body Weight. *Med Sci Sports Exer* 19: 253-259, 1987.

15. McCutcheon MC, Stricha SA, Giese MD, and Nagle FJ. A Further Analysis of the 12-Minute Run Prediction of Maximal Aerobic Power. *Res Q Exer Sport* 61: 280-283, 1990.

16. Safrit MJ, Glaucia Costa M, Hooper LM, Patterson P, and Ehlert SA. The Validity Generalization of Distance Run Tests. *Can J Sport Sci* 13: 188-196, 1988.

Laboratory 9

1. Apostolidis N, Nassis GP, Bolatoglou T, and Geladas ND. Physiological and Technical Characteristics of Elite Young Basketball Players. *J Sports Med Phys Fitness* 44: 157-163, 2004.

2. Bangsbo J. The Physiology of Soccer—With Special Reference to Intense Intermittent Exercise. *Acta Physiol Scand Suppl* 619: 1-155, 1994.

3. Bangsbo J, Iaia FM, and Krustrup P. The Yo-Yo Intermittent Recovery Test: A Useful Tool for Evaluation of Physical Performance in Intermittent Sports. *Sports Med* 38: 37-51, 2008.

4. Bangsbo J, Norregaard L, and Thorso F. Activity Profile of Competition Soccer. *Can J Sport Sci* 16: 110-116, 1991.

5. Billat LV. Interval Training for Performance: A Scientific and Empirical Practice. Special Recommendations for Middle- and Long-Distance Running. Part I: Aerobic Interval Training. *Sports Med* 31: 13-31, 2001.

6. Billat LV. Interval Training for Performance: A Scientific and Empirical Practice. Special Recommendations for Middle- and Long-Distance Running. Part II: Anaerobic Interval Training. *Sports Med* 31: 75-90, 2001.

7. Buchheit M. The 30-15 Intermittent Fitness Test: A New Intermittent Running Field Test for Intermittent Sport Players—Part 1. *Approches Handball* 87: 27-34, 2005.

8. Buchheit M. Illustration of Interval-Training Prescription on the Basis of an Appropriate Intermittent Maximal Running Speed—The 30-15 Intermittent Fitness Test—Part 2. *Approches Handball* 88: 36-46, 2005.

9. Buchheit M. The 30-15 Intermittent Fitness Test: Accuracy for Individualizing Interval Training of Young Intermittent Sport Players. *J Strength Cond Res* 22: 365-374, 2008.

10. Buchheit M. Le 30-15 Intermittent Fitness Test: 10 Year Review. *Myorobie J* 1: 1-9, 2010.

11. Buchheit M. Individualizing High-Intensity Interval Training in Intermittent Sport Athletes With the 30-15 Intermittent Fitness Test. Paper presented as part of NSCA Hot Topic Series. Colorado Springs, CO: NSCA, 2013.

12. Buchheit M, Al Haddad H, Millet GP, Lepretre PM, Newton M, and Ahmaidi S. Cardiorespiratory and Cardiac Autonomic Responses to 30-15 Intermittent Fitness Test in Team Sport Players. *J Strength Cond Res* 23: 93-100, 2009.

13. Buchheit M and Laursen PB. High-Intensity Interval Training, Solutions to the Programming Puzzle: Part I: Cardiopulmonary Emphasis. *Sports Med* 43: 313-338, 2013.

14. Castagna C, Impellizzeri FM, Chamari K, Carlomagno D, and Rampinini E. Aerobic Fitness and Yo-Yo Continuous and Intermittent Test Performances in Soccer Players: A Correlation Study. *J Strength Cond Res* 20: 320-325, 2006.

15. Castagna C, Impellizzeri FM, Rampinini E, D'Ottavio S, and Manzi V. The Yo-Yo Intermittent Recovery Test in Basketball Players. *J Sci Med Sport* 11: 202-208, 2008.

16. Clarke R, Dobson A, and Hughes J. Metabolic Conditioning: Field Tests to Determine a Training Velocity. *Strength Cond J* 38: 38-47, 2016.

17. Covic N, Jeleskovic E, Alic H, Rado I, Kafedzic E, Sporis G, McMaster DT, and Milanovic Z. Reliability, Validity and Usefulness of 30-15 Intermittent Fitness Test in Female Soccer Players. *Front Physiol* 7: 510, 2016.

18. Darrall-Jones J, Roe G, Carney S, Clayton R, Phibbs P, Read D, Weakley J, Till K, and Jones B. The Effect of Body Mass on the 30-15 Intermittent Fitness Test in Rugby Union Players. *Int J Sports Physiol Perform* 11: 400-403, 2016.

19. Duthie G, Pyne D, and Hooper S. Applied Physiology and Game Analysis of Rugby Union. *Sports Med* 33: 973-991, 2003.

20. Elferink-Gemser MT, Visscher C, Lemmink KA, and Mulder TW. Relation Between Multidimensional Performance Characteristics and Level of Performance in Talented Youth Field Hockey Players. *J Sports Sci* 22: 1053-1063, 2004.

21. Halle M. Letter by Halle Regarding Article, "Cardiovascular Risk of High- Versus Moderate-Intensity Aerobic Exercise in Coronary Heart Disease Patients." *Circulation* 127: e637, 2013.

22. Haydar B and Buchheit ML. 30-15 Intermittent Fitness Test-application pour le Basketball. *Pivot* 143: 2-5, 2009.

23. Haydar B, Haddad HA, Ahmaidi S, and Buchheit M. Assessing Inter-Effort Recovery and Change of Direction Ability With the 30-15 Intermittent Fitness Test. *J Sports Sci Med* 10: 346-354, 2011.

24. Heyward VH and Gibson A. *Advanced Fitness Assessment and Exercise Prescription.* 7th ed. Champaign, IL: Human Kinetics, 2014.

25. Jelleyman C, Yates T, O'Donovan G, Gray LJ, King JA, Khunti K, and Davies MJ. The Effects of High-Intensity Interval Training on Glucose Regulation and Insulin Resistance: A Meta-Analysis. *Obes Rev* 16: 942-961, 2015.

26. Krustrup P, Mohr M, Amstrup T, Rysgaard T, Johansen J, Steensberg A, Pedersen PK, and Bangsbo J. The Yo-Yo Intermittent Recovery Test: Physiological Response, Reliability, and Validity. *Med Sci Sports Exerc* 35: 697-705, 2003.

27. Krustrup P, Mohr M, Ellingsgaard H, and Bangsbo J. Physical Demands During an Elite Female Soccer Game: Importance of Training Status. *Med Sci Sports Exerc* 37: 1242-1248, 2005.

28. Krustrup P, Mohr M, Nybo L, Jensen JM, Nielsen JJ, and Bangsbo J. The Yo-Yo IR2 Test: Physiological Response, Reliability, and Application to Elite Soccer. *Med Sci Sports Exerc* 38: 1666-1673, 2006.

29. Léger L and Boucher R. An Indirect Continuous Running Multistage Field Test: The Universite de Montreal Track Test. *Can J Appl Sport Sci* 5: 77-84, 1980.

30. Léger LA and Lambert J. A Maximal Multistage 20 m Shuttle Run Test to Predict V̇O₂max. *Eur J Appl Physiol Occup Physiol* 49: 1-12, 1982.

31. Léger LA, Mercier D, Gadoury C, and Lambert J. The Multistage 20 Metre Shuttle Run Test for Aerobic Fitness. *J Sports Sci* 6: 93-101, 1988.

32. Lemmink KA, Visscher C, Lambert MI, and Lamberts RP. The Interval Shuttle Run Test for Intermittent Sport Players: Evaluation of Reliability. *J Strength Cond Res* 18: 821-827, 2004.

33. Mahar MT, Guerieri AM, Hanna MS, and Kemble CD. Estimation of Aerobic Fitness From 20 m Multistage Shuttle Run Test Performance. *Am J Prev Med* 41: S117-S123, 2011.

34. Pincivero DM and Bompa TO. A Physiological Review of American Football. *Sports Med* 23: 247-260, 1997.

35. Ramos JS, Dalleck LC, Tjonna AE, Beetham KS, and Coombes JS. The Impact of High-Intensity Interval Training Versus Moderate-Intensity Continuous Training on Vascular Function: A Systematic Review and Meta-Analysis. *Sports Med* 45: 679-692, 2015.

36. Rognmo O, Moholdt T, Bakken H, Hole T, Molstad P, Myhr NE, Grimsmo J, and Wisloff U. Cardiovascular Risk of High- Versus Moderate-Intensity Aerobic Exercise in Coronary Heart Disease Patients. *Circulation* 126: 1436-1440, 2012.

37. Sawyer BJ, Tucker WJ, Bhammar DM, Ryder JR, Sweazea KL, and Gaesser GA. Effects of High-Intensity Interval Training and Moderate-Intensity Continuous Training on Endothelial Function and Cardiometabolic Risk Markers in Obese Adults. *J Appl Physiol* 121: 279-288, 2016.

38. Scott TJ, Delaney JA, Duthie GM, Sanctuary CE, Ballard DA, Hickmans JA, and Dascombe BJ. Reliability and Usefulness of the 30-15 Intermittent Fitness Test in Rugby League. *J Strength Cond Res* 29: 1985-1990, 2015.

39. Stickland MK, Petersen SR, and Bouffard M. Prediction of Maximal Aerobic Power From the 20 m Multistage Shuttle Run Test. *Can J Appl Physiol* 28: 272-282, 2003.

40. Strumbelj B, Vuckovic G, Jakovljevic S, Milanovic Z, James N, and Erculj F. Graded Shuttle Run Performance by Playing Positions in Elite Female Basketball. *J Strength Cond Res* 29: 793-799, 2015.

41. Thomas C, Dos'Santos T, Jones PA, and Comfort P. Reliability of the 30-15 Intermittent Fitness Test in Semiprofessional Soccer Players. *Int J Sports Physiol Perform* 11: 172-175, 2016.

42. Weston KS, Wisloff U, and Coombes JS. High-Intensity Interval Training in Patients With Lifestyle-Induced Cardiometabolic Disease: A Systematic Review and Meta-Analysis. *Br J Sports Med* 48: 1227-1234, 2014.

43. Weston M, Taylor KL, Batterham AM, and Hopkins WG. Effects of Low-Volume High-Intensity Interval Training (HIT) on Fitness in Adults: A Meta-Analysis of Controlled and Non-Controlled Trials. *Sports Med* 44: 1005-1017, 2014.

44. Woolford S, Polglaze T, Roswell G, and Spencer M. Field Testing Principles and Protocols. In: Tanner RK and Gore CJ, eds., *Physiological Tests for Elite Athletes*. Champaign, IL: Human Kinetics, 2013, pp. 231-248.

Laboratory 10

1. American College of Sports Medicine. *ACSM's Guidelines for Exercise Testing and Prescription*. 10th ed. Philadelphia: Wolters Kluwer, 2018.

2. Borg GA. Perceived Exertion. *Exer Sport Sci Rev* 2: 131-153, 1974.

3. Bruce RA, Kusumi F, and Hosmer D. Maximal Oxygen Intake and Nomographic Assessment of Functional Aerobic Impairment in Cardiovascular Disease. *Am Heart J* 85(4): 546-562, 1973.

4. Foster C, Jackson AS, Pollock ML, Taylor MM, Hare J, Sennett SM, Rod JL, Sarwar M, and Schmidt DH. Generalized Equations for Predicting Functional Capacity From Treadmill Performance. *Am Heart J* 107: 1229-1234, 1984.

5. Kenney WL, Wilmore JH, and Costill DL. *Physiology of Sport and Exercise*. 6th ed. Champaign, IL: Human Kinetics, 2015.

6. McArdle WD, Katch FI, and Pechar GS. Comparison of Continuous and Discontinuous Treadmill and Bicycle Tests for Max $\dot{V}O_2$. *Med Sci Sports* 5: 156-160, 1973.

7. Noble BJ, Borg GA, Jacobs I, Ceci R, and Kaiser P. A Category-Ratio Perceived Exertion Scale: Relationship to Blood and Muscle Lactates and Heart Rate. *Med Sci Sports Exer* 15: 523-528, 1983.

8. Pollock ML, Foster C, Schmidt D, Hellman C, Linnerud AC, and Ward A. Comparative Analysis of Physiologic Responses to Three Different Maximal Graded Exercise Test Protocols in Healthy Women. *Am Heart J* 103: 363-373, 1982.

9. Robertson RJ and Noble BJ. Perception of Physical Exertion: Methods, Mediators, and Applications. *Exer Sport Sci Rev* 25: 407-452, 1997.

10. Whaley MH, Brubaker PH, Kaminsky LA, and Miller CR. Validity of Rating of Perceived Exertion During Graded Exercise Testing in Apparently Healthy Adults and Cardiac Patients. *J Cardiopulm Rehabil* 17: 261-267, 1997.

Laboratory 11

1. Australian Sports Commission. *Physiological Tests for Elite Athletes*. 2nd ed. Champaign, IL: Human Kinetics, 2013.

2. Bentley DJ, McNaughton LR, Thompson D, Vleck VE, and Batterham AM. Peak Power Output, the Lactate Threshold, and Time Trial Performance in Cyclists. *Med Sci Sports Exer* 33: 2077-2081, 2001.

3. Brooks GA, Fahey TD, and Baldwin KM. *Exercise Physiology: Human Bioenergetics and Its Applications*. Boston: McGraw-Hill, 2005.

4. Coyle EF, Coggan AR, Hopper MK, and Walters TJ. Determinants of Endurance in Well-Trained Cyclists. *J Appl Physiol* 64: 2622-2630, 1988.

5. Coyle EF, Feltner ME, Kautz SA, Hamilton MT, Montain SJ, Baylor AM, Abraham LD, and Petrek GW. Physiological and Biomechanical Factors Associated With Elite Endurance Cycling Performance. *Med Sci Sports Exer* 23: 93-107, 1991.

6. Coyle EF, Martin WH, Ehsani AA, Hagberg JM, Bloomfield SA, Sinacore DR, and Holloszy JO. Blood Lactate Threshold in Some Well-Trained Ischemic Heart Disease Patients. *J Appl Physiol* 54: 18-23, 1983.

7. Dumke CL, Brock DW, Helms BH, and Haff GG. Heart Rate at Lactate Threshold and Cycling Time Trials. *J Strength Cond Res* 20: 601-607, 2006.

8. Faude O, Kindermann W, and Meyer T. Lactate Threshold Concepts: How Valid Are They? *Sports Med* 39: 469-490, 2009.

9. Fry AC, Kudrna RA, Falvo MJ, Bloomer RJ, Moore CA, and Schilling BK. Kansas Squat Test: A Reliable Indicator of Short Term Anaerobic Power. *J Strength Cond Res* 28 (3): 630-635, 2014.

10. Hagberg JM and Coyle EF. Physiological Determinants of Endurance Performance as Studied in Competitive Racewalkers. *Med Sci Sports Exer* 15: 287-289, 1983.

11. Janssen P. *Lactate Threshold Training*. Champaign, IL: Human Kinetics, 2001.

12. Kenney WL, Wilmore JH, and Costill DL. *Physiology of Sport and Exercise*. 6th ed. Champaign, IL: Human Kinetics, 2015.

13. Kindermann W, Simon G, and Keul J. The Significance of the Aerobic-Anaerobic Transition for the Determination of Work Load Intensities During Endurance Training. *Eur J Appl Physiol Occup Physiol* 42: 25-34, 1979.

14. Lindinger MI, Kowalchuk JM, and Heigenhauser GJ. Applying Physicochemical Principles to Skeletal Muscle Acid-Base Status. *Am J Physiol Regul Integr Comp Physiol* 289: R891-R894 (author reply R904-R910), 2005.

15. Martin JS, Fiedenreich ZD, Borges AR, and Roberts MD. Preconditioning With Peristaltic External Pneumatic Compression Does Not Acutely Improve Repeated Wingate Performance Nor Does It Alter Blood Lactate Concentrations During Passive Recovery Compared With Sham. *Appl Physiol Nutr Metab* 40 (11): 1214-1217, 2015.

16. Robergs RA, Ghiasvand F, and Parker D. Biochemistry of Exercise-Induced Metabolic Acidosis. *Am J Physiol Regul Integr Comp Physiol* 287: R502-R516, 2004.

17. Sjodin B and Jacobs I. Onset of Blood Lactate Accumulation and Marathon Running Performance. *Int J Sports Med* 2: 23-26, 1981.

Laboratory 12

1. Abadie BR and Wentworth M. Prediction of 1RM Strength From a 5-10 Repetition Submaximal Test in College Aged Females. *J Exer Physiol* (online) 3: 1-5, 2000.

2. American College of Sports Medicine. *ACSM's Guidelines for Exercise Testing and Prescription.* 8th ed. Philadelphia: Lippincott Williams & Wilkins, 2010.

3. American College of Sports Medicine. *ACSM's Health-Related Physical Fitness Assessment Manual.* 3rd ed. Baltimore: Lippincott Williams & Wilkins, 2010.

4. Acevedo EO and MA Starks. *Exercise Testing and Prescription Lab Manual.* Champaign, IL: Human Kinetics, 2003.

5. Adams GM. *Exercise Physiology Laboratory Manual.* 3rd ed. Boston: McGraw-Hill, 1998.

6. Adams KJ, Swank AM, Berning JM, Sevene-Adams PG, Barnard KL, and Shimp-Bowerman J. Progressive Strength Training in Sedentary, Older African American Women. *Med Sci Sports Exer* 33: 1567-1576, 2001.

7. Adams V, Jiang H, Yu J, Mobius-Winkler S, Fiehn E, Linke A, Weigl C, Schuler G, and Hambrecht R. Apoptosis in Skeletal Myocytes of Patients with Chronic Heart Failure Is Associated With Exercise Intolerance. *J Am Coll Cardiol* 33: 959-965, 1999.

8. Arvandi M, Strasser B, Meisinger C, Volaklis K, Gothe RM, Siebert U, Ladwig KH, Grill E, Horsch A, Laxy M, Peters A, and Thorand B. Gender Differences in the Association Between Grip Strength and Mortality in Older Adults: Results From the KORA-Age Study. *BMC Geriatr* 16: 201, 2016.

9. Baechle TR, Earle RW, and Wathen D. Resistance Training. In: Baechle TR and Earle RW, eds., *Essentials of Strength Training and Conditioning.* 3rd ed. National Strength and Conditioning Association. Champaign, IL: Human Kinetics, 2008, pp. 381-412.

10. Barnard KL, Adams KJ, Swank AM, Mann E, and Denny DM. Injuries and Muscle Soreness During the One Repetition Maximum Assessment in a Cardiac Rehabilitation Population. *J Cardiopulm Rehabil* 19: 52-58, 1999.

11. Berger RA. Relationship Between Dynamic Strength and Dynamic Endurance. *Res Q* 41: 115-116, 1970.

12. Bompa TO and Haff GG. *Periodization: Theory and Methodology of Training.* 5th ed. Champaign, IL: Human Kinetics, 2009.

13. Brown HL. *Lifetime Fitness.* 3rd ed. Scottsdale, AZ: Gorsuch Scarisbrick, 1992.

14. Bryzychi M. Assessing Strength. *Fitness Manage* June: 34-37, 2000.

15. Bryzychi M. Strength Testing: Predicting a One Rep Max From Reps to Fatigue. *J Phys Educ Rec Dance* 64: 88-90, 1993.

16. Chapman P, Whitehead JR, and Binkert RH. The 225-lb Reps-to-Failure Test as a Submaximal Estimation of 1RM Bench Press Performance in College Football Players. *J Strength Cond Res* 12: 258-261, 1998.

17. Cummings B and Finn KJ. Estimation of a One Repetition Maximum Bench Press for Untrained Women. *J Strength Cond Res* 12: 262-265, 1998.

18. Epley B. Poundage Chart. In: *Boyd Epley Workout.* Lincoln, NE: University of Nebraska, 1985.

19. Faigenbaum AD, Westcott WL, Loud RL, and Long C. The Effects of Different Resistance Training Protocols on Muscular Strength and Endurance Development in Children. *Pediatrics* 104: 1-7, 1999.

20. Filippin LI, Teixeira VN, da Silva MP, Miraglia F, and da Silva FS. Sarcopenia: A Predictor of Mortality and the Need for Early Diagnosis and Intervention. *Aging Clin Exp Res* 27: 249-254, 2015.

21. Haff GG, Jackson JR, Kawamori N, Carlock JM, Hartman MJ, Kilgore JL, Morris RT, Ramsey MW, Sands WA, and Stone MH. Force-Time Curve Characteristics and Hormonal Alterations During an Eleven-Week Training Period in Elite Women Weightlifters. *J Strength Cond Res* 22: 433-446, 2008.

22. Haff GG, Stone MH, O'Bryant HS, Harman E, Dinan CN, Johnson R, and Han KH. Force-Time Dependent Characteristics of Dynamic and Isometric Muscle Actions. *J Strength Cond Res* 11: 269-272, 1997.

23. Heyward VH. *Advanced Fitness Assessment and Exercise Prescription.* 6th ed. Champaign, IL: Human Kinetics, 2010.

24. Kallman DA, Plato CC, and Tobin JD. The Role of Muscle Loss in the Age-Related Decline of Grip Strength: Cross-Sectional and Longitudinal Perspectives. *J Gerontol* 45: M82-M88, 1990.

25. Kell RT, Bell G, and Quinney A. Musculoskeletal Fitness, Health Outcomes and Quality of Life. *Sports Med* 31: 863-873, 2001.

26. Kim PS, Mayhew JL, and Peterson DF. A Modified YMCA Bench Press Test as a Predictor of 1 Repetition Maximum Bench Press Strength. *J Strength Cond Res* 16: 440-445, 2002.

27. Kraemer WJ and Fry AC. Strength Testing: Development and Evaluation of Methodology. In: Maud PJ and Foster C, eds., *Physiological Assessment of Human Fitness.* 2nd ed. Champaign, IL: Human Kinetics, 1995, pp. 115-138.

28. Kraemer WJ, Ratamess NA, Fry AC, and French DN. Strength Training: Development and Evaluation of Methodology. In: Maud PJ and Foster C, eds., *Physiological Assessment of Human Fitness.* 2nd ed. Champaign, IL: Human Kinetics, 2006, pp. 119-150.

29. Kraemer WJ, Vingren JL, Hatfield DL, Spiering BA, and Fragala MS. Resistance Training Programs. In: *ACSM's Resources for the Personal Trainer.* Baltimore: Lippincott Williams & Wilkins, 2007, pp. 372-403.

30. Kravitz L, Akalan C, Nowicki K, and Kinzey SJ. Prediction of 1 Repetition Maximum in High-School Power Lifters. *J Strength Cond Res* 17: 167-172, 2003.

31. Kuramoto AK and Payne VG. Predicting Muscular Strength in Women: A Preliminary Study. *Res Q Exer Sport* 66: 168-172, 1995.

32. Lander JE. Maximum Based on Reps. *Natl Strength Cond Assoc* 6: 60-61, 1985.

33. Mathiowetz V, Weber K, Volland G, and Kashman N. Reliability and Validity of Grip and Pinch Strength Evaluations. *J Hand Surg Am* 9: 222-226, 1984.

34. Mayhew JL, Ball TE, Arnold ME, and Bowen JC. Relative Muscular Endurance Performance as a Predictor of Bench Press Strength in College Men and Women. *J Appl Sport Sci Res* 6: 200-206, 1992.

35. Mayhew JL, Johnson BD, Lamonte MJ, Lauber D, and Kemmler W. Accuracy of Prediction Equations for Determining One Repetition Maximum Bench Press in Women Before and After Resistance Training. *J Strength Cond Res* 22: 1570-1577, 2008.

36. Mayhew JL, Prinster JL, Ware JS, Zimmer DL, Arabas JR, and Bemben MG. Muscular Endurance Repetitions to Predict Bench Press Strength in Men of Different Training Levels. *J Sports Med Phys Fitness* 35: 108-113, 1995.

37. Mayhew JL, Ware JS, Cannon K, Corbett S, Chapman PP, Bemben MG, Ward TE, Farris B, Juraszek J, and Slovak JP. Validation of the NFL-225 Test for Predicting 1RM Bench Press Performance in College Football Players. *J Sports Med Phys Fitness* 42: 304-308, 2002.

38. Montoye HJ and Lamphiear DE. Grip and Arm Strength in Males and Females, Age 10 to 69. *Res Q* 48: 109-120, 1977.

39. O'Conner B, Simmons J, and O'Shea P. *Weight Training Today*. St. Paul: West, 1989, pp. 26-33.

40. Reynolds JM, Gordon TJ, and Roberts RA. Prediction of one repetition maximum strength from multiple repetition maximum testing and anthropometry. *J Strength Cond Res* 20: 584-592, 2006.

41. Sale DG. Testing Strength and Power. In: MacDougall JD, Wenger HA, and Green HJ, eds., *Physiological Testing of the High-Performance Athlete*. Champaign, IL: Human Kinetics, 1991, pp. 21-106.

42. Schlicht J, Camaione DN, and Owen SV. Effect of Intense Strength Training on Standing Balance, Walking Speed, and Sit-to-Stand Performance in Older Adults. *J Gerontol A Biol Sci Med Sci* 56: M281-M286, 2001.

43. Shaw CE, McCully KK, and Posner JD. Injuries During the One Repetition Maximum Assessment in the Elderly. *J Cardiopulm Rehab* 15: 283-287, 1995.

44. Spring T, Franklin B, and Dejong A. Muscular Fitness and Assessment. In: *ACSM's Resource Manual for Guidelines for Exercise Testing and Prescription*. 8th ed. Baltimore: Lippincott Williams & Wilkins, 2010, pp. 332-348.

45. Stone MH and O'Bryant HS. *Weight Training: A Scientific Approach*. Edina, MN: Burgess, 1987.

46. Stone MH, Sands WA, Pierce KC, Carlock J, Cardinale M, and Newton RU. Relationship of Maximum Strength to Weightlifting Performance. *Med Sci Sports Exer* 37: 1037-1043, 2005.

47. Stone MH, Sands WA, Pierce KC, Ramsey MW, and Haff GG. Power and Power Potentiation Among Strength-Power Athletes: Preliminary Study. *Int J Sports Physiol Perf* 3: 55-67, 2008.

48. Stone MH, Stone ME, and Sands WA. *Principles and Practice of Resistance Training*. Champaign, IL: Human Kinetics, 2007.

49. Tucker JE, Pujol TJ, Elder CL, Nahikian-Nelms ML, Barnes JT, and Langenfeld ME. One-Repetition Maximum Prediction Equation for Traditional College-Age Novice Females. *Med Sci Sports Exerc* 38: S293, 2006.

50. Warburton DE, Gledhill N, and Quinney A. Musculoskeletal Fitness and Health. *Can J Appl Physiol* 26: 217-237, 2001.

51. Ware JS, Clemens CT, Mayhew JL, and Johnston TJ. Muscular Endurance Repetitions to Predict Bench Press and Squat Strength in College Football Players. *J Strength Cond Res* 9: 99-103, 1995.

52. Wathen D. Load Assignment. In: T.R. Baechle, ed., *Essentials of Strength Training and Conditioning*. Champaign, IL: Human Kinetics, 1994, pp. 435-446.

Laboratory 13

1. Adams GM. *Exercise Physiology Laboratory Manual*. 3rd ed. Boston: McGraw-Hill, 1998.

2. Beam WC and Adams GM. *Exercise Physiology Laboratory Manuel*. 6th ed. Boston: McGraw-Hill, 2011.

3. Beattie K, Carson BP, Lyons M, and Kenny IC. The Relationship between Maximal-Strength and Reactive-Strength. *Int J Sports Physiol Perform* 12: 548-553, 2017.

4. Bobbert MF, Gerritsen KG, Litjens MC, and Van Soest AJ. Why Is Countermovement Jump Height Greater Than Squat Jump Height? *Med Sci Sports Exer* 28: 1402-1412, 1996.

5. Bogdanis GC, Nevill ME, Boobis LH, and Lakomy HK. Contribution of Phosphocreatine and Aerobic Metabolism to Energy Supply During Repeated Sprint Exercise. *J Appl Physiol* 80: 876-884, 1996.

6. Bompa TO and Haff GG. *Periodization: Theory and Methodology of Training*. 5th ed. Champaign, IL: Human Kinetics, 2009.

7. Bosco C and Komi PV. Mechanical Characteristics and Fiber Composition of Human Leg Extensor Muscles. *Eur J Appl Physiol Occup Physiol* 41: 275-284, 1979.

8. Bosco C, Luhtanen P, and Komi PV. A Simple Method for Measurement of Mechanical Power in Jumping. *Eur J Appl Physiol* 50: 273-282, 1983.

9. Breivik SL. Artistic Gymnastics. In: Winter EM, Jones AM, Davison RCR, Bromley PD, and Mercer TH, eds., *Sport and Exercise Physiology Testing Guidelines:*

Volume I—Sport Testing. London: Routledge, 2007, pp. 220-231.

10. Carlock J, Smith A, Hartman M, Morris R, Ciroslan D, Pierce KC, Newton RU, and Stone MH. The Relationship Between Vertical Jump Power Estimates and Weightlifting Ability: A Field Test Approach. *J Strength Cond Res* 18: 534-539, 2004.

11. Cormie, P, McBride JM, and McCaulley GO. Validation of Power Measurement Techniques in Dynamic Lower Body Resistance Exercises. *Journal of Applied Biomechanics* 23: 103-118, 2007.

12. Dotan R and Bar-Or O. Load Optimization for the Wingate Anaerobic Test. *Eur J Appl Physiol Occup Physiol* 51: 409-417, 1983.

13. Evans JA and Quinney HA. Determination of Resistance Settings for Anaerobic Power Testing. *Can J Appl Sport Sci* 6: 53-56, 1981.

14. Fry AC and Kraemer WJ. Physical Performance Characteristics of American Collegiate Football Players. *J Appl Sport Sci Res* 5: 126-138, 1991.

15. Gissis I, Papadopoulos C, Kalapotharakos VI, Sotiropoulos A, Komsis G, and Manolopoulos E. Strength and Speed Characteristics of Elite, Subelite, and Recreational Young Soccer Players. *Res Sports Med* 14: 205-214, 2006.

16. Gore CJ, ed. *Physiological Tests for Elite Athletes.* Champaign, IL: Human Kinetics, 2000.

17. Haff GG, Carlock JM, Hartman MJ, Kilgore JL, Kawamori N, Jackson JR, Morris RT, Sands WA, and Stone MH. Force-Time Curve Characteristics of Dynamic and Isometric Muscle Actions of Elite Women Olympic Weightlifters. *J Strength Cond Res* 19: 741-748, 2005.

18. Haff GG, Kirksey KB, Stone MH, Warren BJ, Johnson RL, Stone M, O'Bryant HS, and Proulx C. The Effects of Six Weeks of Creatine Monohydrate Supplementation on Dynamic Rate of Force Development. *J Strength Cond Res* 14: 426-433, 2000.

19. Haff GG, Stone MH, O'Bryant HS, Harman E, Dinan CN, Johnson R, and Han KH. Force-Time Dependent Characteristics of Dynamic and Isometric Muscle Actions. *J Strength Cond Res* 11: 269-272, 1997.

20. Harman E and Garhammer J. Administration, Scoring, and Interpretation of Selected Tests. In: Baechle TR and Earle RW, eds., *Essentials of Strength Training and Conditioning.* 3rd ed. National Strength and Conditioning Association. Champaign, IL: Human Kinetics, 2008, pp. 249-292.

21. Harman EA, Rosenstein MT, Frykman PN, Rosenstein RM, and Kraemer W. Estimation of Human Power Output From Vertical Jump. *J Appl Sport Sci Res* 5: 116-120, 1991.

22. Hawkins SB, Doyle TL, and McGuigan MR. The Effect of Different Training Programs on Eccentric Energy Utilization in College-Aged Males. *J Strength Cond Res* 23: 1996-2002, 2009.

23. Hespanhol JE, Neto LGS, De Arruda M, and Dini CA. Assessment of Explosive Strength-Endurance in Volleyball Players Through Vertical Jump Testing. *Rev Bras Med Esporte* 13: 160e-163e, 2007.

24. Hoffman JR. *Norms for Fitness, Performance, and Health.* Champaign, IL: Human Kinetics, 2006.

25. Hoffman JR. *Physiological Aspects of Sport Training and Performance.* Champaign, IL: Human Kinetics, 2002.

26. Hori N, Newton RU, Andrews WA, Kawamori N, McGuigan MR, and Nosaka K. Does Performance of Hang Power Clean Differentiate Performance of Jumping, Sprinting, and Changing of Direction? *J Strength Cond Res* 22: 412-418, 2008.

27. Hori N, Newton RU, Nosaka K, and McGuigan MR. Comparison of Different Methods of Determining Power Output in Weightlifting Exercises. *Strength and Cond J* 28: 34-40, 2006.

28. Housh TJ, Cramer JT, Weir JP, Beck TW, and Johnson GO. *Physical Fitness Laboratories on a Budget.* Scottsdale, AZ: Holcomb-Hathaway, 2009.

29. Inbar O, Bar-Or O, and Skinner JS. *The Wingate Anaerobic Test.* Champaign, IL: Human Kinetics, 1996.

30. Johnson DL and Bahamonde R. Power Output Estimate in University Athletes. *J Strength Cond Res* 10: 161-166, 1996.

31. Kalamen J. *Measurement of Maximum Muscle Power in Man.* Columbus, OH: Ohio State University, 1968.

32. Kawamori N, Crum AJ, Blumert P, Kulik J, Childers J, Wood J, Stone MH, and Haff GG. Influence of Different Relative Intensities on Power Output During the Hang Power Clean: Identification of the Optimal Load. *J Strength Cond Res* 19: 698-708, 2005.

33. Kawamori N, Rossi SJ, Justice BD, Haff EE, Pistilli EE, O'Bryant HS, Stone MH, and Haff GG. Peak Force and Rate of Force Development During Isometric and Dynamic Mid-Thigh Clean Pulls Performed at Various Intensities. *J Strength Cond Res* 20: 483-491, 2006.

34. Kirksey B, Stone MH, Warren BJ, Johnson RL, Stone M, Haff GG, Williams FE, and Proulx C. The Effects of Six Weeks of Creatine Monohydrate Supplementation on Performance Measures and Body Composition in Collegiate Track and Field Athletes. *J Strength Cond Res* 13: 148-156, 1999.

35. Klavora P. Vertical-Jump Tests: A Critical Review. *Strength Cond J* 22: 70, 2000.

36. Komi PV. *Strength and Power in Sport.* 2nd ed. Malden, MA: Blackwell Scientific, 2003.

37. Kraska JM, Ramsey MW, Haff GG, Fethke N, Sands WA, Stone ME, and Stone MH. Relationship Between Strength Characteristics and Unweighted and Weighted Vertical Jump Height. *Int J Sports Physiol Perform* 4: 461-473, 2009.

38. Leard JS, Cirillo MA, Katsnelson E, Kimiatek DA, Miller TW, Trebincevic K, and Garbalosa JC. Validity of Two Alternative Systems for Measuring Vertical Jump Height. *J Strength Cond Res* 21: 1296-1299, 2007.

39. Macdougall JD, Wenger HA, and Green HJ. The Purpose of Physiological Testing. In: Macdougall JD, Wenger HA, and Green HJ, eds., *Physiological Testing of the High-Performance Athlete*. Champaign, IL: Human Kinetics, 1991, pp. 1-6.

40. Margaria R, Aghemo P, and Rovelli E. Measurement of Muscular Power (Anaerobic) in Man. *J Appl Physiol* 21: 1662-1664, 1966.

41. Markwick WJ, Bird SP, Tufano JJ, Seitz LB, and Haff GG. The Intraday Reliability of the Reactive Strength Index (RSI) Calculated From a Drop Jump in Professional Men's Basketball. *Int J Sports Physiol Perform* 10: 482-488, 2015.

42. Maud PJ, Berning JM, Foster C, Cotter HM, Dodge C, Dekonning JJ, Hettinga F, and Lampen J. Testing for Anaerobic Ability. In: Maud PJ and Foster C, eds., *Physiological Assessments of Human Fitness*. Champaign, IL: Human Kinetics, 2006, pp. 77-92.

43. Maud PJ and Foster C, eds. *Physiological Assessments of Human Fitness*. 2nd ed. Champaign, IL: Human Kinetics, 2006.

44. McGuigan MR, Doyle TL, Newton M, Edwards DJ, Nimphius S, and Newton RU. Eccentric Utilization Ratio: Effect of Sport and Phase of Training. *J Strength Cond Res* 20: 992-995, 2006.

45. Medbo JI and Tabata I. Anaerobic Energy Release in Working Muscle During 30 Seconds to 3 Minutes of Exhausting Bicycling. *J Appl Physiol* 75: 1654-1660, 1993.

46. Mirzaei B, Curby DG, Rahmani-Nia F, and Moghadasi M. Physiological Profile of Elite Iranian Junior Freestyle Wrestlers. *J Strength Cond Res* 23: 2339-2344, 2009.

47. Nicklin RC, O'Bryant HS, Zehnbauer TM, and Collins A. A Computerized Method for Assessing Anaerobic Power and Work Capacity Using Maximal Cycle Ergometry. *J Appl Sport Sci Res* 4: 135-140, 1990.

48. Newton RU, Cormie P, and Cardinale M. Principles of Athlete Testing. In: Cardinale M, Newton RU, and Nosaka K, eds., *Strength and Conditioning: Biological and Practical Applications*. Chichester, UK: Wiley-Blackwell, 2011, pp. 255-267.

49. Plowman SA and Smith DL. *Exercise Physiology for Health, Fitness and Performance*. 3rd ed. Baltimore: Lippincott Williams & Wilkins, 2011.

50. Robinson JM, Stone MH, Johnson RL, Penland CM, Warren BJ, and Lewis RD. Effects of Different Weight Training Exercise/Rest Intervals on Strength, Power, and High Intensity Exercise Endurance. *J Strength and Cond Res* 9: 216-221, 1995.

51. Sands WA, McNeal JR, Ochi MT, Urbanek TL, Jemni M, and Stone MH. Comparison of the Wingate and Bosco Anaerobic Tests. *J Strength Cond Res* 18: 810-815, 2004.

52. Sayers SP, Harackiewicz DV, Harman EA, Frykman PN, and Rosenstein MT. Cross-Validation of Three Jump Power Equations. *Med Sci Sports Exer* 31: 572-577, 1999.

53. Skinner T, Newton RU, and Haff GG. Neuromuscular Strength, Power, and Strength Endurance. In: Coombes JS and Skinner T, eds., *ESSA's Student Manual for Health, Exercise, and Sport Assessment*. Australia: Elsevier, 2014, pp. 133-173.

54. Stone MH and O'Bryant HO. *Weight Training: A Scientific Approach*. Edina, MN: Burgess, 1987.

55. Stone MH, Sands WA, Carlock J, Callan S, Dickie D, Daigle K, Cotton J, Smith SL, and Hartman M. The Importance of Isometric Maximum Strength and Peak Rate-of-Force Development in Sprint Cycling. *J Strength Cond Res* 18: 878-884, 2004.

56. Stone MH, Stone ME, and Sands WA. *Principles and Practice of Resistance Training*. Champaign, IL: Human Kinetics, 2007.

57. Vargas NT, Robergs RA, and Klopp DM. Optimal Loads for a 30 s Maximal Power Cycle Ergometer Test Using a Stationary Start. *Eur J Appl Physiol* 115: 1087-1094, 2015.

58. Young W. Laboratory Strength Assessment of Athletes. *N Stud Athletics* 10: 89-96, 1995.

59. Zupan MF, Arata AW, Dawson LH, Wile AL, Payn TL, and Hannon ME. Wingate Anaerobic Test Peak Power and Anaerobic Capacity Classifications for Men and Women Intercollegiate Athletes. *J Strength Cond Res* 23: 2598-2604, 2009.

Laboratory 14

1. American Association of Cardiovascular and Pulmonary Rehabilitation. *Guidelines for Pulmonary Rehabilitation Programs*. 4th ed. Champaign, IL: Human Kinetics, 2011.

2. American College of Sports Medicine. *ACSM's Exercise Management for Persons with Chronic Diseases and Disabilities*. 3rd ed. Champaign, IL: Human Kinetics, 2009.

3. Babcock MA, Pegelow DF, Harms CA, and Dempsey JA. Effects of Respiratory Muscle Unloading on Exercise-Induced Diaphragm Fatigue. *J Appl Physiol* 93: 201-206, 2002.

4. Black LF, Offord K, and Hyatt RE. Variability in the Maximal Expiratory Flow Volume Curve in Asymptomatic Smokers and in Nonsmokers. *Am Rev Respir Dis* 110: 282-292, 1974.

5. Clanton TL, Dixon GF, Drake J, and Gadek JE. Effects of Swim Training on Lung Volumes and Inspiratory Muscle Conditioning. *J Appl Physiol* 62: 39-46, 1987.

6. Crapo RO, Morris AH, and Gardner RM. Reference Spirometric Values Using Techniques and Equipment That Meet ATS Recommendations. *Am Rev Respir Dis* 123: 659-664, 1981.

7. Dempsey JA. Is the Lung Built for Exercise? (JB Wolffe Memorial Lecture). *Med Sci Sports Exer* 18: 143-155, 1986.

8. Dempsey JA, Harms CA, and Ainsworth DM. Respiratory Muscle Perfusion and Energetics During Exercise. *Med Sci Sports Exer* 28: 1123-1128, 1996.

9. Dempsey JA, Sheel AW, Haverkamp HC, Babcock MA, and Harms CA. Pulmonary System Limitations to Exercise in Health (John Sutton Memorial Lecture: CSEP, 2002). *Can J Appl Physiol* 28 Suppl: S2-S24, 2003.

10. Dominelli PB, Render JN, Molgat-Seon Y, Foster GE, Romer LM, Sheel AW. Oxygen Cost of Exercise Hyperpnoea Is Greater in Women Compared With Men. *J Physiol* Apr 15; 593 (8): 1965-1979, 2015.

11. Ferreira SA, Guimaraes M, and Taveira N. Pulmonary Rehabilitation in COPD: From Exercise Training to "Real Life." *J Bras Pneumol* 35: 1112-1115, 2009.

12. Ghanem M, Elaal EA, Mehany M, and Tolba K. Home-Based Pulmonary Rehabilitation Program: Effect on Exercise Tolerance and Quality of Life in Chronic Obstructive Pulmonary Disease Patients. *Ann Thorac Med* 5: 18-25, 2010.

13. Hankinson JL, Crapo RO, and Jensen RL. Spirometric Reference Values for the 6 s FVC Maneuver. *Chest* 124: 1805-1811, 2003.

14. Harms CA, Babcock MA, McClaran SR, Pegelow DF, Nickele GA, Nelson WB, and Dempsey JA. Respiratory Muscle Work Compromises Leg Blood Flow During Maximal Exercise. *J Appl Physiol* 82: 1573-1583, 1997.

15. Harms CA and Dempsey JA. Does Ventilation Ever Limit Human Performance? In: Ward SA, ed., *The Physiology and Pathophysiology of Exercise Tolerance.* New York: Plenum Press, 1996, pp. 91-96.

16. Harms CA, McClaran SR, Nickele GA, Pegelow DF, Nelson WB, and Dempsey JA. Effect of Exercise-Induced Arterial O$_2$ Desaturation on $\dot{V}O_2$max in Women. *Med Sci Sports Exer* 32: 1101-1108, 2000.

17. Harms CA, McClaran SR, Nickele GA, Pegelow DF, Nelson WB, and Dempsey JA. Exercise-Induced Arterial Hypoxaemia in Healthy Young Women. *J Physiol* 507 (Pt 2): 619-628, 1998.

18. Harms CA, Wetter TJ, St Croix CM, Pegelow DF, and Dempsey JA. Effects of Respiratory Muscle Work on Exercise Performance. *J Appl Physiol* 89: 131-138, 2000.

19. Jobin J, Maltais F, LeBlanc P, and Simard C. *Advances in Cardiopulmonary Rehabilitation.* Champaign, IL: Human Kinetics, 2000.

20. Jobin J, Maltais F, Poirier P, LeBlanc PJ, and Simard C. *Advancing the Frontiers of Cardiopulmonary Rehabilitation.* Champaign, IL: Human Kinetics, 2002.

21. Kory RC, Callahan R, Boren HG, and Syner JC. The Veterans Administration-Army Cooperative Study of Pulmonary Function. I. Clinical Spirometry in Normal Men. *Am J Med* 30: 243-258, 1961.

22. Melissant CF, Lammers JW, and Demedts M. Relationship Between External Resistances, Lung Function Changes and Maximal Exercise Capacity. *Eur Respir J* 11: 1369-1375, 1998.

23. Mickleborough TD, Stager JM, Chatham K, Lindley MR, and Ionescu AA. Pulmonary Adaptations to Swim and Inspiratory Muscle Training. *Eur J Appl Physiol* 103: 635-646, 2008.

24. Morris JF. Spirometry in the Evaluation of Pulmonary Function. *West J Med* 125: 110-118, 1976.

25. Powers SK, Dodd S, Criswell DD, Lawler J, Martin D, and Grinton S. Evidence for an Alveolar-Arterial PO$_2$ Gradient Threshold During Incremental Exercise. *Int J Sports Med* 12: 313-318, 1991.

26. Powers SK, Lawler J, Dempsey JA, Dodd S, and Landry G. Effects of Incomplete Pulmonary Gas Exchange on $\dot{V}O_2$max. *J Appl Physiol* 66: 2491-2495, 1989.

27. Powers SK, Martin D, and Dodd S. Exercise-Induced Hypoxaemia in Elite Endurance Athletes. Incidence, Causes and Impact on $\dot{V}O_2$max. *Sports Med* 16: 14-22, 1993.

28. Riario-Sforza GG, Incorvaia C, Paterniti F, Pessina L, Caligiuri R, Pravettoni C, Di Marco F, and Centanni S. Effects of Pulmonary Rehabilitation on Exercise Capacity in Patients With COPD: A Number Needed to Treat Study. *Int J Chron Obstruct Pulmon Dis* 4: 315-319, 2009.

29. St Croix CM, Morgan BJ, Wetter TJ, and Dempsey JA. Fatiguing Inspiratory Muscle Work Causes Reflex Sympathetic Activation in Humans. *J Physiol* 529 (Pt 2): 493-504, 2000.

30. Stocks J and Quanjer PH. Reference Values for Residual Volume, Functional Residual Capacity and Total Lung Capacity. ATS Workshop on Lung Volume Measurements. Official Statement of the European Respiratory Society. *Eur Respir J* 8: 492-506, 1995.

31. Wetter TJ, Harms CA, Nelson WB, Pegelow DF, and Dempsey JA. Influence of Respiratory Muscle Work on $\dot{V}O_2$ and Leg Blood Flow During Submaximal Exercise. *J Appl Physiol* 87: 643-651, 1999.

Laboratory 15

1. Bray GA and Gray DS. Obesity. Part I—Pathogenesis. *West J Med* 149: 429-441, 1988.

2. Friedl KE, DeLuca JP, Marchitelli LJ, and Vogel JA. Reliability of Body-Fat Estimations From a Four-Compartment Model by Using Density, Body Water, and Bone Mineral Measurements. *Am J Clin Nutr* 55: 764-770, 1992.

3. Heyward VH and Wagner DR. *Applied Body Composition Assessment.* 2nd ed. Champaign, IL: Human Kinetics, 2004.

4. Jackson AS and Pollock ML. Generalized Equations for Predicting Body Density of Men. *Br J Nutr* 40: 497-504, 1978.

5. Jackson AS, Pollock ML, and Ward A. Generalized Equations for Predicting Body Density of Women. *Med Sci Sports Exer* 12: 175-181, 1980.

6. McArdle WD, Katch FI, and Katch VL. *Exercise Physiology: Energy, Nutrition, and Human Performance.* Baltimore: Lippincott Williams & Wilkins, 2007.

7. Morrow JR, Jr., Jackson AS, Bradley PW, and Hartung GH. Accuracy of Measured and Predicted Residual Lung Volume on Body Density Measurement. *Med Sci Sports Exer* 18: 647-652, 1986.

8. National Heart, Lung, and Blood Institute. *Clinical Guidelines on the Identification, Evaluation, and Treatment of Overweight and Obesity in Adults: The Evidence Report.* Bethesda, MD: National Institutes of Health, 1998.

9. Pollock ML and Jackson AS. Research Progress in Validation of Clinical Methods of Assessing Body Composition. *Med Sci Sports Exer* 16: 606-615, 1984.

10. Prior BM, Cureton KJ, Modlesky CM, Evans EM, Sloniger MA, Saunders M, and Lewis RD. In Vivo Validation of Whole Body Composition Estimates from Dual-Energy X-Ray Absorptiometry. *J Appl Physiol* 83: 623-630, 1997.

11. Siri WE. Body Composition From Fluid Spaces and Density: Analysis of Methods. 1961. *Nutrition* 9: 480-491 (discussion 480, 492), 1993.

12. Withers RT, Smith DA, Chatterton BE, Schultz CG, and Gaffney RD. A Comparison of Four Methods of Estimating the Body Composition of Male Endurance Athletes. *Eur J Clin Nutr* 46: 773-784, 1992.

Laboratory 16

1. American College of Sports Medicine. *ACSM's Guidelines for Exercise Testing and Prescription.* 10th ed. Philadelphia: Wolters Kluwer, 2018.

2. Thaler MS. *The Only EKG Book You'll Ever Need.* Philadelphia: Lippincott Williams & Wilkins, 2007.

3. Kenney WL, Wilmore JH, and Costill DL. *Physiology of Sport and Exercise.* 6th ed. Champaign, IL: Human Kinetics, 2015.

4. Dubin, D. *Rapid Interpretation of EKGs.* Fort Meyers, FL: Cover, 2000.

ABOUT THE AUTHORS

G. Gregory Haff, PhD, CSCS,★D, FNSCA, ASCC, is an associate professor and the course coordinator for the postgraduate degree in strength and conditioning at Edith Cowan University in Joondalup, Australia. Haff has published more than 80 articles, centering his

Courtesy of Greg Haff

research on performance effects in the areas of strength training, cycling, and nutritional supplementation.

Haff is the president of the National Strength and Conditioning Association (NSCA) and a senior associate editor for the *Journal of Strength and Conditioning Research*. He was the United Kingdom Strength and Conditioning Association (UKSCA) Strength and Conditioning Coach of the Year for Education and Research and the 2011 NSCA William J. Kraemer Outstanding Sport Scientist Award winner. He is a certified strength and conditioning specialist with distinction, a UKSCA-accredited strength and conditioning coach (ASCC), and an accredited Australian Strength and Conditioning Association level 2 strength and conditioning coach.

Additionally, Haff is a national-level weightlifting coach in the United States and Australia. He serves as a consultant for numerous sporting bodies, including teams in the Australian Football League, Australian Rugby Union, Australian Basketball Association, and National Football League.

Charles Dumke, PhD, FACSM, is a full professor in the department of health and human performance at the University of Montana, where he teaches undergraduate and graduate courses. He has taught courses in exercise physiology for over 15 years, first at Appalachian State

Courtesy of Charles Dumke

University and then at the University of Montana. He earned his doctoral degree in kinesiology from the University of Wisconsin at Madison. His areas of interest in exercise science are energy expenditure, fuel utilization, economy of movement, mechanisms of mitochondrial adaptation, and diabetes. He has published more than 100 peer-reviewed articles on these topics. Dumke is a fellow of the ACSM and serves on several national and regional committees.

In his free time, Dumke enjoys competing in triathlons, biking, running, taking on building projects with little know-how, and coaching his son in ball sports. He resides in Missoula, Montana, with his wife, Shannon; son, Carter; and dog, Rastro.